THE NATURAL PHILOSOPHY OF TIME

BY

G. J. WHITROW

Emeritus Professor of the History and Applications of Mathematics in the University of London and Senior Research Fellow of the Imperial College of Science and Technology

Second Edition

CLARENDON PRESS · OXFORD
1980

Oxford University Press, Walton Street, Oxford OX2 6DP

OXFORD LONDON GLASGOW

NEW YORK TORONTO MELBOURNE WELLINGTON

IBADAN NAIROBI DAR ES SALAAM LUSAKA CAPE TOWN

KUALA LUMPUR SINGAPORE JAKARTA HONG KONG TOKYO

DELHI BOMBAY CALCUTTA MADRAS KARACHI

Published in the United States by Oxford University Press, New York
First edition published by Nelson, 1961

British Library Cataloguing in Publication Data

Whitrow, Gerald James
The natural philosophy of time. – 2nd ed.
1. Time
I. Title
115 BD638 79-41145
ISBN 0-19-858212-9

Filmset in Northern Ireland at The Universities Press (Belfast)
and printed in Great Britain by Lowe and Brydone, Thetford.

PREFACE

In an article published in *New Scientist* in 1959, Professor J. L. Synge wrote that, in his view, of all measurements made in physics the measurement of time is the most fundamental, and 'the theory underlying these measurements of time is the most basic theory of all'. He argued that Euclid put us on the wrong track by taking space as the primary concept of science and relegating time to a poor second. The continuing absence of any generally recognized term for the study of time was itself evidence of this curious neglect. Synge proposed that the word 'chronometry' be used to denote that part of science which deals with the concept of time with the same wide scope as 'geometry' deals with space. In making this suggestion, he pointed out that pure, or theoretical, chronometry should be distinguished from applied, or practical, chronometry, (i.e. the technology of clock construction, the astronomical determination of time, dendrochronology, varve analysis, radiocarbon dating, etc).

The first draft of the first edition of the present book had already been completed when Synge's 'timely' article appeared. It was gratifying to discover that the task on which I had been engaged at intervals over the previous five years might help to fill, however inadequately, a recognized gap in the literature of science and natural philosophy.

Since 1961, when the first edition of this book was published by Nelson, there has been a remarkable and widespread increase in interest in all aspects of the concept of time. Early in 1966 an important collection of essays on various aspects of time was published–*The Voices of Time*, edited by J. T. Fraser. The same year, at a meeting of the New York Academy of Sciences, it was agreed to form an International Society for the Study of Time. I was invited to become the first President, and Dr. Fraser became Honorary Secretary. The first, and highly successful, meeting of this interdisciplinary Society was held at the Mathematisches Forschungs institut at Oberwohlfach, in the Black Forest in the Federal Republic of Germany, in September 1969. Other meetings have since been held every three of four years. The proceedings of these conferences have been published by Springer-Verlag. Meanwhile various international colloquia on different aspects of time have been held in several countries. It can, therefore, no longer be claimed that a book such as this 'fills a gap'. Nevertheless, I believe that a comprehensive survey by one author of the current state of scientific knowledge concerning this fundamental concept is still of value.

I should like to thank all those who have sent me copies of their publications, and in particular the Earl of Halsbury, Professor K. G. Denbigh, Professor O. Heckmann, Professor B. Hoffman, Professor W. H. McCrea, Professor G. C. McVittie, Professor Sir Karl Popper, Professor A. G. Walker and Christopher Sykes (my former producer at the B.B.C.) for their encouragement over the years. I am grateful for the help given and forbearance shown by the staff of the Clarendon Press who have been waiting patiently for delivery of the typescript. A special word of thanks is due to my three successive secretaries, Vera van der Bergh, Elsie Bryce, and Elizabeth Phare-Sinha for their efficient and beautiful typing. I am also greatly indebted to Jagna Pindelska, Librarian of the Mathematics Department, Imperial College, for her expertise and assiduity in tracking down any book or periodical that I wished to consult. Finally, as always, I owe more than I can say to the loyal support of my wife, Magda Whitrow, who has read the proofs and compiled the Index.

London
2 April 1979 G.J.W.

Similarly, time by itself does not exist; but from things themselves there results a sense of what has already taken place, what is now going on and what is to come.

Lucretius

We see which way the stream of time doth run.

Shakespeare, *Henry IV*, Part II, Act IV, Sc. 1.

A l'époque vraiment rationelle, on n'aime pas le temps; en général, les hommes ne l'aiment pas beaucoup, mais les philosophes ont une horreur particulière pour lui; ils ont tout fait pour supprimer le temps.

P. Janet

Our knowledge of time as of space owes more to the labours of mathematicians and physicists than to those of professed philosophers.

C. D. Broad

It is impossible to meditate on time and the mystery of the creative process of nature without an overwhelming emotion at the limitations of human intelligence.

A. N. Whitehead

CONTENTS

1

UNIVERSAL TIME

1.1. The 'elimination' of time

The history of natural philosophy is characterized by the interplay of two opposing points of view which may be conveniently associated with the names of Archimedes and of Aristotle, those intellectual giants of antiquity whose writings were of decisive importance for the late medieval and Renaissance founders of modern science. Archimedes is the prototype of those whose philosophy of physics presupposes the 'elimination' of time,† i.e. of those who believe that temperoral flux is not an intrinsic feature of the ultimate basis of things. Aristotle, on the other hand, is the forerunner of those who regard time as fundamental, since he insisted that there are real 'comings-into-being' and that the world has a basic temporal structure.‡

Archimedes was the founder of the science of hydrostatics and the author of the first important treatise on statics. What Euclid did for the art of the stonemason, Archimedes did for the practical and intuitive knowledge of generations of technicians who had operated simple machines like the balance and the lever. He laid bare the theoretical basis of this knowledge and, following the example of Euclid, presented it as a logically coherent system. His treatise *On the Equilibrium of Planes* is an outstanding example of a scientific discipline based on rigorous argument from apparently incontrovertible premises. It attained the ideal, so earnestly sought in our own day by Einstein and others, of reducing a branch of physics to a branch of geometry, but the concept of time played no part in it.

Aristotle's treatment of physical problems was very different. The metaphysical principle that every change requires a cause was fundamental to his thought. For example, Book VII of the *Physics* opens with the statement, 'Everything that is in motion must be moved by something', a postulate which had to be rejected before an effective science of dynamics could be formulated. Nevertheless, however erroneous Aristotle's principles may now seem, the fact that they were adhered to for so

† The term is due to Emile Meyerson (1930).

‡ Earlier and more shadowy thinkers who could be taken as protagonists of these two viewpoints are Parmenides and Heraclitus. Parmenides maintained that ultimate physical reality is timeless, whereas the central doctrine of Heraclitus was that the world is the totality of *events* and not of *things*. (For a modern analysis of Aristotle's philosophy of nature, see Randall (1960).)

long shows that they must have appeared to be just as 'self-evident' as the axioms and postulates of Euclid and Archimedes. The essential difference was that, whatever the mathematicians may themselves have thought, they were in fact concerned with abstract limiting cases, whereas Aristotle was an empiricist whose exclusive concern was the actual physical universe *as he conceived it*—and hence subject to all the shortcomings of that limited conception. Indeed, before modern physics could arise Aristotle's physics had to be overthrown and the method of Archimedes adapted to the study of motion.

Nevertheless, with all its faults, Aristotle's physics was in one vital respect superior to that of Archimedes. The certainty and lucidity of Archimedes' principles are largely due to the fact that they are gathered, so to speak, from the surface of phenomena and not dug out from the depths. His logically perfect treatise on statics was, in fact, less profound and less rich in promise of fruitful developments than the immature and disorderly work of Aristotle. The reason for this is clear: Archimedes avoided the problems of motion; Aristotle faced them. In the natural philosophy of Archimedes laws of nature are laws of equilibrium and temporal concepts play no part, whereas for Aristotle nature was 'a principle of motion and change' and could not be understood without an analysis of time.

Although the peculiarly fundamental nature of time in relation to ourselves is evident as soon as we reflect that our judgments concerning time and events in time appear themselves to be 'in' time, whereas our judgments concerning space do not appear in any obvious sense to be in space, physicists have been influenced far more profoundly by the fact that space seems to be presented to us all of a piece, whereas time comes to us only bit by bit. The past must be recalled by the dubious aid of memory, the future is hidden from us, and only the present is directly experienced. This striking dissimilarity between space and time has nowhere had a greater influence than in physical science based on the concept of measurement. Free mobility in space leads to the idea of the transportable unit length and the rigid measuring rod. The absence of free mobility in time makes it much more difficult for us to be sure that a process takes the same time whenever it is repeated. Consequently, as Einstein (1954) remarked, 'It is a characteristic of thought in physics . . . that it endeavours in principle to make do with "space-like" concepts *alone*, and strives to express with their aid all relations having the form of laws'. It is true that in the term 'space-like' he included the concepts of *time* and *event* as they featured in his theory, but he believed it to be more natural 'to think of physical reality as a four-dimensional existence, instead of, as hitherto, the *evolution* of a three-dimensional existence'. Thus, in order to study the temporal aspect of nature effectively, men have strained their ingenuity to devise means whereby the peculiar characteristics of time are either ignored or distorted. (Indeed, this is evident even at the level of ordinary conversation when we speak of 'a short space of time', as if an interval of time could be regarded as an interval of space.) Great achievements in physical science have been made by relentlessly pursuing this paradoxical policy.

There is nothing specifically modern or revolutionary in the tendency to subordinate the temporal to the spatial. As long ago as 1872, in his famous address *On the Limitations of Natural Science*, Emile du Bois-Reymond made the sweeping statement that 'The cognition of nature is the reduction of all changes in the physical world to the motions of atoms governed by forces independent of time'. A quarter of a century earlier, Helmholtz in his lecture *On the Conservation of Energy* maintained that 'The task of physical science is finally to reduce all phenomena of nature to forces of attraction and repulsion, the intensity of which is dependent only upon the mutual distance of material bodies. Only if this problem is solved are we sure that nature is conceivable'. In similar vein Poinsot declared in his *Éléments de Statique* that 'In perfect knowledge we know but one law—that of constancy and uniformity. To this simple idea we try to reduce all others, and it is only in this reduction that we believe science to consist'.

Going back to the eighteenth century, we find that Lavoisier's conception of chemistry was based on the postulate that in every chemical transformation there is a conservation of 'matter': 'It is upon this principle that is based the entire art of making experiments in chemistry'. The chemical equation was an expression of a principle of identity, of preservation, of time-elimination—in short a statement that, despite appearances, basically *nothing has happened.* Thus the philosopher of science Emile Meyerson (1930, p. 238) concluded that 'Science in its effort to become "rational" tends more and more to suppress variation in time'.

In mathematical physics, Lavoisier's contemporary Lagrange was the forerunner of Minkowski and Einstein in affirming that time could be regarded as a fourth dimension of space. He realized that the time variable of rational mechanics based on Newton's laws of motion does not point in a unique direction and that, in principle, all motions and dynamical processes subject to these laws are reversible, just like the axes of geometrical co-ordinate systems. Moreover, the origin of Newtonian time can be as freely and arbitrarily chosen as can the origin of a Cartesian co-ordinate system. By regarding physical time as a fourth dimension of space, Lagrange all but eliminated time from dynamical theory.

The 'elimination' of time from natural philosophy is closely correlated with the influence of geometry. Archimedes' theory of statical phenomena was almost entirely geometrical (the non-geometrical elements in it are not immediately apparent—for example, the implicit assumption that the turning effect about the fulcrum of a number of weights spaced along one arm of a lever will be the same as if they were all collected at their centre of gravity). Galileo's great achievements in dynamics were largely due to his successful exploitation of the device for representing time by a geometrical straight line. The primary object of Einstein's profound researches on the forces of nature has been well epitomized in the slogan 'the geometrization of physics', time being completely absorbed into the geometry of a hyperspace. Thus, instead of ignoring the temporal aspect of nature as Archimedes did, post-Renaissance mathematicians and physicists have sought to explain it away in terms of the spatial, and in this they have been aided by philosophers, notably the idealists.†

1.2. Trend and symmetrical time

If the concept of time in physics is to be subordinated to that of space, we must somehow circumvent the asymmetry of past and future which characterizes our temporal experience. Although this has become increasingly difficult, the most determined attempts have been made to attain this end.

Despite the achievements of Lavoisier and Lagrange, evidence of trend in nature could not be ignored by the nineteenth century founders of thermodynamics. By the middle of the century it was clear that two distinct principles were involved in the theory of heat. On the one hand, in any closed system, that is to say in any system that can be isolated from the rest of the universe, the total quantity of energy is constant. In other words, the quantity of heat that disappears in such a system is equivalent to the amount of other kinds of energy that appears, and *vice versa.* This law of conservation of energy (First Law of Thermodynamics) therefore asserts the invariance of the total *quantity* of energy in a system that is not interacting with its surroundings. On the other hand, the Second Law of Thermodynamics concerns the *quality* of this energy, that is the amount of energy available in the system for doing useful work. It

† A similar attitude has also been observed among biologists. Some years ago J. Z. Young (1938) felt obliged to draw attention to the fact that 'emphasis on the direction of biological activities is curiously unpopular among some biologists, the word 'teleological' being (wrongly) applied to it as a label of undiscriminating reproach. Yet it is impossible for anyone concerned with living things to ignore this point of view'.

determines the direction in which thermodynamic processes occur and expresses the fact that, although energy can never be lost, it may become unavailable for doing mechanical work. This law, as formulated by Rudolph Clausius and William Thomson (afterwards Lord Kelvin), was a refinement and generalization of the hypothesis that heat cannot, of itself, pass from a colder to a hotter body. Whereas we can get work out of the heat of a body which is hotter than surrounding bodies, we cannot get work out of a body which is cooler than surrounding bodies. Clausius (in 1850 and 1854) and Thomson (in 1851) based their independent formulations of this law on the much neglected investigations of Sadi Carnot (1796–1832), who in his pioneer *Réflexions sur la puissance motrice du feu*, published in 1824, was concerned with the fact that, although energy (as we now call it) is conserved, it may nevertheless be rendered unavailable for doing mechanical work.

In 1865 Clausius restated Carnot's principle in terms of the concept of entropy. He derived this word from the Greek *τροπή* (a transformation). He defined it differentially: a change of entropy is equal to heat received divided by the temperature at which the heat is received, provided that this heat is insufficient to produce any appreciable change of temperature, which is measured on the absolute scale introduced by Kelvin. The Second Law of Thermodynamics can then be stated in the following form: the entropy of an isolated, or closed, system never diminishes. Every reversible change occurring in a closed system will leave its total entropy unaltered, for the gain of entropy in one part of the system will be balanced by its loss in the other part, but every irreversible change in the system will increase its entropy. Irreversible changes occur when heat passes of its own accord from one part of a system to another at lower temperature. In general, any spontaneous change in the physical or chemical state of a system will lead to an increase of its entropy.†

Clausius argued that this law conflicted with the customary view that the general state of the world remains invariable, transformations in one direction at particular places and times being counterbalanced by transformations in the reverse direction at other places and times. Although the First Law of Thermodynamics might be taken to confirm this view, the Second Law contradicts it explicitly. 'From this it results that the state of the universe must change more and more in a determined direction'. (Meyerson 1930, p. 265.)

It is remarkable that no-one before Carnot seems to have had any real understanding of this principle and its implications. Even Heraclitus

† No such changes will occur when the entropy is a maximum, i.e. when no increase of entropy can occur without changing the conditions of the system. The system will then be in a state of stable thermodynamic equilibrium.

believed that his eternal flux was a cyclic process. Carnot's principle was accepted only with great reluctance, and repeated attempts were made to escape from its cosmological consequences. The idea of a continual change of the universe in the same direction until complete thermal equilibrium has been attained was repugnant to many scientists. Emile Meyerson has drawn attention to typical instances. For example, Haeckel (1900) declared that 'If this theory of entropy were true, we should have a "beginning" corresponding to the assumed "end" of the world. Both ideas are quite untenable in the light of our monistic and consistent theory of the eternal cosmogenetic process; both contradict the law of substance... The second thesis of the mechanical theory of heat contradicts the first and so must be rejected'. He argued that Carnot's principle can only be applied to 'distinct processes', but in the world at large 'quite other conditions obtain'. Similarly, the chemist Arrhenius (1910) wrote that 'If the ideas of Clausius were true, this thermal death ought to have been realized in the infinite time of the existence of the world'. Moreover, we cannot suppose that there was a beginning, since energy cannot be created. Consequently 'this is totally incomprehensible to us'. Commenting on these forthright declarations, Meyerson (1930, p. 269), maintained that the point of view of both Haeckel and Arrhenius is determined by an *a priori* consideration anterior to all experience. 'What shocks them is that on Carnot's principle we must suppose ourselves to be at a precise moment of a continuous development'.

Boltzmann attempted to evade the cosmological consequences of Carnot's principle by speculating on the possibility of there being regions of the universe in which thermal equilibrium has been attained and others in which time runs in the opposite sense to that in our stellar system. He believed that for the whole universe the two directions of time are indistinguishable, just as in space there is neither upper nor lower. More recently, at the British Association discussion in 1931 on 'The Evolution of the Universe', Sir Oliver Lodge claimed that too much attention had been paid to the Second Law of Thermodynamics and that 'the final and inevitable increase of entropy to a maximum is a bugbear, an idol, to which philosophers need not bow the knee'.

It was at this meeting that E. A. Milne put his finger on a logical gap in the proof that the entropy of the universe as a whole automatically tends to a maximum. He argued that in establishing the Second Law of Thermodynamics the following additional axiom is required: Whenever a process occurs in the universe, it is possible to divide the universe into two portions such that one of the portions is entirely unaffected by the process. This axiom would, however, automatically exclude world-wide processes. Nevertheless, he was careful to point out that we cannot say that the entropy of the universe is *not* increasing, for every local irreversible process causes such an increase. What we can say is that we have no means for assessing change of entropy for the whole universe, since we can calculate such a change for 'closed systems' with something outside them but the universe *ex hypothesi* has nothing (physical) outside it.

One of the boldest and most thoroughgoing attempts to deny the existence of any *objective* temporal trend in the physical universe was made in 1930 by the distinguished physical chemist, G. N. Lewis (1930). He maintained that the idea of 'time's arrow', to use Eddington's graphic phrase, is almost entirely due to the phenomena of consciousness and memory and that throughout the sciences of physics and chemistry 'symmetrical' time suffices. He claimed that nearly everywhere in these sciences the ideas of unidirectional time and unidirectional causality have been purged, as if the physicist were aware that these ideas introduce an irrelevant 'anthropomorphic element'. Moreover, in the cases where these ideas have been invoked he believed that they have always been used to support some false doctrine: for example, that the universe is actually 'running down'. Instead, the statistical interpretation of thermodynamics leads to the conclusion that if the universe is finite then the precise present state of the universe has occurred in the past and will recur in the future, and in each case within finite time.

For the simple, but typical, case of three identifiable molecules in a closed cylinder with a middle wall provided with a shutter, he proved that the entropy of the general *unknown* distribution of these molecules is greater than that of any known distribution, for example, a particular pair on the left and the remaining one on the right. He showed that an increase of entropy occurs when, after trapping any one *known* distribution, we open the shutter. If however, we start with the shutter open, with all eight distributions occurring one after another, and then close the shutter, so that the system is then trapped in a particular distribution, no change of entropy occurs. Consequently, he argued, increase in entropy comes only when a *known* distribution goes over into an *unknown* distribution, and the loss which characterizes an irreversible process is *loss of information.* He therefore concluded that gain in entropy always means loss of information and nothing more. 'It is a subjective concept', he wrote, 'but we can express it in its least subjective form, as follows. If, on a page, we read the description of a physico-chemical system, together with certain data which help to specify the system, the entropy of the system is determined by these specifications. If any of the essential data are erased, the entropy becomes greater; if any essential data are added, the entropy becomes less. Nothing further is needed to show that the irreversible process neither implies one-way time, nor has any other temporal implications. Time is not one of the variables of pure thermodynamics'.

Lewis also analysed the role of time in optical and electromagnetic phenomena. In his view the laws of optics are entirely symmetrical with respect to the emission and absorption of light. If time is imagined as reversed, emitting and absorbing objects exchange roles but the optical laws are unchanged. Radiation from a particle seems, however, to be in direct conflict with his idea of temporal symmetry, and he admitted that the emission of a continuous spherical shell of energy is essentially irreversible. All parts of the shell move outward until they meet absorbing bodies, but some parts will not meet such bodies perhaps for years, whereas others may do so within minute fractions of a second. The exact physical reversal arrangement whereby a number of bodies located at vastly different distances would each emit the appropriate amount of energy at the appropriate time and in the appropriate direction, so that in the neighbourhood of the particle all these particles of radiation would combine to form a continuous contracting spherical shell. Nevertheless, instead of being embarrassed by considerations of this character, Lewis welcomed them, arguing that the concept of symmetrical time immediately leads to the conclusion that the basic radiation process must be one in which a single emitting particle sends its energy to only one absorbing particle—in other words, a process analogous to Einstein's theory of the photon.

In the case of electromagnetic theory, it is immediately seen that Maxwell's equations, like the equations of classical mechanics, are unchanged when the time direction is reversed. How then was it possible to derive the old one-way theory of radiation from equations which are themselves compatible with symmetrical time? The derivation depends on the fact that of two symmetrical solutions which arise in the mathematical analysis, only the *retarded potential* is admitted as physically significant. 'In the whole history of physics', wrote Lewis, 'this is the most remarkable example of the suppression by physicists of some of the consequences of their own equations because they were not in accord with the old theory of unidirectional causality'. Instead, he maintained that if *advanced potentials* had been used, and retarded potentials had been discarded, we should have had an electromagnetic theory of light in equally good agreement with empirical facts, but in the interpretation of which we should have been led to regard the absorbing particle as the active agent, sucking in energy from all parts of space in a spherical shell contracting with the velocity of light. He claimed that no satisfactory quantum electrodynamics could be developed until the retarded and advanced potentials were used simultaneously and symmetrically.

The theory of the equilibrium of matter and radiation at constant temperature was shown by Lewis to depend on a principle which was first used in a limited way by Boltzmann, but which Lewis deduced as a universal law from his idea of temporal symmetry. This law, now usually known as the *principle of detailed balancing*, asserts that *every* process of transformation occurring in an enclosed system in thermodynamical equilibrium is capable of direct reversal, and transformations in the two directions occur with equal frequency. The gain by any process balances the loss by the converse process, so that any statistical enumeration, however detailed, of the processes of change occurring in a system in equilibrium at constant temperature would remain true if the direction of time were reversed. Consequently, in any system in equilibrium, 'time must lose the unidirectional character which plays so important a part in the development of the time concept'.†

1.3. Irreversible phenomena

The arguments advanced by Lewis in support of his theory of symmetrical time are ingenious and powerful. Nevertheless, as Miss Cleugh (1937, p. 165) has shrewdly remarked, 'The ghost of time cannot be permanently laid'. Cogent as is much of Lewis's reasoning, it leaves out of account many important factors.

Thus, the particular argument by which he endeavoured to circumvent his own contention that the emission of a continuous spherical shell of radiation is essentially an irreversible process cannot be extended to other types of spherical wave. For, as has been pointed out by K. R. Popper (1956, 1957, 1958), the absence of the converse phenomenon of waves closing in isotropically to a precise point of annihilation is not a peculiarity of light and electromagnetic radiation, but is equally true of other types of phenomena, for example water waves generated by a disturbance at a particular place. We cannot explain away the apparent absence of temporal symmetry in these other cases by appeals to the corpuscular nature of the phenomena in question. Instead, we are compelled to acknowledge their essential irreversibility.

This condition has been disputed by Hinckfuss (1975). In rejecting Popper's argument that the expansion of a wave-like process from a point

† A good account of Lewis's views and their reception has been given by R. H. Stuewer (1976). He points out that Lewis 'took great pleasure in the role of scientific *provocateur* that he chose for himself'. Incidentally, although Einstein introduced the idea of the photon in 1905, it was Lewis who gave it its name, in 1926.

is irreversible, he considers only the electromagnetic case and asserts that, in fact, the conditions prevailing at any point of space at any time (in particular, the electric and magnetic field strengths there) depend on radiation that has converged on that point from every direction. This remark in no way disposes of Popper's argument, for Hinckfuss overlooks an essential feature of that argument—the coherence condition. The temporal converse to a wave-like process that spreads out spherically symmetrically from a point would depend on a large number of sources scattered throughout space being able to emit radiation in such a way that it all converges on the same point at the same time with the same wavelength and intensity from every direction. Popper argues that the improbability of such a situation occurring is so great that it must be regarded as virtually impossible. Hinckfuss objects to any such argument for irreversibility being based on improbability because we may in fact be inhabiting a very unusual part of the universe if our solar system is the product of highly improbable initial conditions. This criticism is irrelevant, for the essential difference between an outgoing isotropic monochromatic wave and its precise temporal converse has nothing to do with whether we inhabit an unusual corner of the universe or not. It simply depends on the difference between a single, easily realized initial condition, namely that radiation is emitted isotropically from one source, and an indefinitely large number of initial conditions concerning the emission of radiation, with the necessary temporal and directional coherence and equality of wavelength, from a multitude of sources scattered *throughout space*. The artificiality and complexity of the second case as compared with the simplicity of the first means that its occurrence is virtually impossible, and therefore Popper is justified in regarding isotropic radiation and other wave-like processes from a point as indicators of a preferred direction of time.

A remarkably simple but fascinating example of irreversibility was described by E. A. Milne in 1932. He noted that any swarm of non-colliding particles moving uniformly in straight lines, if contained in a finite volume at some particular initial instant, will eventually, that is at some finite later time, be an expanding system, even if it were originally a contracting one. On the other hand, an expanding system of uniformly moving particles will never of its own accord become a contracting one. Although Milne considered a swarm of particles (with a cosmological analogy in mind), it is sufficient for our purposes to consider only two particles. If initially they are approaching each other, then eventually they will be found to be moving apart, but if initially they are moving apart, they will continue to move apart and will never approach each other. Thus, the simplest possible kinematic situation automatically reveals the irreversibility of time.

This conclusion has, however, been rejected by Gold (1958, 1962), who argues that, if the particles can be thought of as becoming indefinitely far removed from each other, they can also be thought of as starting indefinitely far apart. This criticism is irrelevant, since the particles can be thought of as initially at finite distances apart when they are set in motion. The essential point is that, if the particles all move uniformly in straight lines and continue to do so throughout, approach precedes recession but recession never precedes approach. Nevertheless, it is implicitly assumed that the spatial extension of the universe is either infinite or very large compared with the distances apart of the particles at all times considered, so that we do not see the particles converging after making a complete circuit of the universe. Since the argument depends on this assumption, it is to that extent incomplete.

Another common physical phenomenon which indicates asymmetry as between past and future is that of impact. For, although a perfectly elastic collision can be regarded as temporally reversible, we cannot regard inelastic collisions in the same way, particularly collisions which destroy relative motion, for example the impact of a falling stone with the ground. Reversal of time in this case would produce a wholly mysterious phenomenon in which a previously inert stone would suddenly begin to rise spontaneously at great speed. Unlike the phenomenon of impact which is immediately intelligible irrespective of the cause, if any, of the stone's original motion, the reverse phenomenon would be inexplicable, for, even if we introduced the concept of a repulsive force, we would still be baffled by the fact that the stone begins to move at one instant rather than another.

Moreover, within the field of optical and electromagnetic phenomena there are definite indications of temporal asymmetry to which Lewis does not refer. For example, in his account of Einstein's pioneer theory of the emission and absorption of light by molecules, E. T. Whittaker (1953) has explicitly drawn attention to the fact that 'as there is spontaneous emission, but not spontaneous absorption, there is asymmetry as between past and future'. Moreover, as R. H. Stuewer (1976) has pointed out, Lewis's deep-rooted conviction that time is symmetrical led him at first to reject and later to mistrust Einstein's concept of stimulated emission, which eventually proved to be fundamental for the invention of the laser.

Lewis's discussion also completely ignored the observer and his conditions of perception. Thus, he paid no attention to the fact that we can only perceive incoming light but not outgoing light. Consequently, if time were reversed and the stars attracted light from us instead of emitting light to us, they would be invisible. The part of the universe which we see must have its past-future relations, at least in so far as the emission of light is concerned, concordant with our own.

Norbert Wiener (1961) has analysed the hypothetical situation in which we are confronted with another being whose time runs in the sense opposite to ours. Communication with such a being would be impossible. Any signal which he might send us would have as consequences from his point of view events which were its antecedents from ours. These antecedents would seem to us to be the natural explanation of his signal, without presupposing an intelligent being to have sent it. If he drew us a square, we should not see the square being drawn, but instead we should observe its gradual disappearance line by line. This other being would have exactly similar ideas concerning us, and so Wiener concludes that within any world with which we can communicate the direction of time must be uniform.

1.4. Evolution

Another outstanding omission from Lewis's discussion of time is the absence of any mention of processes associated with 'long' intervals of time, i.e. of processes extending over many millions of years. These are the very processes which, when scientists first came to study them in detail, caused men generally to question their age-old belief that the general state of the world remains more or less invariable. Astronomy and palaeontology were the sciences primarily concerned, but the line of thought is comparatively recent. Indeed, until the philosopher Immanuel Kant began to speculate on the evolution of the Milky Way some two hundred years ago, astronomy appeared to be the science *par excellence* of symmetrical time. Similarly, until the nineteenth century the concept of biological evolution made little impact on man's way of thinking about the world.

The idea of the irreversibility of organic evolution has been called Dollo's law after the Belgian palaeontologist Louis Dollo (1857–1931) who drew attention to the evidence for it in the fossil record (Dollo 1893) nearly a century after Girard Soulavie had come to the conclusion that the stratigraphical ordering of the rocks (in his case, the Tertiary rocks of the Paris basin) can be regarded as a chronological ordering. Direction in evolution is, however, a more subtle concept than appears at first sight. Micromutations are believed by geneticists to be the starting points for biological evolutionary changes. The inherited information that is copied from one generation to the next is occasionally erroneously transmitted, and since the copying process is blind the error is perpetuated and the heriditary material is changed. Laboratory studies have shown, however, that many mutations are reversible, often with a frequency comparable with that of the original mutation and are therefore 'undirected'. For example, in *Drosophila* most mutations can be reversed and the reverse mutation appears to reconstitute the original gene. Moreover, where there is a difference in mutation frequency in the two directions the greater frequency is usually in the direction opposite to that which we believe must have been the past evolutionary one (Muller 1939). Does this imply that, in principle, evolution is reversible? On the contrary, the reversibility of mutation means that mutation alone cannot be responsible for sustained evolutionary trends. The process of mutation gives rise to the inherited variations that make evolution possible but it is not evolution itself. Moreover, most mutations that have been studied by geneticists appear to be unfavourable and many are actually lethal. It would therefore appear that evolution must have occurred 'in the face of a blizzard of predominantly unfavourable mutations' (Fisher 1932). Also it has been shown that for mutations to dominate the trend of evolution it would be necessary to postulate mutation rates far greater than those that are known to occur.

According to the now prevalent neo-Darwinian view, the irreversibility of the phylogenetic process is due to the interactions of organisms and their environments. This functional relationship is called *adaptation.* Although it is difficult to give a precise definition of this technical term, it essentially denotes any aspect of the organism that promotes its welfare, or the general welfare of the species to which it belongs, in the environment it usually inhabits. 'Welfare' means the organism's success in obtaining food, avoiding predators, and generally surviving and satisfying its whole range of biological needs. The welfare of the species includes that of its members and also the maintenance and increase of the population. Adaptations are, in short, the goal-directed features of living things. The mechanism of adaptation is natural selection. This is regarded as an automatic or self-produced process by which those organisms survive that are able to reproduce efficiently in the environment they inhabit.

One of the more serious difficulties that has been cited from time to time against the hypothesis that natural selection is the essential driving force of biological evolution is its failure to account for alleged examples of non-advantageous development, notably in the case of plants. Compared with animals they are passive organisms and might be expected to show comparatively little evolutionary development. On the contrary, flowering plants, the most recent and highly evolved, show a vastly greater number of species than mammals. Darwin himself was aware of the problem when he wrote to Hooker in 1879 that 'The rapid development, as far as we can judge, of all the higher plants within recent geological times is an abominable mystery' (Darwin 1903). Indeed, in plants the principal differences (for example, of calyx or flower arrangement) *appear* to have no advantage in the struggle for existence. J. C. Willis has drawn attention to the remarkable multiplicity of form in the family of water plants known as Postodemaceae (about 40 genera and 160 species) which grow in conditions of remarkable uniformity on smooth water-worn rocks. 'It would seem', he wrote, 'as if, in cases like this, if perhaps not in most, *evolution must go on, whether there be any adaptational reason for it or not*' (Willis 1940). He therefore argued that natural selection—which he suggested might more appropriately be called 'natural elimination'— is not the driving force of evolution but a regulative force determining whcthcr a given form shall survive.

The modern concept of selection, however, is not primarily a process of elimination but rather of differential reproduction influenced by favourable genetic factors that were unknown to Darwin. Although it is still recognized that natural selection can eliminate unfavourable mutations, the modern theory does not insist that it is always effective nor that it can eliminate all unfavourable mutations immediately, but those that are favourable will tend to spread through the population increasingly in successive generations.

As regards the possibility of non-Darwinian evolution, there has been much controversy in recent years concerning the role of what is called 'genetic drift' which is due to chance mutations that are neutral from the point of view of natural selection. Thoday (1975) has argued that, although in general the establishment of unconditionally neutral alleles is mere evolutionary 'noise', this non-Darwinian form of evolution may in future combinations or conditions cease to be neutral and yield an additional form of progressive evolution.

Other forms of evolutionary mechanism to which attention has been directed are (i) trial-and-error learning and (ii) differentiation in cellular development, leading to highly specialized types which cannot revert to their original undifferentiated form (Simpson, Pittendrigh, and Tiffany 1957). Both processes are either self-correcting or self-reinforcing and become progressively more difficult to reverse. The principal difference between natural selection and learning concerns their respective time scales; the influence of natural selection on the evolution of organic forms is normally effective only after millions of years, whereas the influence of learning on patterns of behaviour can be extremely rapid. With man, trial-and-error learning, culminating in modern scientific method, has become the decisive factor controlling social development. The fact that this is happening is in accord with the general trend of past biological evolutionary 'progress', characterized by increasing control by the organism of its environment and increasing independence of environmental charge—for example, homeothermy in birds and mammals.

The general conclusion of paleontologists is that the phylogenetic process has definite features of irreversibility, for, although the fossil remains of both vertebrates and invertebrates indicate that structures or functions once gained can be lost, structures that are lost can seldom be regained (Needham 1938). One of the best known examples is the pseudo-dentition of *Odontopteryx*, an eocene bird. Instead of regaining its lost teeth, its beak and lower jaw were serrated in saw-like form. Although some small-scale reversals have been reported (Kurtèn 1963), generally speaking evolution is the consequence of many variations occurring in a definite order and for it to be reversible there would have to be a highly improbable recurrence of specific variations acting in an inverse sense to those that brought about the original transformations. Moreover, evolution necessarily occurs on the basis of what has already happened, that is to say of previous evolution. If there is an evolutionary sequence of organisms A, B and C, in the sense that B evolved from A and then C from B, the fact that C differs from A makes it unlikely that B could at some later stage result from it. Consequently, it would appear highly probable that the overall evolutionary process is essentially irreversible and that 'the Ammonites, the Dinosaur and the Lepidodendron are gone beyond recall' (Blum 1962, p. 201).

Nevertheless, the question has been raised by J. Maynard Smith (1972) whether there is any *law* which plays a similar role in biology to the Second Law of Thermodynamics in physics, thereby assigning a definite arrow of time to evolutionary processes. In other words, if we could make observations on the members of a species at two epochs millions of years apart, is there a criterion that would enable us to decide which epoch was the earlier? Maynard Smith points out that, although evolutionary change may have tended to be in the direction of increasing complexity, not all species become more complex. Instead, he concludes that it is the property of multiplication which leads us to assign a direction to biological processes, but he finds it difficult to decide whether evolution as a whole has a direction.

The criterion of multiplication can, however, be applied generally, since living organization undoubtedly tends to increase the amount of matter within itself—the *biomass*. The total biomass of fish in the sea probably exceeds that of any previous type of marine life. Similarly, the total number of birds in the world (order of one hundred thousand million) represents a biomass smaller than that of mammals (Young 1950). It has been argued by A. J. Lotka (1945) that the 'direction' of evolution is controlled by the basic principle that the collective efforts of living organisms tend to maximize both their energy intake from the sun and the loss of free energy by dissipating processes occurring within them and in decaying dead organisms. Thus, the total energy flux throughout the biomass tends to increase, birds and mammals turning over energy faster than lower classes of vertebrates. Irrespective of whether we are justified in regarding this as a universal *law* of biological evolution, the evidence for it supports the view that the evolutionary process as a whole has a definite temporal direction.

Further confirmation of this conclusion comes from current ideas concerning the origin of life. Most biologists believe that life was 'spontaneously' generated from inanimate matter at some time in the remote past and that this process has not been repeated for a very long time. Since the work of Pasteur, belief in the possibility of continuing spontaneous generation has been generally abandoned. Indeed, if spontaneous generation continued, the whole evolutionary process could always be repeating itself, subject to the effect of prevailing environmental conditions, and in these circumstances it might be difficult to obtain any clear evidence of evolution (Blum 1962, p. 173). The problem of the origin of life, however, cannot be properly investigated without taking into account the terrestrial conditions that made it possible.

Terrestrial life is believed to have originated when the earth's atmosphere was devoid of oxygen. Various lines of evidence indicate that the present atmosphere is not primordial but has evolved in the course of the earth's history. The primitive atmosphere is thought to have been generated by volcanic action. Since no free oxygen is released in this way, all the oxygen now in the earth's atmosphere is thought to be due to photosynthesis by living organisms. Indeed, if there had been any oxygen at the stage when the complicated molecules that characterize animate matter, particularly the nucleic acids and proteins, were first formed, they would have been broken down again since they are not stable in the presence of oxygen. Under existing conditions their formation out of inorganic substances would not be possible without the co-operation of living organisms. Originally, these molecules could only have arisen in an oxygen-free atmosphere (Berkner and Marshall 1964). The existence of living organisms is therefore evidence that irreversible evolutionary change is not confined to them alone but also characterizes their environment, particularly the earth's atmosphere.

In contrast to the one-way process of biological evolution, the surface history of the earth seems, at first sight, to be cyclic. Nevertheless, when studied on a sufficiently long time scale this also reveals evidence of trend. The upthrust of land masses, in particular mountain building, is now believed to be due to the collisions of huge slowly moving 'plates' giving rise to continental drift of the kind first suggested, in a general way, early this century by the German meteorologist Alfred Wegener. The driving force of this relentless surface motion of the Earth is now believed to be ocean-floor spreading caused by the upthrust of molten material from the earth's hot interior. Since heat is continually being radiated by the earth into outer space, a continual source of internal heat must be postulated to maintain the flow. This problem was studied in the latter part of the last century by Kelvin, who calculated that on the basis of the known rate of loss of terrestrial heat the earth's surface must have been molten some forty million years ago, and that therefore this must be the upper limit to the ages of the rocks. Palaeontologists and students of biological evolution objected strongly to Kelvin's result, and at the turn of the century an entirely unsuspected new source of terrestrial heat was discovered in the phenomenon of radioactivity. It was soon realized that radioactive elements are widely distributed in the earth's crust and that radioactive transformations generate heat. It is now known that these deposits are almost sufficient to maintain a steady balance of heat generation and loss to outer space. Consequently, the earth's surface could have remained approximately within the present temperature range for thousands of millions of years. Also the earth's interior could have been maintained at a relatively high temperature for an equally long period and can continue for even longer, Indeed, W. D. Urry has calculated that the Earth will not have lost its entire stock of atomic fuel until 150,000 million years have elapsed (Hawkes 1952). During the last 500 million years, since Cambrian times, there is no evidence of any marked diminution of crustal or volcanic activity, and this is in accord with calculations indicating that heat production during this period has not diminished by more than about 4%. Nevertheless, despite this immense lengthening of the period during which the pattern of the past surface history of the earth can be continually repeated, the fact remains that, on the basis of present knowledge, a general trend towards a steady state is inevitable, with all the continents finally submerged beneath a world-wide ocean.

When we come to consider the radiation of energy from the sun and stars, we are again confronted by one-way processes. It is true that we no longer accept the suggestion of Kelvin and Helmoltz that the sun maintains its enormous output by a process of steady contraction whereby gravitational energy is converted into electromagnetic energy, and are therefore no longer confronted with the conclusion that the sun can continue to radiate for only some twenty million years. We now believe that the release of nuclear energy is responsible for solar radiation. The sun's heat is thus maintained by the conversion of matter into radiation. The process can continue steadily for thousands of millions of years, but in the absence of any known compensating process it cannot continue indefinitely.

More generally, in the universe at large this process is repeated on a prodigious scale, so that localized sources are continually dissipating energy into the depths of space. The bearing of this phenomenon on the problem of the spatial extent of the universe was considered by Olbers† (1823) who speculated on the fact that the background brightness of the sky is finite. For the present discussion, however, the significant problem is the temporal history of the universe. The mere fact that stars and galaxies are visible to us seems to imply that they are not eternal and that they have an evolutionary history, unless there is some unknown process providing them with inexhaustible sources of energy. Thus, even if the dark material of the universe, whether diffuse or concentrated, could in principle exist for ever, it appears that the overall appearance of the universe must ultimately change—that its present 'bright' appearance must have had a beginning and will ultimately come to an end. The only escape from this conclusion is either to postulate the creation of new stars and energy-radiating sources or to assume, as previously suggested, that the stars are inexhaustible sources.

Expert opinion is now unanimous in regarding thermonuclear processes as the origin of stellar energy. Consequently, the most highly luminous stars are thought to be comparatively short lived. Assuming the validity of Eddington's law that luminosity is a function of mass, it follows that, since the rate of loss of mass in the thermonuclear process is very small, a star like the sun will tend to radiate energy at a constant rate. It is thought that it may have been doing so for the past four or five thousand million years. In marked contrast, if Rigel had been shining at the time when the coal seams were being laid down, some two hundred million years ago, then it could not be shining now. We are thus led to believe that it began to shine long after life appeared on the earth's surface. Indeed, there are some stars which appear to have been shining for less than a million years. If such stars began to radiate so recently, it is not unlikely that new stars are being formed in the Milky Way even now.

This consideration clearly has an important bearing on our problem of temporal change in the overall appearance of the stellar system. If new stars are continually being created, it might be thought that the heavens could continue indefinitely to preserve the same general aspect, as was maintained by Aristotle. Nevertheless, a difficulty remains. By what process are new stars created? The most plausible suggestion is that they are formed by the gravitational condensation of diffuse matter. This suggestion finds some support from the fact that in the extragalactic nebulae we find that the regions in which large quantities of diffuse dark matter occur are also the regions in which the comparatively short-lived

† Olbers' Paradox can be traced back to Halley a hundred years earlier, and also to J.-P. Loys de Chéseaux, in 1744 (Jaki 1969).

high-luminosity stars abound. For the indefinite continuance of the process an inexhaustible supply of diffuse matter is essential. The most generally acceptable mechanism by which such matter could be generated is through the violent explosion of novae and supernovae. Nevertheless, unless the entire star disintegrates this mechanism will not supply an unending source of interstellar matter and the cycle cannot continue indefinitely. Thus it seems that our stellar system, the Milky Way, must like its component stars, have an evolutionary history too.

The Milky Way is, however, only one stellar system among myriads and so we must next consider the grand system of all such stellar systems, the system of galaxies. The constituents of this system appear to have their individual evolutionary trends, but what of the system as a whole? Are new galaxies still in process of creation, as we believe new stars to be? The most plausible mechanism of generation is the condensation of diffuse matter which still appears to exist in between the galaxies belonging to some large clusters. This, however, would appear to be a wasting asset. If so, then the system of galaxies must change and thus display a definite evolutionary trend.

Both on the terrestrial scale and on the celestial scale there is abundant evidence of temporal trend *in* the universe when sufficiently long intervals of time are considered. Nevertheless, this evidence alone does not compel us to believe that there must be a temporal trend *of* the universe. For, even if all the large-scale processes of nature are themselves irreversible, it has been claimed that the universe as a whole need not necessarily have an evolutionary history, either because its general appearance is always the same or because it progresses through an endless sequence of identical cycles. When, however, we come to consider this problem of unidirectional time in relation to the whole physical universe, and not merely to individual objects within the universe, we are confronted by deeper issues than those so far considered.

1.5. The cosmological origin of time

Cosmological problems play a peculiar role in modern physics. The scientific revolution which reached its climax in the seventeenth century is generally believed to have owed its success to the fact that natural philosophers like Galileo ceased to speculate about the world as a whole and confined their attention to definite limited problems in which specific objects and processes were regarded in isolation from their environment. Descartes criticized Galileo on this very score. While agreeing with him in his protests against the Scholastics and in his belief that problems of physics should be studied with the aid of mathematics, Descartes maintained that Galileo 'is continually wandering from the point and does not explain any matter thoroughly; which goes to show that ... without having considered the first causes of nature, he has merely looked for the causes of certain particular facts, building thus without any foundation' (Beck 1952). The great disadvantage of the Cartesian attitude to physical research is that it is based on the principle that before we can know anything we must, in a general sense, know everything. The Galilean attitude, on the other hand, is based on the principle of isolation and piecemeal inquiry. Because it works by deliberately ignoring many factors, it can often afford to be ignorant of them. It was primarily for this reason that Newtonian physics ultimately superseded Cartesian. Newton knew no more than Descartes what the mechanism of gravitation was, but he succeeded in side-stepping the question, whereas Descartes could not.

The treatment of other basic ideas of classical physics provides further evidence of the success, and at the same time of the limitations, of this policy. Newtonian space is absolute, but the problem of its identification is skilfully evaded by the principle of Newtonian relativity. Thus, although the problem of space was realized to be a cosmological one, a specific technique was devised for evading cosmological considerations. The analogous treatment of energy is particularly relevant for its remarkable similarity to that of time. The usefulness of the concept depends on the idea of potential energy. Classical physics cannot provide a complete rule for its measurement, but it evades this difficulty by concentrating on problems in which we only need to know differences of this energy. Similarly, in classical physics there is no complete rule for assigning epochs to events, but in practice this does not matter because only time differences are needed. Thus the origins or zero points of measurement, of both potential energy and time, are freely chosen by the investigator; in other words, they are purely conventional. These conventions can therefore be regarded as means of circumventing the classical physicist's ignorance of the natural zero level of potential energy and the natural origin of time. By putting his questions in such a way that knowledge bearing on these factors becomes irrelevant, the physicist takes a methodological short cut, but in so doing is in danger of falling into the philosophical fallacy of believing that the very factors which he has neglected are *ipso facto* non-existent. In fact, the method of isolation and convention both sharpens and narrows our questioning, thereby imposing limitations upon it.

Mathematically, the origin—if any—of time is projected to 'minus infinity', which means that in practice it is irrelevant and only time differences matter. This irrelevance of the origin of time is directly associated with the fact that the time variable does not appear explicitly in the mathematical formulation of the fundamental laws of physics. Indirectly, it is also associated with the fact that the laws of classical mechanics are reversible and do not distinguish between past and future. In classical mechanics there is no special epoch which can serve as a fundamental point of reference, with respect to which earlier and later can be distinguished. The Second Law of Thermodynamics suggests the possibility of a terminal limiting point in the future, but, as we have seen, the cosmological application of this law is a disputed hypothesis. The difficulty does not absolve us, however, from the duty of considering the problem of a natural origin of time.

In a famous lecture on cosmogony in 1871, Helmholtz (1881) argued that men of science were not only entitled to, but ought to, investigate whether 'on the supposition of an everlasting uniformity of natural laws, our conclusions from present circumstances as to the past . . . imperatively lead to an impossible state of things; that is, to the necessity of an infraction of natural laws, of a beginning which could not have been due to processes known to us'. As Helmholtz rightly emphasized, this question is no idle speculation for it concerns the extent to which existing laws are valid.

On this point there has been, and still is, considerable confusion. A natural origin of time is often assumed to be an epoch of universal creation. Such an epoch would certainly be an origin of physical time, but the idea of a temporal origin can, and in fact does, arise most readily in physics as a limit imposed on our extrapolation into the past of the laws of nature. Whether or not such a limit is to be regarded as an epoch of world creation is, strictly speaking, a metaphysical or theological question that lies outside science proper. We can divide the laws of physics into two groups according as their indefinite extrapolation into the past is, in principle, possible or impossible. Laws falling into the second group are all comparatively recent discoveries, for example the Rutherford–Soddy law of radioactive decay.

According to this law, the number of atoms of a given concentration of a naturally radioactive element, such as uranium-238, which will decay in a given short interval of time $\mathrm{d}t$ is proportional to the number N of atoms of this element present at the beginning of the interval, the constant of proportionality λ being independent of physical conditions such as temperature and pressure. Thus,

$$\mathrm{d}N/\mathrm{d}t = -\lambda N$$

where the reciprocal of λ is a time interval characterizing the particular element concerned. We find that, in fact,

$$1/\lambda = \alpha/\ln 2$$

where α is the half-period. The law does not isolate a particular origin of time, but it is immediately seen that it limits the past history of the material deposits to which it applies. For, if we were to attempt to extrapolate the law back into an infinite past, we should find that N itself would have then been infinite.

Strictly speaking, however, it is not the indefinite extrapolation into the past of the *law* of radioactive decay which is immediately called in question but the extrapolation of the *application* of this law to a given radioactive source. There must have been a temporal origin of the source, and other radioactive sources may have been in existence earlier. Nevertheless, there is an important difference between the law of radioactive decay and the law of gravitation, for the latter does not of itself impose any temporal limitation on its application to a given system of bodies.

The concept of a natural limitation on the extrapolation into the past of an actual physical law arises in connection with the hypothesis of the expansion of the universe. It has been found that the spectra of the extragalactic nebulae are shifted towards the red, the more distant galaxies exhibiting the greater shifts. The evidence is compatible with, and most naturally explained by, the hypothesis that these stellar systems are receding from the Milky Way. The distribution of these systems in the sky, when corrected for the effect of obscuring matter inside the Milky Way, is found to be roughly isotropic, and is generally regarded as convincing evidence that the grand system of extragalactic stellar systems forms the framework of the whole physical universe. Moreover, it is thought that these constituent systems are not only receding from the Milky Way, but also from each other. If these ideas are valid, then it would seem that the universe as a whole cannot be in a steady state, but must be expanding. Consequently, there would be an evolutionary history of the universe itself and not merely of objects within the universe; also, there might be a finite range of past time since the system began to expand from its most condensed state, and if so there would be a natural origin of time.

These conclusions do not, however, follow automatically from the hypothesis that the observed spectral shifts are due to the Doppler effect associated with motion in the line of sight, for we only know the spectral shifts of the galaxies as we observe them at the present day and it is possible that in the distant past a terrestrial observer would have observed different shifts, and as a result we might be able to reconcile the hypothesis of the recession of the galaxies with the possibility of an indefinitely long measure for the range of past time. Three different ways of coming to this conclusion have in fact been suggested: either the universe has always been an expanding system but an infinite time has elapsed since expansion began, or the universe alternately expands and contracts concertina fashion, this process having neither beginning nor end, or, although the galaxies continually recede from each other, the universe *as a whole* is in a 'steady state' and does not change with lapse of time.

The first alternative is generally regarded as leading to a fictitious eternity of past time, all actual physical events that have already occurred being confined to a finite stretch of past time. The infinity concerned is thus purely a feature of mathematical formalism and not the correlative of an infinite sequence of actual happenings.

The second alternative could, in the opinion of some cosmologists, lead to a genuine infinity of past time and the consequent abolition of a natural origin, since time would consist of an endless sequence of similar cycles. However, in order to reconcile this idea with the apparently finite life histories of individual stars and galaxies, it must be assumed that in each cycle stars and galaxies are created anew from the material remains of the previous cycle. Although this hypothesis of endless cycles of expansion and contraction avoids the difficult concept of world creation, it must be regarded as sheer speculation since we do not have, and probably never can have, any empirical evidence for it.†

The third alternative has been exemplified by what was for some years one of the most widely accepted cosmological theories of modern times. The first attempts to discard the idea that the systematic correlation of extragalactic red shifts with distance necessarily implies that the universe of mutually receding galaxies as a whole is expanding, and therefore has an evolutionary history, were made by abandoning the assumption that the spectral displacements concerned are Doppler shifts. However, no alternative explanation of these displacements proved generally acceptable to astronomers and cosmologists, and so in 1948 a new hypothesis was devised with the object of reconciling the modern idea of the mutual recession of the galaxies with the traditional assumption that the universe *as a whole* does not change with time. This association of ideas that had previously been regarded as incompatible was based on the hypothesis that, as old galaxies recede from one another, new ones are continually being formed to fill the ever-increasing gaps that would otherwise result. For such a process to continue without intermission so that the universe is truly eternal, its general appearance not changing with time, it was

† Moreover, in such a concertina-type universe there is a difficulty concerning the continual increase, from cycle to cycle, of world entropy (if this concept is allowable) and hence its continual decrease as we extrapolate backwards (see p. 343; also see the discussion of Kant's first antinomy on pp. 27–32).

postulated† that new matter, in the form of atoms of hydrogen, is continually being created *ex nihilo* at an invariable rate uniformly throughout space. In due course these atoms come together under their mutual gravitational attraction to form new stars and galaxies, each with a definite life history, although the system as a whole is changeless and therefore eternal. This meant that on the grand scale time was effectively 'eliminated'.

For more than a decade and a half this concept of a steady-state universe had a strong appeal for many students of cosmology, but since 1965 it has been generally abandoned following the discovery of the cosmic background radiation (see p. 306) and other empirical evidence relating to radio galaxies and quasars that cannot be reconciled with it. Instead, the belief is now widespread that the universe has an evolutionary history of its own, with an origin at a finite epoch in the past. It is, therefore, now generally recognized that on the grand scale time *cannot* be 'eliminated'.

1.6. Time and the universe

Despite the many attempts to dissociate the concept of time from that of the universe, it has long been assumed that these two concepts stand in a peculiarly intimate relation to each other, irrespective of whether or not there is a unique natural origin of time. As C. D. Broad (1921, p. 334) has remarked, 'It is commonly believed that when the analysis is made into moments and momentary events, all the events in the history of the world fall into their places in a single series of moments'. In other words, it has been commonly assumed that physical time is essentially unidimensional and that there is a unique time sequence associated with the world as a whole. The former assumption arises from man's psychological awareness of a definite before-and-after sequence of events in his own immediate conscious experience. The latter is an immense extrapolation from that experience to the world as a whole.

The intimate association of the universe and time was discussed by Plato in the *Timaeus.* In Plato's cosmology the universe was fashioned by a divine artificer, the Demiurge, imposing form and order on primeval matter and space, which were regarded as originally in a state of chaos. The Demiurge was, in effect, the principle of reason which imposed order on chaos reducing it to the rule of law. The pattern of law was provided by ideal geometrical shapes. These were eternal and in a perfect state of absolute rest. 'But to bestow this attribute altogether upon a created thing was impossible, so he bethought him to make a moving image of eternity,

† See for example Bondi (1952).

and while he was ordering the universe he made of eternity that abides in unity an eternal image moving according to number, even that which we have named time' (Archer-Hind 1888). According to this view, time and the universe are inseparable. Time, unlike space, is not regarded as a pre-existing framework into which the universe is fitted, but is itself produced by the universe, being an essential feature of its rational structure. Unlike the ideal model ('eternity') on which it is based, the universe is subject to change. Time, however, is that aspect of change which bridges the gap between the universe and its model, for being governed by a *regular* numerical sequence it is a 'moving image of eternity'. This moving image manifests itself in the motions of the heavenly bodies. It came into existence simultaneously with the construction of the heavens, and if the heavens should ever be dissolved then time would pass away too. Thus, a further analogy of the moving image to the eternal is that the created heavens have been, are, and will be in all time. As C. D. Broad (1921, p. 343) has pointed out, this view is similar to that held by Spinoza who believed that things as they really are for 'reason' are timeless, but this timelessness cannot be grasped by 'imagination' which represents it confusedly as duration through endless time.

Whereas Plato's analysis of time was based on the hypothesis that time and the universe are inseparable, Aristotle, in his analysis, did not begin with a broad world view such as we find in the *Timaeus*. Not only did Aristotle regard the Platonic identification of time with the uniform revolution of the universe as unsatisfactory, but he maintained that time must not be identified with motion in general, for motion (which for him meant not merely locomotion, but physical change of any kind) can be 'faster' or 'slower', or indeed uniform or non-uniform, and these terms are themselves defined by time, whereas time cannot be defined by itself. Nevertheless, although time is not *identical* with motion, it seemed to him to be *dependent* on motion—to be something pertaining to motion, for 'we apprehend time only when we have marked motion, marking it by "before" and "after"; and it is only when we have perceived "before" and "after" in motion that we say that time has elapsed' (Hardie and Gaye 1930). What then is the precise relation between time and motion? Aristotle thought that time is a kind of number—the numerable aspect of motion. In justification of this view he argued that 'we discriminate the more or less by number, but more or less movement by time'.† Thus, in his view, time is a numbering process founded on our perception of 'before' and 'after' in motion: 'Time is the number of motion with respect

† The Greeks did not have the modern concept of velocity. For them, rate of motion signified the time to cover a given distance.

to earlier and later'. In other words it is that aspect of motion which makes possible the enumeration of successive states.

Although Aristotle was more careful than his predecessors in distinguishing between time and motion, he maintained that the relation between the two was reciprocal. 'Not only do we measure the movement by the time, but also the time by the movement, because they define each other. The time marks the movement, since it is number; and the movement the time'. An obvious difficulty of this point of view is that motion can lapse or cease, but time cannot. Aristotle attempted to face this difficulty by arguing that time is also the measure of rest—indirectly—since rest is the privation of motion.

The outstanding example of motion that continues unceasingly is that of the heavens, and despite the fact that Aristotle did not base his discussion explicitly on cosmological considerations, he was profoundly influenced by the cosmological view of time. In particular, he appears to have been swayed by the definition formulated by the Pythagorean Archytas of Tarentum, a contemporary of Plato, who said that time is the number of a certain movement and is the interval appropriate to the nature of the universe. The Pythagoreans believed in eternal recurrence and the interval to which Archytas referred was probably the 'Great Year', that is the interval after which all celestial phenomena were thought to repeat themselves. Consequently, although Aristotle began by specifically rejecting any close association between time and a particular motion in favour of one between time and motion in general, he too ultimately came to the conclusion that there was a peculiarly close correlation between time and the circular motion of the heavens, which was for him the perfect example of uniform motion. Rectilinear motion could not be 'continuous', that is continually uniform, unless it was motion in an *infinite* straight line, and Aristotle did not believe in the possibility of such a line. The primary form of motion was therefore motion in a circle, for this alone could continue uniformly and eternally, and time must be primarily the measure of such motion. For Aristotle, time was therefore a circle,† at least in so far as it was measured by 'the

† Throughout Greek thought (and likewise in other ancient cosmologies, for example Hindu, Maya, etc.) time was regarded as essentially periodic, because the universe was thought to be cyclic. F. M. Cornford (1937) points out that the origin of the circular image of time 'is borrowed from the revolving year—*annus, anulus,* the ring'. He also draws attention to a remark of Proclus who said explicitly that time is not like a single straight line of unlimited extent in both directions, but is limited and circumscribed. This may be contrasted with Locke's statement that 'duration is but as it were the length of one straight line extended *in infinitum*'(Locke 1690). Proclus refers to the Great Year as recurring again and again. 'It is in that way that Time is unlimited'. For, 'The motion of Time joins the end to the beginning and this an infinite number of times'. Consequently, the idea of a cyclic universe did not imply a truly cyclic view of time, as discussed in §1.9, but only the

circular movement', by which he meant the circular motion of the heavens. Thus, Aristotle's conception of time was ultimately no less cosmological than Plato's. Time, in his view, was not the numerable aspect of any particular kind of motion, for 'there is the same time everywhere at once'.

The idea of world-wide time was presupposed by Kant in his celebrated discussion of time in the formulation of the first of the four antinomies of pure reason. Indeed, Kant came to the central problem of his *Kritik der reinen Vernunft* by considering whether the universe could have had an origin in time or not.† He believed that there were indisputable arguments for rejecting both alternatives, and he therefore concluded that our idea of time is inapplicable to the universe itself but is merely a part of our mental apparatus for imagining or visualizing the world. It is essential to our experience of things in the world, but we get into trouble if we apply it to anything which transcends all possible experience, in particular to the universe as a whole.

I shall dispute Kant's conclusion, because I do not believe that his antinomy exhausts all possibilities of associating time and the universe. First, let us consider his proof that the world cannot have existed for an infinite time. If we assume that the world has no beginning in time, then up to every given moment an eternity has elapsed and there has occurred in the world an infinite series of successive states of things. 'Now', argued Kant, 'the infinity of a series consists in the fact that it can never be

---

periodic repetition of the various states of the universe. Thus, according to S. Sambursky (1959) the Stoics, who regarded the universe as a dynamic continuum, understood by the cosmic cycle that 'the cosmos, although subject to continual metabolism, never dies and that its immortality is only another expression of the infinite extension of time, of the never-ceasing succession of events'. Similarly, the ancient atomists, notably the Epicureans, who believed that worlds composed of the same indestructible elementary particles were continually being destroyed and re-created, also seem to have regarded time in much the same way. The Stoics even went so far as to adopt a theory of eternal recurrence that involved repetition of *everything that happens* over and over again to all eternity. (For an authoritative account of the 'myth of the eternal return', see Eliade (1959).) The rise of Christianity, with its central doctrine of the Crucifixion as a *unique* event in time, was the cardinal factor causing men to think of time as linear progression rather than cyclic repetition. The first philosophical theory of time inspired by the Christian revelation was that of St. Augustine who rejected the traditional concept of a cyclic universe and instead maintained that time is the measure by human consciousness of the irreversible and unrepeatable 'rectilinear' movement of history (*Confessions*, Book XI). It is significant that the most famous proponent in modern times of eternal recurrence was the famous anti-Christian philosopher Nietzsche.

† Kant was essentially concerned with the question of the finitude or otherwise of the universe in time and not of time itself. He assumed that different times are parts, or delimitations, of a single underlying time and that this single underlying time is unlimited. For a recent discussion of Kant's theory of time see Al-Azm (1967).

completed by successive synthesis. It thus follows that it is impossible for an infinite world-series to have passed away, and that a beginning of the world is therefore a necessary condition of the world's existence' (Kemp Smith 1934). This argument as stated by Kant is imprecise by modern standards. Kant did not clearly distinguish between the question of whether there was, or was not, a first event in the world's history and the question of whether the total duration of the world's past time is finite or infinite. Kant's discussion of the hypothesis that the world had no beginning in time is formulated as an argument against the idea that the past sequence of successive states of things is, in the language of modern mathematics, an open set with no first member. Whether or not an infinite measure is to be associated with it depends on the choice of temporal unit. Kant's idea of successive 'states of things' presupposes that there is a unique time sequence for the whole universe, but his argument applies to any set of discrete events in temporal sequence, e.g. successive oscillations of a fundamental natural process.†

It is widely assumed nowadays that Kant's argument can be readily disposed of by appealing to the modern theory of infinite sets and sequences. In my view, Kant's argument cannot be automatically rejected in this way, since all reference to temporal concepts has been purged from the modern theory of infinite sets and sequences which are not thought of as being produced in time. Kant's argument, however, concerns successive actions or events occurring in time. It does not conflict with the idea that infinite sets and sequences are legitimate objects of thought, but it rejects the possibility of the actual occurrence of an infinite sequence of past events.

This problem has had a long history‡ and opinion concerning the possibility of an infinite past of the universe has been divided. Thus, influenced by Aristotle as well as by Christian theology, St. Thomas Aquinas (1225–1274) maintained 'That the world has not always existed is to be held by faith alone and cannot be demonstratively proved' (Aquinas). In his opusculum *De Aeternitate contra Murmurantes*, in which he criticized those whom he designated as 'murmurers against Aristotle', he argued at some length that it was logically possible for the universe to have been created by God out of nothing and yet to have existed from all eternity as Aristotle believed. The medieval followers of St. Augustine rejected this view. Prominent among those who argued against it was St.

† Kant's question must be distinguished from the purely mathematical analysis of time into an infinity of durationless instants which is discussed in Chapter 4.

‡ It can be traced back at least as far as Joannes Philoponos of Alexandria in the sixth century A.D. For a scholarly historical account see Craig (1979).

Thomas's contemporary St. Bonaventure (1221–1274). Of the various arguments that he formulated against the idea of the eternity of the world the most compelling is that the infinite cannot be bridged: if the universe has no beginning an infinite number of celestial revolutions must have taken place and therefore the present day could not have been reached (Gilson 1938).

The possibility of an infinite past was also rejected by the redoubtable Richard Bentley in the sixth of his Boyle Sermons in 1692. He argued that the world cannot be eternal and there cannot have been an infinite number of past revolutions of a planet about the sun. 'For, consider the *present* revolution of the Earth . . . God almighty, if he so pleaseth, may continue this motion to perpetuity in infinite revolutions to come; because futurity is inexhaustible, and can never be spent or run out by *past and present* moments. But then, if we look backwards from this present revolution, we may apprehend the impossibility of infinite revolutions on that side; because all are already past, and so were once actually *present* and consequently are finite . . . For surely we cannot conceive a preteriteness (if I may say so) still backwards *in infinitum*, that never was present, as we can an endless futurity that never will be present. So that one is potentially infinite, yet nevertheless the other is actually finite' (Bentley 1838).

Bentley's point was that an infinite sequence of past events is not analogous to an infinite sequence of future events. To say that a sequence of future events is infinite means that each event in the sequence will have a successor. Following Bentley, we can describe such a sequence as 'potentially infinite'. Correspondingly, if a sequence of past events is infinite, each event in the sequence must have had a predecessor. If this is all that is meant by a sequence of past events being infinite, so that an infinite past sequence is exactly analogous to an infinite future sequence, then Bentley was wrong, and so was Kant when he argued that an elapsed infinitity of successive events is a self-contradictory concept. However, before we can accept this conclusion or reject it, we must consider the question further.

Bertrand Russell, for example, has criticized Kant on the grounds that classes, or sets, which are infinite in number are given all at once by the definining property of their members, so that there is no question of completeness or of 'successive synthesis'. Moreover, when Kant says that an infinite sequence can 'never' be completed by successive synthesis all that he has a right to say is that it cannot be completed in a finite time. "Thus what he really proves is at most, that if the world had no beginning, it must already have existed for an infinite time" (Russell, 1926).

Kant's argument and Russell's objections to it have been examined recently by Zwart (1976). He attacks Russell's contention that Kant only had the right to say that an infinite sequence (of events) cannot be completed in a finite time "for to say that something will not occur in a finite time is exactly the same as to say that it will never occur". Nevertheless, Zwart agrees with Russell that Kant's argument is fallacious, for he maintains that the proposition that an infinite number of past events in sequence could not possibly have occurred is only true if it is assumed in advance that there was a *first* event. Instead, Zwart concludes that the situation as between past and future events is completely symmetrical, the present being an event somewhere in a sequence of events that has neither beginning nor end.

Similarly, in a reply to my formulation of the case against the possibility of an infinite sequence of past events (Whitrow 1978), Popper has asserted against Kant that we can regard the past time, or elapsed time, and the future time, or impending time, as symmetrical with respect to infinity. Both may be regarded as infinite sequences of temporal units and therefore as actual infinities (Popper 1978). Popper believes that the opposite impression is due to viewing past time both as infinite and as having a beginning (in the infinitely distant past); as he rightly objects, these two hypotheses contradict each other.

There are, however, significant differences between past and future infinite sequences which should not be overlooked. For example, in the case of past events we can, at least in principle, be presented with records, or traces, of their occurrence. This is not the case with future events, for whereas any genuine trace that has been left to us of the occurrence of a particular past event was made when that event occurred, there can be no corresponding trace available now of any future event made at the time of its occurrence. An infinite sequence of past events could, in principle, have bequeathed to us an infinite number of traces, whereas no such embarrassing possibility can be associated with the concept of an infinite sequence of future events.

The crucial difference, however, between past and future sequences of events is that all the events in a past sequence, whether finite or infinite in number, have actually occurred and consequently an infinite sequence of past events cannot be potentially infinite but must be actually infinite. The difficulty in regarding any infinite sequence of future events as other than potentially infinite can be illustrated by what Bertrand Russell has called the 'Tristram Shandy paradox' (Russell 1937, p. 358). Tristram Shandy, in Sterne's famous novel, on finding that it took him two years to write an account of the first two days of his life lamented that material for his biography would thus accumulate faster than he could deal with it, so that he could never come to an end. "Now I maintain", argued Russell, "that if he lived forever, and not wearied of his task, then even if his life had continued as eventually as it began, no part of his biography would remain unwritten". This statement is perfectly correct so long as 'living forever' signifies living for a potentially infinite sequence of days, for since Tristram Shandy writes in a year the events of a day the events of the nth day will be written in the nth year, and since any assigned day will be the nth day it will eventually be written about. Nevertheless, Tristram Shandy too was right in lamenting that he would never finish his autobiography, for this would be true even if he lived forever, since as time goes on not only does he get no nearer his goal but, on the contrary, it continually recedes. Only if Tristram Shandy could actually live through an infinite number of days would the time arrive when all the days he had lived would have been written about. However, this could never happen because at no stage in the sequence of days, one after another, could the number that he had lived through cease to be finite.

If it were possible, in principle, to 'live through' an infinite sequence of future events, this sequence would have to comprise events, such as the completion of Tristram Shandy's autobiography, that would be separated from the present by an infinite number of intermediate events. This too can never happen because at no stage in the succession of events from the present can the number that have occurred, however large, become actually infinite.

A potentially infinite sequence of future events can be enumerated as 1, 2, 3, . . . and so on indefinitely. Similarly, it has been argued that an infinite sequence of past events can be associated with the sequence of negative integers ending with -1 and that this demolishes Kant's objection to the possibility of an infinite sequence of past events. However, we can only enumerate the events in such a sequence by counting backwards, that is by beginning with -1 instead of ending with it. This is the reverse of the way in which the events would actually occur and yields only a potentially infinite sequence.

Since *all* the events of a past sequence have already occurred, an infinite past sequence would *in this respect* be like an infinite future sequence of events that could, in principle, be 'lived through'. We have rejected the latter because it would imply the actual occurrence of events that are separated from the present by an infinite number of intermediate events, and this cannot happen. If an infinite sequence of past events could have occurred and therefore, in principle, could have been experienced, i.e. 'lived through', would not this too imply that events must have occurred in this sequence that are separated from the present by an infinite number of intermediate events? If, however, this were so, then how could the present event in the sequence have been attained? It should be noted that in formulating these questions we have not assumed the existence of a *first* event.

I do not believe that these questions can be answered merely by asserting that an infinite sequence of past events is an open set with no first member just as a potentially infinite future sequence is an open set with no last member and that any event in either sequence is separated from the present by only a finite number of intermediate events. If it is accepted that the number of future events that occur in sequence following the present can never become actually infinite because at each stage the number that will have occurred must remain finite, how is it possible for a corresponding sequence of past events culminating in the present to be infinite? The conclusion that this is not possible and that any sequence of discrete past events must necessarily be finite has not, however, been generally accepted in recent years.† Nevertheless, I am not satisfied that Kant's thesis has been disproved.

Let us now turn to Kant's counter-argument that the world cannot have had a beginning. Kant's analysis is a valid proof that the world cannot have had a beginning *in* time. That is to say, it cannot have existed for only a finite part of an infinite past. For, he argued, if it had it must have been preceded by an empty time. No coming-to-be is possible, however, in a completely empty time, since no part of such a time can be distinguished from any other part and 'no part of such a time possesses as compared with any other, a distinguishing condition of existence rather than of non-existence; and this applies whether the thing is supposed to arise of itself or through some other cause' (Kemp Smith 1934). In other words, the moment before the world began would have contradictory properties: it would be like all other moments of empty time and also unlike them because of its immediate temporal relation to an event of the world.

Although Kant's second argument is a valid reason for rejecting the idea that the universe was created *in* time, *we are not compelled to accept his conclusion that the two arguments together imply that time does not pertain to the universe.* Instead, we are free to accept the answer previ-

† It may be argued that this conclusion cannot possibly be accepted because it would seem that an apparently empirical result has been obtained by a purely logical argument and this is impossible. This objection, however, cannot be sustained, because the argument presupposes that there are temporal successions in the universe and this is an empirical assumption. For another example of a synthetic statement about time that seems to be necessarily true *a priori*, see that due to Pap discussed in Section 1.9.

ously given by Plato, and also by St. Augustine,† that the world and time coexist. However puzzling it may seem, the concept of a first moment of time is not a self-contradictory concept, for it may be defined as the first event that happened—for example, the initial 'explosion' of an expanding universe out of a primeval point-like singularity of infinite density. There was no time before that.‡

Kant's argument against the possibility of there having been a first event depends essentially on the idea—which he was striving to demolish—that time is something which exists of its own accord. It is well known that his analysis resulted from his reflections on the natural philosophy of Newton, who believed not merely in the existence of universal time (involving world-wide simultaneity) but elevated this concept to the status of an entity existing in its own right independently of actual physical events.

1.7. Absolute time

'Absolute, true and mathematical time', wrote Newton, 'of itself, and from its own nature, flows equably without relation to anything external' (Cajori 1934, p. 6). This famous definition which appears at the beginning of the *Principia* has been one of the most criticized, and justly so, of all Newton's statements. It reifies time and ascribes to it the function of flowing. If time were something that flowed then it would itself consist of a series of events in time and this would be meaningless. Moreover, it is equally difficult to accept the statement that time flows 'equably' or uniformly, for this would seem to imply that there is something which controls the rate of flow of time so that it always goes at the same speed. However, if time can be considered in isolation 'without relation to anything external', what meaning can be attached to saying that its rate of flow is *not* uniform? If no meaning can be attached even to the possibility of non-uniform flow, then what significance can be attached to specifically stipulating that the flow is 'equable'?

† In a famous passage (*De Civitate Dei*, Book XI, Chapter VI) St. Augustine was led to ask, 'seeing therefore that God, whose eternity alters not, created the world and time, how can he be said to have created the world in time, unless you will say that there was something created before the world whose course time doth follow?' And he answered: 'verily the world was made with time and not in time, for that which is made in time is before some time and after some time'.

‡ The idea of time itself having either a beginning or an end is one that many people automatically reject. As has been pointed out, however, by von Wright (1968), if *everything* should eventually come to a complete standstill then time itself would come to an end. Refusal to admit this conclusion presupposes that there is, in principle, someone who can 'contemplate this dead world *in time*', and this, he argues, contradicts the idea of *everything* being at a complete standstill. A similar argument applies to refusal to admit the idea of time having a beginning.

Newton was not a philosopher in the modern professional sense of the term, and so it is perhaps not surprising that he gave no critical analysis of his definitions but generally contented himself with their practical use. What is surprising, however, is that his definition of absolute time has no practical use! In practice we can only observe events and use processes based on them for the measurement of time. The Newtonian theory of time assumes, however, that there exists a unique series of moments and that events are distinct from them but can occupy some of them. Thus temporal relations between events are complex relations formed by the relation of events to the moments of time which they occupy and the before-and-after relation subsisting between distinct moments of time.†

Why did Newton introduce this complicated metaphysical concept? Two reasons can be advanced: one physical and the other mathematical. Physically, Newton must have regarded the concept as the essential correlative of absolute space and absolute motion. It is well known that he had definite empirical evidence which he interpreted as a conclusive argument in favour of his belief in absolute motion. This evidence was dynamical. 'True motion is neither generated nor altered but by some force impressed upon the body moved, but relative motion may be generated and altered without any force impressed upon the body' (Cajori 1934, p. 10). The actual effects by which Newton believed that absolute motion could be distinguished from relative were the centrifugal forces associated with motion in a circle. 'For there are no such forces in a circular motion purely relative, but in a true and absolute circular motion they are greater or less according to the quantity of the motion. If a vessel, hung by a long cord, is so often turned about that the cord is strongly twisted, then filled with water and held at rest together with the water, thereupon by the sudden action of another force it is whirled about the contrary way, and while the cord is untwisting itself the vessel continues for some time in this motion, the surface of the water will at first be plain, as before the vessel began to move; but after that the vessel, by gradually communicating its motion to the water, will make it begin sensibly to revolve and recede by little and little from the middle, and ascend to the sides of the vessel, forming itself into a concave figure (as I have experienced); and the swifter the motion becomes, the higher will

† Unfortunately, since the advent of the theory of relativity it has become a common habit for the adjectives 'universal' and 'absolute' to be regarded as synonymous when applied to time. Strictly speaking, the former signifies 'world-wide', whereas the latter should be used only for the Newtonian concept that time is independent of events. In Newton's view, time is *both* universal and absolute. On the other hand, the modern concept of 'cosmic time' (see Chapter 6) is universal but *not* absolute. (Nevertheless, because of relativity, 'universal' cannot mean 'with respect to all possible frames of reference'.)

the water rise, till at last, performing its revolutions in the same times with the vessel, it becomes relatively at rest in it.'

This experimental evidence shows that after the pail begins to spin there is at first relative motion between the water and the pail which gradually diminishes as the water takes up the motion of the pail. Newton pointed out that when the *relative* motion was greatest it produced no effect on the surface of the water, but that as it diminished to zero and the rotational motion of the water increased the surface became more and more concave. He interpreted this as evidence that rotational motion is absolute. Consequently, it is not necessary to refer to any other body to attach a definite physical meaning to saying that a particular body rotates, and from this he argued that time, as well as space, must be absolute.

Mathematically, Newton seems to have found support for his belief in absolute time by the need, in principle, for an ideal rate-measurer. He pointed out that, although commonly considered equal, the natural days are in fact unequal. 'It may be', he wrote, 'that there is no such thing as an equable motion, whereby time may be accurately measured. All motions may be accelerated and retarded, but the flowing of absolute time is not liable to any change. The duration or perseverance of the existence of things remains the same, whether the motions are swift or slow, or none at all; and therefore this duration ought to be distinguished from what are only sensible measures thereof'. Newton regarded the moments of absolute time as forming a continuous sequence like that of the real numbers and believed that the rate at which these moments succeed each other is a variable which is independent of all particular events and processes.

An argument which has been used by Bertrand Russell (1937, p. 265) in favour of the absolute theory of time turns on the relation of time to position. When the time is given, the position of a material particle is uniquely determined, but when the position is given then there may be a number, indeed there may be an infinity, of corresponding moments. Thus, the relation of time to position is not one–one but may be many–one. From this consideration he claimed that the time sequence must form an independent variable existing in its own right, and that the correlation of events is made possible only through their prior correlation with moments of absolute time.

Despite Newton's advocacy and Russell's erstwhile support (which he later abandoned), the absolute theory of time has found little favour with philosophers. That moments of absolute time can exist in their own right is now generally regarded as an unnecessary hypothesis. Events are simultaneous not because they occupy the same moment of time, but simply because they happen together. As Gunn (1929) has so forcefully remarked, 'They correlate themselves because they coexist, and they have no need of an entity "moment of absolute time" to do this, but rather because they happen, we speak of a moment, and this moment is not a temporal entity existent in its own right, it is simply the class of co-existent events themselves. We derive time from events, not vice versa'. For the temporal correlation of events which do not coexist, it is sufficient to postulate that there is a linear sequence of states of the universe, each of which is the class of events simultaneous with a given event, and that these states have the simple before-and-after relation.

1.8 Relational time

The theory that events are more fundamental than moments—which do not exist in their own right, but are classes of events defined by the concept of simultaneity—is usually known as the relational (or relative) theory of time. It was formulated by Leibniz† who opposed it to Newton's absolute theory. Leibniz's theory was founded on his principles of sufficient reason, identity of indiscernibles, and pre-established harmony.

According to the first of these principles, nothing happens without there being a reason why it should be thus rather than otherwise. 'Truths of reasoning', Leibniz wrote, 'are necessary and their opposite is impossible; truths of fact are contingent and their opposite is possible.... But there must also be a sufficient reason for contingent truths or truths of fact' (Leibniz 1925). A particular form of this rather ill-defined general principle is that symmetry of causes must persist in the symmetry of effects. For example, as Leibniz himself pointed out in the second of his five letters to Clarke, the defender of Newton, 'Archimedes, wishing to proceed from mathematics to physics in his book *On Equilibrium* was compelled to make use of a particular case of the great principle of sufficient reason; he takes it for granted that if there is a balance in which everything is the same on both sides, and if, further, two equal weights be hung on the two ends of the balance, the whole will remain at rest. This is because there is no reason why one side should go down rather than the other' (Leibniz 1934, p. 194).

Leibniz applied this principle to time in a famous passage of his third letter. 'Suppose someone asks why God did not create everything a year sooner; and that the same person wants to infer from that that God did something for which He cannot possibly have had a reason why He did it

† Essentially the same view was held by Lucretius, *De rerum natura*, Book I, lines 455–65. In the translation by R. E. Latham (1951) the passage runs: 'Similarly, time by itself does not exist; but from things themselves there results a sense of what has already taken place, what is now going on and what is to ensue. It must not be claimed that anyone can sense time by itself apart from the movement of things or their restful immobility'.

thus rather than otherwise, we should reply that his inference would be true if time were something apart from temporal things, for it would be impossible that there should be reasons why things should have been applied to certain instants rather than to others, when their succession remained the same. But this itself proves that instants apart from things are nothing, and that they only consist in the successive order of things; and if this remains the same, the one of the two states (for instance that in which the Creation was imagined to have occurred a year earlier) would be nowise different and could not be distinguished from the other which now exists' (Leibniz 1934, p. 200).

According to the principle of the identity of indiscernibles, which Leibniz deduced from his principle of sufficient reason, it is impossible that there should exist things which differ *sole numero*, or only because they are two, and are otherwise completely similar. In his fourth letter to Clarke he wrote, 'To suppose two things indiscernible is to suppose the same thing under two names. Thus the hypothesis that the universe should have originally had another position in time and place from that which it actually had, and yet all the parts of the universe should have had the same position with regard to one another as that which they have in fact received, is an impossible fiction' (Leibniz 1934, p. 204).

Leibniz's monads† are mutually independent but, in order that they should form one universe, each mirroring the whole course of the universe from its own point of view, the famous principle of pre-established harmony stipulated that the states of all monads at every instant correspond with each other. Leibniz illustrated this principle by the simile of the two clocks which may be made to keep perfect time with one another in three different ways. They may be physically connected, as in Huygens's experiments in which two pendulums hung on a bar of wood were set swinging out of time with one another but ultimately swung in harmony as the result of the mutual transference of vibrations through the wood. Alternatively, they could be kept in time by the continual intervention of a skilled workman. Finally, they may have been so perfectly constructed that they keep time without either mutual influence or external assistance. The last possibility corresponds to the pre-established harmony.‡

Thus, in Leibniz's theory neither space nor time can exist in their own right independently of bodies, except as ideas in the mind of God. Space

† Leibniz's monads are atoms endowed, in varying degrees, with the power of perception.

‡ This principle survives in the current belief that all similarly constituted atoms in precisely similar external circumstances, e.g. the same gravitational fields, states of motion, etc. with respect to the observer, have identical physical properties including the frequencies and wavelengths of the photons associated with transitions between the corresponding states of different energy levels.

is the order of coexistences, and time is the order of succession of phenomena. This order is the same for all monads, for since each of the latter mirrors the whole universe they must necessarily keep pace with one another. Consequently, in so far as the temporal aspect of the universe is concerned, Leibniz's principle of harmony is equivalent to the postulate of universal time. He was very clear on the question of the temporal origin of the universe. 'It is a similar, that is to say an impossible, fiction to suppose that God had created the world several million years sooner. Those who incline towards such kinds of fiction will be unable to reply to those who are in favour of the eternity of the world. For since God does nothing without a reason, and since there is no reason assignable why He did not create the world sooner, it will follow either that He created nothing at all, or that he produced the world before any assignable time, which is to say that the world is eternal. But when we show that the beginning, whatever it was, is always the same thing, the question why it was not otherwise ceases to exist. If space and time were something absolute, that is to say if they were something other than certain orders of things, what I an saying would be a contradiction. But since this is not the case, the hypothesis is contradictory, that is to say it is an impossible fiction'. (Leibniz 1934, p. 206).

Leibniz does not seem to have formulated any detailed criticism of Newton's strongest argument for absolute time which was based, as we have seen, on his belief that rotational motion is absolute. The first attack on this interpretation of the rotating bucket experiment was made by Berkeley, whose whole philosophy was founded on the rejection of abstract general ideas and, in particular, on the rejection of absolute space and time as objective realities existing independently of our perception. In his essay *De Motu*, published in 1721, he showed that the crucial point in Newton's argument was his implicit assumption that the experiment would yield the same result if it were performed in empty space, whereas in fact the pail was at first rotating and then at rest relative to the earth. Its motion was only apparently, and not truly, circular because it is necessary to take into account the rotation of the earth about its axis, the revolution of the earth about the sun, and so on. He concluded that the phenomena cited by Newton merely indicated rotation relative to the other bodies of the universe and that it is not necessary to introduce the idea of absolute rotation. The same point was made by Mach, in the latter part of the nineteenth century, in his classic *Science of Mechanics*. He remarked that the only experimental test that could be imagined for disproving the idea that rotational motion is relative (with respect to the universe as a whole) would be to compare Newton's experiment as he performed it with one in which the bucket is left undisturbed and the universe is made to rotate around it. This test is impossible to carry out and so we are not compelled to accept Newton's interpretation of his experiment (Mach 1960). Consequently, his case for absolute time collapses.†

1.9. Cyclic time

We have previously remarked that, on the basis of the relational theory of time, we can correlate events which do not coexist if we postulate that there is a linear sequence of states of the universe, each of which is the class of events simultaneous with a given event, and that these states have the simple before-and-after relation. We must now consider the following objection to the relational definition of a moment as a particular state of the universe which was formulated by Russell (1901). He maintained that it is not *logically* absurd to imagine the separate occurrence of two apparently identical states of the universe. However, if we define a moment as a particular state of the universe, we should then be faced with the logical absurdity that two moments could be both different and identical.

Fortunately, this contradiction can be resolved without appealing to Newton's concept of absolute time, for, if a state of the universe is defined as the class of all simultaneous events, two states which are not simultaneous cannot be identical *in all respects.* However, this resolution of the difficulty involves explicit recognition of time as fundamental: the date becomes an essential characteristic of an event.‡ States of the universe will then, strictly speaking, be non-recurrent.

Russell's argument turned on the essential distinction between the ideas of a cyclic universe and of cyclic time. The former leads to the concept of periodic universal time (compare the concept of the Great Year discussed on p. 26), whereas the latter implies that time is closed like a ring. This concept of cyclic time has been rightly condemned by M. F. Cleugh (1937, p. 225). Of the assertion that the 'same' event can recur over and over again she writes, 'This is rubbish. The same *content* may or may not recur—that is a different question. This is implicit in the very word "recur".' It may perhaps be objected that when two states of the universe are exactly the same in every particular except date it is mere pedantry to insist on calling them 'two' and that they are really identical. However, as

† In asserting that time is relational we do not necessarily imply that it depends only on material events. It may depend on mental events as well.

‡ In a careful analysis of the relational theory of time by Schuster (1961), it is shown that time cannot be asymmetrical (if A precedes B then B cannot precede A) and transitive (if A precedes B and B precedes C then A precedes C) unless the temporal position of an event is one of its characteristics.

she points out, it is *not* nonsense to insist on the distinction between cycles of things and cycles of events.

In making this sharp distinction, it may seem that we are surreptitiously assuming that, after all, time is independent of things and exists in its own right, i.e. is absolute. We take our stand, however, with Miss Cleugh in maintaining that, even if we regard time as relational and hence pertaining to the universe, it is not nonsense to assert that the event of the universe passing a particular stage once would be a different event from the event of passing it again.† Indeed, we can go further and assert that, if time were truly cyclic, there would be no difference between the universe going through a single cycle of events and through a sequence of identical cycles, for any difference would necessarily imply that time is not cyclic. i.e. that there is a basic non-cyclic time to which the different cycles could be related and distinguished one from another. Moreover, the same argument would also apply to the initial and final events of a single cycle, for if they were identical there would be no sense in regarding them as occurring separately. In other words, if there is no basic non-cyclic time to refer to, we cannot distinguish a 'circular' sequence of states of the universe from a 'rectilinear' one.

The necessarily non-cyclic nature of time which is expressed in the statement 'No event precedes itself' has been cited by the philosopher Arthur Pap (1963) as an example of an *a priori* truth that is not strictly analytic. Pap argues that this statement cannot be a logical truth, since the verb 'to precede' is not in the vocabulary of logic but designates something we find in the external world, a temporal relation. In other words the statement has the form: For any *x*, *not-(xRx)*, and since some statements of that form are false (e.g. if we substitute for *R* the phrase 'is at the same place as'), the statement in question is not a formal, or logical, truth. It might yet be strictly analytic if we could analyze it into a statement that could be expressed in purely logical terms (e.g. as the statement that 'all uncles have siblings' can be revealed as analytic by defining an uncle as a male who has a sibling who is a parent). However,

† If, in accordance with Einstein's theory of relativity, a body moving through the universe has an individual time which differs from the universal time of the world as a whole, then we can, in certain circumstances, imagine the possibility of such a body describing a closed path in time. In this case the same *event* would recur. The relevant circumstances are discussed in Chapter VI (pp. 305–6) and rejected because, in principle, an observer travelling with the body could influence his own past at a later date. (Incidentally, the whole point of the fable about the man who was granted his wish to relive the past hour of his life—which included the expression and automatic granting of the wish itself—turns on endless *repetition*, and this presupposes that time goes on relentlessly, that is, that there is a difference between living the events of the hour once and reliving them again and again. In short, the same hour was *not* relived but the actions which filled it were simulated in all succeeding hours.)

as Pap points out, *precedence seems to be a simple relation that does not admit of further analysis*. If so, he argues, we have here a *synthetic* statement that is necessarily true *a priori*. It cannot be a logical truth since it is about an empirically given relation and therefore has empirical content, whereas logical truths do not contain descriptive terms essentially and in that sense are not 'about' the world of experience. Nevertheless, it is not an empirical statement, because the truth or falsehood of any such statement necessarily depends on facts of experience. Pap concludes by remarking that, although a being who had no experience of temporal succession could not possibly understand the meaning of the term 'to precede', it does not follow that the assertion we make about this non-logical relation when we say that it is irreflexive and asymmetrical (and, for that matter, transitive) is subject to the test of experience.

1.10. The scale of time

In formulating his much criticized definition of absolute time, Newton not only stated that 'all things are placed in time as to order of succession' but also that another name for it is 'duration'. 'Relative, apparent and common time', he pointed out, 'is a measure of duration by means of motion', although he admitted that it might well be that there is no 'equable motion, whereby time may be accurately measured' (Cajori 1934, p. 6). We thus see that Newton referred explicitly to both of the characteristic properties of physical time: its order and its rate. In his view they are distinct: the temporal order of events (the before-and-after sequence) does not of itself determine the duration of time between one event and another, nor the rate at which events succeed each other. Instead, these are determined by the respective moments of absolute time with which the events are correlated and the rate of 'flow' of this time.

On the other hand, by defining time as the order of succession of phenomena, Leibniz appears to have overlooked both its durational aspect and the associated problem of its continuity. Successive pictures on a roll of film may inform us of the temporal order of events in, say, the growth of a plant, but they tell us nothing about the rate at which the plant grows. Leibniz's definition refers, however, to successive states of the whole universe. From the practical point of view, the difference between his definition and Newton's may be summed up in the statement that according to Newton the universe *has* a clock, whereas according to Leibniz it *is* a clock. Thus, in Leibniz's view, the concept of the rate of growth of a plant would have significance only in relation to the whole universe, which is itself 'mirrored' in each monad.

So far, in discussing universal time, we have concentrated mainly on the question of its nature—whether absolute or relational—and on the question of whether it has a natural zero or origin. In considering the question of duration, however, we are now confronted with the further problems of determining a satisfactory unit of measurement and of constructing a significant scale of time. As regards these problems, Newton's definition gives us no more assistance than does that of Leibniz. Moreover, both of these great thinkers seem to evade rather than to take account of, let alone resolve, the fundamental antinomy: that, whereas the concept of spatial measurement does not conflict in any way with that of spatial order, despite the sharp distinction which geometers have learned to draw between the metrical and the topological, the concept of succession clashes with the concept of duration.

This clash of concepts has led to the formulation of paradoxes concerning time and its measurement which have puzzled many modern philosophers as well as the great thinkers of antiquity. The fleeting aspect of time gave rise long ago to the question of its reality. Thus, in his book *Against the Physicists*, Sextus Empiricus argued that the past no longer exists and the future does not yet exist, and therefore, at most, only the present can exist. The present, however, must be either indivisible or divisible. If it is indivisible it will neither have a beginning whereby it is joined on to the past nor an end whereby it is joined on to the future, for that which has a beginning and end is not indivisible. Moreover, since it has neither beginning nor end it will not have a middle, and having neither beginning nor middle nor end he argued that it will not exist at all. On the other hand, if present time is divisible it is divided into existent or non-existent times. That which is divided either into non-existent times will not itself exist, but if time is divided into existent times it will no longer as a whole be present.

This argument, which is similar to others discussed in Chapter 4, turns on the difficulties associated with the division of time into parts. In practice, the measurement of time has tended to depend as far as possible on spatial concepts. That this is an ancient device is revealed by etymology. Thus, in Greek and Latin, for example, we find that the words *τέμενος*, *tempus*, and *templum* all signified bisection or intersection; for carpenters two crossing beams constituted a *templum*. This division of space into quarters, for example the east, was reproduced in the division of the day, for example morning. Thus, despite the leading role played by temporal phenomena in the development of the idea of the universal cosmic order, the concept of spatial division became the basis of measurement. Consequently, the universal natural scale of time, of which the movements of the heavenly bodies were the clearest visible image, ultimately came to be represented *geometrically* as a one-dimensional locus. This geometrical line was implicitly assumed to be uniquely graduated, and in Newton's view it was independent of phenomena.

If, however, we adopt a purely relative measure of time in terms of a specific sequence of particular phenomena, we obtain a scale which may be adequate for the temporal ordering of all phenomena but not for the metrical comparison of different intervals of time. Indeed, one could imagine an endless variety of clocks of this type. Given three successive events A, B. and C, the temporal intervals between A and B and between B and C, respectively, might be judged of equal duration according to one such clock and of unequal duration according to another. Indeed, if one clock were represented mathematically by the variable t and another by the variable τ, the relationship between the two could be of the type

$$\tau = f(t)$$

where $f(t)$ denotes *any* monotonically increasing function of t. To arrive at a unique measure of duration, we require some universal criterion which will enable us to get rid of the arbitrary function f and replace it by a function with the property that to equal intervals of τ correspond equal intervals of t. Such a function is necessarily linear, i.e. of the form

$$\tau = at + b$$

where a and b are constants, and yields an effectively unique measure of duration, since the constant b is irrelevant for this purpose and the constant a depends solely on the interval which we choose as *numerical* unit, for example a second or a year. Moreover, the conversion factor from one such numerical unit to another does not change with lapse of time.

To obtain a universal criterion of this type, neither Newton's definition of time nor that given by Leibniz will suffice. Nor, in the long run, will it be sufficient to base our definition of time on the observed motions of the heavenly bodies. With modern refinements of observational technique, we know that the moon's revolutions are not strictly uniform but subject to a very small secular acceleration, that minute irregularities can be detected in the diurnal rotation of the earth, and so on. Greater accuracy in the measurement of time can be obtained by means of atomic and molecular clocks. Implicit in these developments is the hypothesis that all atoms of a given element behave in exactly the same way, irrespective of place and epoch. The ultimate scale of time is therefore a theoretical concomitant of our concept of universal laws of nature. This was recognized last century, long before the advent of modern ultra-precise timekeeping, in particular by Thomson and Tait in their famous treatise on *Natural Philosophy*. In discussing the law of inertia, they pointed out that it could be stated in the following form: the times during which any particular body not compelled by force to alter the speeds of its motions passes through equal spaces are equal. In this form, they said, the law expresses our convention for measuring time (Thomson and Tait 1890).

More generally, Poincaré argued that, in calculating the secular acceleration of the moon, for example, astronomers invoke the fundamental laws of Newtonian physics and consequently assume that time should be defined in such a way that these laws can be maintained.† Poincaré was puzzled by the fact that we have no direct intuition of the equality of two intervals of time, so that, although we can know that one event is anterior to another, we cannot assign the same precise meaning to saying by how much it is anterior, except by invoking some definition of duration which has a certain degree of arbitrariness. Therefore, he argued that, since different ways of defining time would lead to different 'languages' for describing the same experimental facts, time should be so defined that the fundamental laws of physics, in particular the equations of mechanics, are as 'simple' as possible. He concluded that 'there is not one way of measuring time more true than another; that which is generally adopted is only more *convenient.* Of two watches, we have no right to say that the one goes true, the other wrong; we can only say that it is advantageous to conform to the indications of the first' (Poincaré 1929).

Poincaré appears, however, to have overlooked the possibility that the customary 'simple' formulations of distinct fundamental physical laws may entail different scales of 'uniform time'. Thus, we have no *a priori* guarantee that the time scale implied, for example, by the usual formulation of the law of radioactive decay of uranium-238 is identical with that implied by the law of inertia, the law of gravitation, etc. To presuppose, as we usually do, that in the application of these different laws to the physical universe, the same universal scale of time is involved, is *not* a mere question of convention, for it depends upon the hypothesis, which may or may not be true, that there is a unique basic rhythm of the universe.‡

† The astronomer's point of view has been put very clearly by G. Clemence (1952): an 'invariable measure of time' is a measure that leads to no contradictions between the observations of celestial bodies and the rigorous theories of their motions. Clemence states explicitly that this measure of time is, in fact, defined by the accepted laws of motion. He also points out that any angle that is a known continuous function of time and that can be measured independently of distance is suitable for a measure of time. It is not even necessary that it should increase uniformly with time, but only that an adequate theory of its motion be available. (Incidentally, the pendulum is inadequate because we have no adequate theory of the disturbances—due to imperfections of suspension, gravity variation, etc.—to which it may be subjected.)

‡ The suggestion that some of the 'constants' of nature occurring in fundamental physical laws may vary over long periods of time, which is equivalent to modifying this hypothesis, is discussed in Chapter 7, pp. 361–4.

References

Al-Azm, S. J. (1967). *Kant's theory of time.* Philosophical Library, New York.

Aquinas, St. Thomas, *Summa theologica*, la, xlvi, 2.

Archer-Hind, R. D. (1888). *The Timaeus of Plato*, p. 199. Macmillan, London.

Arrhenius, S. (1910). *L'évolution des mondes* (trans. J. Segny), p. 206. Béranger, Paris.

Beck, L. (1952). *The method of Descartes*, p. 242. Clarendon Press, Oxford.

Bentley, R. (1838). *Sermons preached at Boyle's Lecture, etc.* (ed. A. Dyce), p. 134. Macpherson, London.

Berkner, L. V., and Marshall, L. C. (1964). The history of oxygen concentration in the earth's atmosphere. *Discuss. Faraday Soc.* **37,** 122.

Blum, H. F. (1962). *Time's arrow and evolution.* Princeton University Press, Princeton, N.J.

Bondi, H. (1952). *Cosmology*, Chapter XII. Cambridge University Press, Cambridge.

Broad, C. D. (1921). Time. In *Encyclopaedia of religion and ethics* (ed. J. Hastings), vol. 12. Clark, Edinburgh.

Cajori, F. (1934). *Newton's Principia* (trans. A. Motte). California University Press, Berkeley, Calif.

Clemence, G. (1952). *Am. Sci.* **40,** 267.

Cleugh, M. F. (1937). *Time,* Methuen, London.

Cornford, F. M. (1937). *Plato's cosmology*, p. 104. Routledge and Kegan Paul, London.

Craig, W. L. (1979). *The kalām cosmological argument.* Macmillan, London.

Darwin, C. (1903). *More letters* (ed. F. Darwin and J. C. Seward), vol. II, p. 20. John Murray, London.

Dollo, L. (1893). *Bull. Soc. Belge Géol. Pal. Hydr.* **7,** 164.

Einstein, A. (1954). *Relativity: the special and the general theory* (transl. R. W. Lawson), p. 141. Methuen, London.

Eliade, M. (1959). *Cosmos and history: the myth of the eternal return.* (transl. W. R. Trank). Harper, New York.

Fisher, R. A. (1932). *Sci. Prog.* **27,** 281.

Gilson, E. (1938). *The philosophy of St. Bonaventure* (transl. E. Threthowan and F. J. Sheed), p. 192. Sheed and Ward, London.

Gold, T. (1958). In *La structure et l'évolution de l'univers* (ed. R. Stoops), p. 95. Institut International de Physique Solvay, Brussels.

—— (1962). *Am. J. Phys.* **30,** 403.

Gunn, J. A. (1929). *The problem of time*, p. 323, Allen and Unwin, London.

Haeckel, E. (1900). *The riddle of the universe* (transl. J. McCabe), p. 253. Watts, London.

Hardie, R. P. and Gaye, R. K. (1930). In *The works of Aristotle* (ed. W. D. Ross), vol. II, *Physics,* Book IV, 219a. Clarendon Press, Oxford.

Hawkes, L. (1952). *Geology and time, Abbott Memorial Lecture*, p. 14. University of Nottingham.

Helmholtz, H. (1881). *Popular lectures on scientific subjects* (Second Series) (transl. E. Atkinson), p. 144. Kegan Paul, London.

HINCKFUSS, I. (1975). *The existence of space and time*, p. 134. Clarendon Press, Oxford.

JAKI, S. L. (1969). *The paradox of Olbers's paradox.* Herder and Herder, New York.

KEMP SMITH, N. (1934). *Immanuel Kant's Critique of Pure Reason* (trans. N. Kemp Smith), p. 217. Macmillan, London.

KURTÈN, B. (1963). *Soc. Sci. Fenn. Commun. Biol.*, **24,** 3.

LATHAM, R. E. (1951). *Lucretius: on the nature of the universe*, pp. 40–1. Penguin Books, Harmondsworth.

LEIBNIZ, G. W. (1925). *The Monadology* (transl. R. Latta), p. 236 Clarendon Press, Oxford.

—— (1934). *Philosophical Writings* (trans. M. M.). Dent, London.

LEWIS, G. N. (1930). *Science*, **71,** 569.

LOCKE, J. (1690). *Essay concerning human understanding.* Book II, chap. 15, Para. 11.

LOTKA, A. J. (1945). *Hum. Biol.* **17,** 167.

MACH, E. (1960). *Science of mechanics* (transl. T. J. McCormack), (6th edn.), p. 284. Open Court, Chicago.

MAYNARD SMITH, J. (1972). Time and the evolutionary process. In *The study of time* (ed. J. T. Fraser, F. C. Haber and G. Müller), p. 203, Springer-Verlag, Berlin, Heidelberg, New York.

MEYERSON, E. (1930). *Identity and reality* (trans. K. Loewenberg). Allen and Unwin, London.

MULLER, H. J. (1939). *Biol. Rev. Camb. phil. Soc.* **14,** 261.

NEEDHAM, J. (1938). *Biol. Rev. Camb. phil. Soc.* **13,** 225.

OLBERS, H. M. W. (1823). *Berliner astronomisches Jahrbuch für das Jahr 1826*, p. 10. Späthen, Berlin.

PAP, A. (1963). *An introduction to the philosophy of science*, pp. 97–8, Eyre and Spottiswoode, London.

POINCARÉ, H. (1929). The value of science. In *The foundations of science* (transl. G. B. Halsted), p. 228. Science Press, New York.

POPPER, K. R. (1956). *Nature (Lond.)* **177,** 538.

—— (1957). *Nature (Lond.)* **179,** 1297.

—— (1958). *Nature (Lond.)* **181,** 402.

—— (1978). *Brit. J. Phil. Sci.*, 29, 47–8.

RANDALL, J. H., JR. (1960). *Aristotle.* Columbia University Press, New York.

RUSSELL, B. (1901). *Mind* **10,** 296.

—— (1926). *Our knowledge of the external world* (revised edn.), p. 161. Allen and Unwin, London.

—— (1937). *The principles of mathematics* (2nd edn.). Allen and Unwin, London.

SAMBURSKY, S. (1959). *Physics of the Stoics*, p. 107. Routledge and Kegan Paul, London.

SCHUSTER, M. M. (1961). *Rev. Metaphys.* **15,** 209.

SIMPSON, G. G., PITTENDRIGH, C. S., and TIFFANY, L. H. (1957). *Life: an introduction to biology*, p. 336. Routledge and Kegan Paul, London.

STUEWER, R. H. (1976). G. N. Lewis on detailed balancing, the symmetry of time and the nature of light, *Hist. Stud. phys. Sci.* **6,** 469–511.

THODAY, J. M. (1975). *Nature* (*Lond.*) **255,** 675–7.

THOMSON, W., and TAIT, P. G. (1890). *Natural philosophy*, Part 1, p. 241. Cambridge University Press, Cambridge.

WHITROW, G. J. (1978). *Br. J. phil. Sci.* **29,** 39–45.

WHITTAKER, E. T. (1953). *A history of the theories of aether and electricity: the modern theories (1900–1926*), p. 198. Nelson, London.

WIENER, N. (1961). *Cybernetics*, p. 35. M.I.T. Press, Cambridge, Mass.

WILLIS, J. C. (1940). *The course of evolution*, p. 21. Cambridge University Press, Cambridge.

VON WRIGHT, G. H. (1968). *Time, change and contradiction*, p. 17. Cambridge University Press, Cambridge.

YOUNG, J. Z. (1938). Evolution of the nervous system. In *Evolution* (ed. G. R. de Beer), p. 180. Clarendon Press, Oxford.

—— (1950). *The life of vertebrates*, p. 409. Clarendon Press, Oxford.

ZWART, P. J. (1976). *About time: a philosophical inquiry into the origin and nature of time*, p. 242. North-Holland, Amsterdam.

2

HUMAN TIME

2.1. Time and the mind

Despite its intimate association with the universal world-order, the idea of time is closely associated with the human mind. This was clearly realized by Aristotle. If only soul, or intellect, is able to count, then 'Whether if soul did not exist time would exist or not is', he admitted, 'a question that may fairly be asked'. (Hardie and Gaye 1930). He thought that without soul there would be no time, but only that of which time is an attribute, namely motion, that is if it is possible to imagine motion existing without a soul as its motive force.† Aristotle did not purse this train of thought any further, because he believed that when we investigate nature and the role of time, we do so as beings with souls for whom time is that aspect of motion in virtue of which its is measurable. Moreover, in his view, our minds must necessarily conform to the world-order, which therefore controls both our perception of time and the process by which we calculate or measure it. For him all motion is ultimately referred to the uniform circular motion of the first heaven, or sphere of fixed stars, for its time.

In late antiquity Aristotle's analysis was submitted to searching criticism by Plotinus and, above all, by St. Augustine, who pointed out that if we regard motion as measured in terms of time and time in terms of motion then we come perilously near to circularity of definition. 'Do I measure, O my God, and know not what I measure'? According to him, time and motion must be even more carefully distinguished from one another than they were by Aristotle. In particular, time must not be correlated with the motion of the heavenly bodies, for, if the heavens should cease to move but a potter's wheel continue to run round, it would still be possible to measure its revolution. Even though one would not be inclined to maintain that each revolution constituted a day, one could hardly deny that in some way it represented the passage of time. Similarly, when at the prayer of Joshua, the Sun stood still, time nevertheless continued, 'For, if a body be sometimes moved, sometimes stand still,

† Unlike Democritus, who believed that atoms move automatically, Aristotle seems to have taken a more animistic point of view, but in fact his ideas were more sophisticated. His concept of *psyche* (customarily translated as 'soul', but not to be confused with the Pythagorean, Christian, or Cartesian concepts) signified the living body's natural purposive function. The relation of the living organism to its *psyche* was like that of the flute to flute playing (see J. H. Randall 1960).

then we measure, not his motion only, but his standing still too by time... Time then is not the motion of a body' (Pusey 1907).

Dissatisfied, therefore, with Aristotle's close association of time with motion, St. Augustine turned to the mind rather than the physical order for the ultimate source and standard of time.† 'It is in thee, my mind', he cried, 'that I measure times'. He arrived at his solution of the problem by one of the most acute analyses in the history of the subject. Instead of appealing to motion, *with its spatial associations*, he considered purely temporal phenomena—auditory rather than visual—such as the reading of a poem and the sounding of a voice. 'Thus measure we the spaces of stanzas by the spaces of the verses, and the spaces of the verses by the spaces of the feet, and the spaces of the feet by the spaces of the syllables, and the spaces of the long by the spaces of the short syllables; not measuring by pages', he significantly comments, 'for then we measure spaces, not times'. Nevertheless, we do not yet arrive at a fundamental unit or scale of time, 'because it may be that a shorter verse pronounced more fully may take up more time than a longer pronounced hurriedly'. This consideration, however, suggests to him that time is a protraction or distension 'but of what I know not; and I marvel if it be not of mind itself'. He then considers the problem of measuring the time taken by a voice in making a single sound and is faced by a characteristic conundrum concerning the apparently conflicting concepts of succession and duration. Clearly, before the sound begins we cannot measure the time it is going to take, nor after it has sounded can we measure the time it has taken, for it is then no more. Can we then measure it in the present, while it is being sounded? This will not be possible, he points out, so long as the present is regarded as truly momentary and without duration. Therefore, any stretch of time, however short, necessarily involves something of either past or future. St. Augustine thus came to the conclusion that we can measure times only if the mind has the power of holding within itself the impression made on it by things as they pass by even after they have gone. 'This it is which still present I measure, not the things which pass by to make this impression. This I measure when I measure times. Either then this is time or I do not measure times'. Although St. Augustine failed to explain how the mind could be an accurate chronometer for the

† The idea that time exists *per se* (absolute time) does not seem to have been considered by thinkers in antiquity, except that:

(i) according to Strato of Lampsacus, the pupil of Aristotle, 'day and night and year are not Time nor part of Time, but are respectively light and darkness and the revolution of Sun and Moon; instead, Time is the quantity in which these exist' (Diels 1882);

(ii) according to Galen, as reported by a tenth-century writer Ibn Abi Said (Pines 1955), 'motion does not produce time for us; it only produces for us days, months, and years. Time, on the other hand, exists *per se*, and is not an accident consequent upon motion'.

external order of physical events, he must be regarded as the great pioneer of the study of *internal* time.†

Following the publication of Newton's *Principia*, the empirical philosophers Locke, Berkeley, and Hume considered the origin of the notion of time and all agreed that it was the succession of ideas in the mind, but they too failed to make clear how this is related to the time used in physics. Berkeley complained that whenever he attempted to frame a simple idea of time, abstracted from the succession of ideas in his mind, he was 'lost and embrangled in inextricable difficulties'. He believed that the duration of any finite spirit was measured by 'the number of ideas succeeding each other in that spirit or mind'.‡ He paid no attention, however, to the problems of the uniformity and universality of time, and in the *First Dialogue between Hylas and Philonous* the latter suggests that ideas may succeed each other twice as fast in one mind as in another (Berkeley 1713). Whether or not Zawirski (1936) is justified in complaining that 'Berkeley semble rejeter non seulement le temps absolu du Newton, mais aussi le temps du sens commun', there is no doubt that both he and Hume failed to give any explanation of the distinction which we make between the temporal order of our ideas and that of the external objects which we claim to know by means of them.

This crucial point was grasped by Kant. He believed that time is the form of 'intuition' appropriate to our internal sense, so that we only conceive our states of mind as being in time in introspection, but that they are not really in time. Although Kant considered that all knowledge begins with experience, he did not regard the concept of time (or of space) as derived from experience. 'For neither coexistence nor succession would ever come within our perception, if the representation of time were not presupposed as underlying them *a priori*. Only on the presupposition of time can we represent to ourselves a number of things as existing at one and the same time (simultaneously) or at different times

† Although St. Augustine regarded time as being in the soul, or mind, he believed that time is inconceivable apart from the universe. Time, in his view, implied both a universe containing motion and change and a soul, or mind, that exists in its own right (Callahan 1948).

‡ Berkeley owed this idea to Locke (1690, p. 107), but it can be traced back to Hobbes. The common sense objection to it was clearly stated by Hume's contemporary and critic Thomas Reid (1785), 'I am rather inclined to think that the very contrary is the truth. When a man is racked with pain, or with expectation, he can hardly think of anything but his distress and the more his mind is occupied by that sole object, the longer the time appears. On the other hand, when he is entertained with cheerful music, with lively conversation and brisk sallies of wit, there seems to be the quickest succession of ideas, but the time appears shortest'.

The influence of different activities on the subjective rate of time has been studied by Loehlin (1959). His experimental investigation of what he called the 'activity factor' led him to conclude that the amount of 'information' actively coped with by the subject can be an important influence on the apparent rate of time (see also Michon 1967). For further discussion see p. 62.

(successively)' (Kemp Smith 1934). Although Kant was an enthusiastic disciple of Newton, he denied that time had any claim to absolute reality. In his view, the concept of time 'does not inhere in objects but merely in the subject which intuits them'. In other words, time (like space) pertains essentially to the functioning of mind and not to things-in-themselves. But whereas it is only the *mediate* condition of the appearances of external objects (which we also represent as being in space), it is the *immediate* condition of those of our inner sense, which we represent to ourselves as being solely in time.

2.2. The prehistoric origin of the idea of time

Towards the end of the nineteenth century Kant's idea of time as a necessary condition of our experience of the physical world was submitted to increasing criticism by psychologists. In a brilliant essay, *La Genèse de l'Idée de Temps*, published in 1890, two years after his death, M. J. Guyau turned from the formal problem posed by Kant to consider the actual development of the concept of time. He regarded time not as a prior condition, but as a consequence, of our experience of the world, the result of a long evolution. He argued that it was essentially a product of human imagination, will, and memory. In direct opposition to the English associationist and evolutionist school led by Herbert Spencer, who regarded the idea of time as the source of the idea of space, he maintained that, even though we may use the one in order to measure the other, nevertheless they are utterly distinct ideas with their own characteristics. Moreover, the idea of space originally developed before the idea of time. In the period of primitive mental confusion the succession of ideas did not automatically give rise to the idea of their succession, whereas movements in all directions made in response to natural desires gave rise to the idea of space as the natural mode of representation of simultaneous sensations coming from different parts of the organism. The idea of events in their temporal order came after the idea of objects in their spatial order, for the latter is related to the perceptions or *presentations* themselves whereas the former depends on reproductive imagination or *representation*. The ultimate origin of the idea of time lies in our perception of differences and resemblances. Both are necessary, for too great a diversity in the succession of images presented to the mind is just as ineffective as is too small a diversity, since each new image will occupy our entire field of awareness to the exclusion of all that has gone before. Consequently, a certain degree of continuity and regularity in the heterogeneous flood of sensations is a necessary condition for the idea of time to arise. Therefore, argued Guyau, it cannot be a purely *a priori* concept.

Despite its cogency, I do not believe that this argument provides a complete refutation of Kant's idea that time is 'the form of our inner sense, that is, of the intuition of ourselves and of our inner state', although we cannot agree with him that it is nothing but this. For Guyau assumed that the human mind has the *power*, apparently not possessed by animals, to *construct* the idea of time from our recognition, or awareness, of certain features characterizing the data of experience. Even if Kant threw no light on the origin of this power, since he regarded our idea of time as an invariable mental framework with no evolutionary history, at least he realized that a peculiarity of the human mind was involved.

We owe to Guyau some penetrating suggestions on the way in which this power of the mind has developed. He rejected Herbert Spencer's naïve assumption that the idea of time was derived from a primitive awareness of temporal sequence. On the contrary, he maintained that in the primitive state of mental confusion there was no clear conception either of simultaneity or of succession. Guyau suggested that the idea of time arose when man became conscious of his reactions towards pleasure and pain and of the succession of muscular sensations associated with these reactions. When a baby is hungry it cries and stretches out its arms towards food—'voilà la germe de l'idée d'avenir'. Every need implies the possibility of satisfaction, and the aggregate of these possibilities leads to our concept of 'the future'. 'Un être qui ne désirerait rien, qui n'aspirerait à rien, verrait se fermer devant lui le temps... L'*avenir* n'est pas *ce qui vient vers nous*, mais *ce vers quoi nous allons*' (Guyau 1890, p. 32). The psychological origin of the concept of time is therefore to be found in the conscious realization of the distinction between desire and satisfaction. The sense of purpose and associated effort is the ultimate source of the ideas of cause and effect, but it was only by a series of scientific abstractions that men eventually arrived at the concepts of a uniform temporal sequence and a definite causal process.

In Guyau's view, the concept of time was always intimately associated with that of space. The future was in effect that which lay ahead and was sought for, whereas the past lay behind and was no longer in view. 'En somme, la *succession* est un abstrait de *l'effort moteur* exercé dans *l'espace*; effort qui, devenu conscient, est *l'intention*' (Guyau 1890, p. 36). The idea of purpose was associated with some direction in space and thus with motion. Time can thus be regarded as an abstraction of motion which is itself associated with a sequence of sensations of muscular effort and resistance experienced along a line from the point of space where one is originally to another point where one wishes to be.

Thus, whereas spatial conceptions may have originated when man became fully conscious of, and reflected upon, his movements, Guyau argued that temporal concepts must be traced back to the feelings of effort and fatigue associated with these movements. As has since been stressed by Janet (1928), however, man had to acquire the ability to discriminate between the respective sensations of commencing, continuing, and terminating an action. In recent years it has become clear that all man's mental abilities are potential capacities which he can only realize in practice by learning how to use them. For, whereas animals inherit various particular patterns of sensory awareness—known as 'releasers' because they function as automatic initiators of specific types of action—man inherits only one releaser, that made by a mother's smile to her infant.† He must therefore learn to construct all his other patterns of awareness from his own experience.† Consequently, our ideas of space and time, which according to Kant function as if they were releasers, must instead be regarded as mental constructs which have to be learnt.

Guyau's hypothesis that the original source of man's idea of time was an accumulation of sensations which produced an internal perspective directed towards the future is supported by the current opinion of anthropologists that the great development of the prefrontal lobes of the brain of *homo sapiens* may have been intimately associated with his growing power of adjustment to future events. For, although Neanderthal man may have shown some rudimentary concern for the future, since he appears to have buried his dead, the emergence of modern man has been correlated with a strongly increased tendency to look forward§, the principal evidence being the sudden development of tools which, unlike the primitive Neanderthal handaxes, were used to make a wide variety of other tools (barbed harpoons, fish hooks, eyed needles, etc.) for future use.

A vital step in the development of man's appreciation of time was his discovery that this forward perspective could be viewed retrospectively: that, during our life, there is formed within our minds a kind of deposit of that which was originally present in our thoughts and feelings. The gradual development of coherent memory, as well as of coherent thought, was presumably closely connected with the transition from evocative

† Some anthropologists, however, believe that our understanding of other facial expressions depends on innate releasing mechanisms (Eibl—Eibesfeldt 1970).

‡ This is in accord with the general character of the later stages of evolutionary advance, for whereas even the highest invertebrates (insects) rely mainly on inherited patterns, there is an increasing tendency as we go up the evolutionary scale for the higher vertebrates to depend on patterns which are learnt from individual experience. Incidentally, it may be observed that the increasing need, with evolutionary advance, to incorporate individual experience in mental structure is reflected in the general tendency towards prolongation of infancy and of the total span of life.

§ In this context it is interesting to note that children grasp the idea of the future sooner than that of the past (Stern 1930). According to Paul Fraisse (1964, p. 181), as soon as a child frees himself from the confusion of the past and the future, the latter plays a far larger part than the former in his conscious perspectives. If he does turn to his past it is only to locate himself in relation to others and he attaches little importance to it.

'speech', pointing forward in time, to descriptive 'speech', pointing backward in time. This was dependent on man's recognition of long-enduring objects to which names could be given and must have been an extremely difficult step.

The hypothesis that the prehistoric origin of the idea of time was intimately associated with that of language† is supported by the fact that, although time has come to be represented primarily in terms of spatial images derived from vision,‡ it is actually far more closely connected with hearing, the principal sense involved in the development of speech. Fundamental for both time and speech, particularly primitive speech, is rhythm. Rhythm involves repetition, the function of which is to secure that it should be recognized. Moreover, it has a natural tendency to kinaesthetic stimulation and self-perpetuation. This may be explained by the fact that the nervous system puts itself into a state of expectancy and is therefore ready for the appropriate discharge at the right moment. A highly developed sense of rhythm enabled a tribe to function with precision as a single unit in both war and hunting.

2.3. The historical development of the idea of time

Primitive man's intuition of time was dominated by his sense of rhythm rather than by the idea of continual succession. There was no explicit sense of time itself but only of certain temporal associations which divide time into intervals similar to bars in music. Even with the rise of civilization it appears that primary importance was attached to simultaneity rather than to succession. Specific religious and sacrificial acts had to be performed on specific occasions, often associated with particular phases of the Moon or solar solstices, and only on these occasions. Even in medieval Europe the first stages in the development of the mechanical clock seem to have been influenced by the monastic demand for accurate determination of the hours when the various religious offices should be said, rather than for any desire to register the passage of time.

Indeed, for long the aspects of time which were of primary significance for the human mind were not duration, trend, and irreversibility, but repetition and simultaneity. These were the characteristic features of what

† This hypothesis has been investigated in detail by the Swiss mathematician and philosopher of science Ferdinand Gonseth (1972) who, in discussing the role of time in the development of language, has argued that the emergence of the conjugated-verb system of tenses that we find in Indo-European languages indicates *a consciousness of universal time as opposed to the time of the self.*

‡ The Chinese and Japanese have even used the sense of smell as a means of telling the time! Fire and incense time-keepers appear to have been used since the sixth century A.D. (Bedini 1963). Some years ago a 200 year old incense clock was displayed in a Tokyo department store. Small pieces of incense, stuck in the top, were burned down one by one, each giving off a different aroma and enabling those with sensitive noses to tell the approximate time (see *The Illustrated London News*, 233, No. 6213, 5 July 1958, p. 17).

has been called 'mythical time'. In primitive thought we find innumerable examples of the belief that an object or an act is 'real' only in so far as it imitates or repeats an ideal prototype. We are therefore, presented with the paradoxical situation that in his first conscious awareness of time man instinctively sought to transcend, or abolish, time.† In particular, every ritual sacrifice was believed to reproduce an initial divine sacrifice and coincide with it. As Mircea Eliade (1954) has shown by numerous examples, the life of archaic man was characterized by the repetition of archetypal acts and the unceasing rehearsal of the same primordial myths, so that he tended to live in a continual present.‡

† In an article on 'The Origin of Religion' S. G. F. Brandon (1959) argued that it was primarily the mental and emotional tension resulting from man's discovery that every living creature is born and dies which led him intuitively to try to circumvent the relentless flux of time, for example by the 'ritual perpetuation of the past'. Professor Brandon maintains that *religion originally stemmed from human consciousness of the temporal process*, 'the menace of which was focussed in death but alleviated by the promise of new life manifest in the phenomenon of birth'. He cites archaeological support for this view, namely, whereas upper palaeolithic man buried his dead and equipped them with weapons, tools, ornaments, and even food (which often must have been in short supply for the living), and also fashioned primitive figurines symbolic of maternity and thus of the promise of new life, there is no evidence that he attained the concept of deity (or displayed any interest in celestial phenomena).

‡ This conclusion is confirmed by the behaviour of surviving primitive races. For example, Evans-Pritchard (1937), in his famous investigation of witchcraft, oracles and magic among the Azande, found from their behaviour, as distinct from their expressed patterns of thought, that there exists for them no sharp distinction between present and future. The future depends on the disposition of mystical forces that can be tackled here and now. Thus, when a man consults an oracle which says he will fall sick in the coming month, he believes that this is bound to happen unless the oracle can be changed, but, whereas we would say that there is only a change in the prediction of the oracle, the Azande believes that there is a real change in his condition since his future is already part of time. The Azande cannot explain these matters, but 'content themselves with believing and enacting them'.

In his well-known study of the Hopi Amerindians, Benjamin Lee Whorf (1956) found that their language contains 'no words, grammatical forms, constructions or expressions that refer directly to what we call "time", or to past, present or future, or to enduring or lasting'. A similar attitude towards time has been revealed by tribal African children, as P. M. Bell (1975) discovered when he spent a year teaching in Uganda. They not only found it difficult to be punctual but also to judge how long something took to happen. 'For example, a two-hour bus journey I was told had taken only 10 minutes; by other children 20 minutes or six hours. They had no idea. Nor were these children backward in an educational sense'. Similar behaviour has been exhibited by Australian aboriginal children. Although they have been found, on the whole, to be of similar mental capacity to white children, they find it extremely difficult to tell the time by the clock. According to C. Ralling (1959), 'They will read off the hands and face of a clock as a memory exercise, but to relate it to the time of day seems to involve a mental gap that few of them manage to jump. The reason is that their lives, unlike ours, are not dominated by time'.

Of course, most primitive peoples have some idea of time and some method of reckoning, usually based on astronomical observations. For example, the Australian aborigine will fix the time for a proposed action by placing a stone in, say, the fork of a tree so that the sun will strike it at the agreed hour. Nevertheless, it is significant that Rousseau, who extrolled the 'noble savage', detested time and clocks. When he threw away his watch, he thanked heaven because he would no longer need to know what time of day it was (*Rousseau juge de Jean-Jacques*, second dialogue, *see* Cassirer 1945).

This view has been reinforced by recent attempts to interpret the cultural peculiarities of ancient civilizations. For example, faced with the characteristic phenomena of ancient Egyptian civilization—the divinity of Pharaoh, his entombment in a pyramid, the burial of cats and dogs, the mummification of the dead—Henri Frankfort rejected Spengler's view that Egyptian civilization was an embodiment of conscious concern with the future, and argued instead, and I believe far more plausibly, that the Egyptians had very little sense of history or of past and future. 'For they conceived the world as essentially static and unchanging. It had gone forth complete from the hands of the Creator. Historical incidents were, consequently, no more than superficial disturbances of the established order, or recurring events of never-changing significance. The past and the future—far from being a matter of concern—were wholly implicit in the present; and ... the divinity of animals and kings, the pyramids, mummification as well as several other seemingly unrelated features of Egyptian civilization—its moral maxims, the forms peculiar to its poetry and prose—can all be understood as a result of a basic conviction that only the changeless is truly significant' (Frankfort 1956).

Our knowledge of the highly developed calendars, king lists, and annals of ancient civilizations does not conflict with this view. As a distinguished French Assyriologist has remarked, 'We must accept the fact that the inhabitants of Mesopotamia in antiquity did not regard history in the same light as the modern world, at least intermittently, does so. They were mainly interested in themselves, and were content to leave matters relatively indefinite' (Contenau 1954). Even for the Greeks significant history was, on the whole, essentially contemporary history. Moreover, time as recorded by their sundials, sand-clocks, water-clocks, etc. 'resembled the irregular flowing of a river rather than an exactly graduated measuring rod' (Leach 1954). When we encounter examples of abstract speculation concerning immense aeons of time, notably by the ancient Hindus and the Mayas,† we find that, underlying the orgy of calculation, time was regarded merely as the eternal repetition of the cosmic rhythm.

† Of all ancient peoples, the Maya priests developed the most elaborate and accurate astronomical calendar, and thereby gained enormous influence over the masses. Indeed, the calendar-correction formulae obtained by the astronomer-priests at Copan in the sixth and seventh centuries A.D. were even more accurate than our present leap-year *correction* which was not introduced until 1582 by Pope Gregory XIII. Our correction is 0·03 days per century too long, whereas the corresponding ancient Maya correction was 0·02 days too short (Morley 1947).

Unlike the Greeks, whose philosophy was dominated by the assumption that perfect knowledge was essentially geometrical, the Mayas were obsessed with the idea of time. Every stela and altar was erected to mark the passage of time and was dedicated at the end of a period. The divisions of time were pictured as burdens carried on the backs of a hierarchy of divine bearers (the personification of the numbers by which the different

Generally, in primitive societies and in most ancient civilizations, change was not conceived as a continuous process spread out in time but as discontinuous and abrupt. The principal transitions in nature were regarded as occurring suddenly, but inevitably, in a cycle with a definite rhythm. Similarly, man's journey through life was visualized as a sequence of distinct stages punctuated by sudden crises and transitions. These prompted *les rites de passage*—the ritual ceremonies that, as van Gennep (1909) was the first to note, differ only in detail from one culture to another but are in essence universal.†

It was indeed a long step from the inhomogeneity of mythological time, with its specific holy days and lucky and unlucky secular days, to the homogeneity of physical time as conceived by modern civilized man.

Nevertheless, the primitive idea of time as rhythmical repetition became the basis for its division, and ultimately for its measurement. One of the oldest and most widespread conscious expressions of this idea is to be found in myths concerning the moon, many of the most ancient civilizations—for example that of Ur—being based on moon worship. The lunar phases presented a vivid example of eternal repetition and provided a more obvious unit of time than the solar year. Also in the Indo-European languages we find that most words for month and moon derive from the same root *me*, producing in Latin, for example, *menis* and *metior*, 'to measure'. Again, in the religion of the ancient Egyptians the scribe of the gods who endowed mankind with the arts of writing and counting was Thoth, the moon god, who as the divider and measurer of time was the lord over exact and immutable measurement.

Although restricted, the ancient conception of time was thus of immense significance for the growth of civilization. Increasing emancipation of human thought from the domination of immediate sensory impressions involved closely related developments in man's awareness of time and in his ideas of the universe. Whereas primitive man tended to visualize all natural processes purely subjectively and to regard them as being at the mercy of arbitrary demonic forces which could be influenced by magic,

---

periods—days, months, years, etc.—were distinguished). There were momentary pauses at the end of each prescribed period when one god with his burden succeeded another with his. Nevertheless, the Mayas never attained the idea of time as the journey of one bearer and his load. Moreover, in their prophecies, past, present, and future tended to become one. The divine bearers in relays carried time forward on its unending journey, but at the same time events moved in a cycle represented by the recurring spells of duty for each god in the succession of bearers. Days, months, years, etc. were all members of relay teams marching through eternity. By calculating which gods would be marching together on a given day, the priests could determine the combined influence of all marchers, and thus forecast the fate of mankind (Thompson 1956).

† The idea that man's journey through life is a sequence of distinct stages is exemplified by Shakespeare's 'seven ages of man'.

civilized man was inclined to direct his thought more and more to the contemplation of a universal world-order. In this greatest of all revolutions in human thought the heavenly bodies played a fundamental role. No longer considered solely in terms of their immediate physical effects, they came to be regarded as the constant measures of time ensuring the proper synchronization of events. Thus, out of man's primeval awareness of rhythm and periodicity there eventually emerged the abstract idea of world-wide uniform time.

Nevertheless, this concept, like that of physical space, was not clearly formulated in terms of mathematics until the 'scientific revolution' of the seventeenth century. For, even in so culturally and technologically advanced a civilization as the Chinese, different intervals of time tended, as a rule, to be thought of as separate discrete units, so that time was effectively discontinuous. The universe was regarded as a vast organism undergoing a cyclic pattern of alternation, with now one and now another component taking the lead, the idea of *succession* being subordinate to that of *interdependence.* Just as space was decomposed into regions, time was split up into eras, seasons and epochs. In other words, time was 'boxed'† (Granet 1934, Needham 1956). Indeed, man was aware of different times long before he formulated the idea of time itself. Even the Mayas never seem to have grasped the idea of time as the journey of one bearer and his load. Despite the influence of Christianity with its emphasis on the Incarnation as a unique event not subject to repetition, the theory of cycles and astral influences was accepted by most western thinkers down to the seventeenth century. It is true that many of the superstitions about 'unlucky days' were vigorously combated by the medieval church, although with little success. However, as Keith Thomas (1971) has reminded us, the Church itself had endowed each date in the year with some symbolic significance, and the ecclesiastical calendar reinforced the conviction that time was uneven in quality. In other words, for most men magical time had not been superseded by homogeneous scientific time. Moreover, throughout the whole medieval period there was a conflict between the cyclic and linear concepts of time, and it was the scientists and scholars influenced by astronomy and astrology who

† However, in his stimulating and important Henry Myers Lecture on 'Time and eastern man', Joseph Needham (1965, especially pp. 8–9) has provided powerful evidence that the Chinese were far from being a people who took no account of time. 'In so far as the traditional natural philosophy was committed to thinking of time in separate compartments or boxes perhaps it was more difficult for a Galileo to arise who should uniformise time into an abstract geometrical co-ordinate, a continuous dimension amenable to mathematical handling'. However, Needham goes on to argue that in the Chinese invention of mechanical clockwork there is no indication of 'inhibition by any ideas of sharp boundaries between stretches of time', since the clocks were continually looked after and 'ticked away continually decade by decade'.

emphasized the cyclic concept, whereas the linear concept was fostered by the mercantile class with the rise of a money economy. So long as power was concentrated in the ownership of land, however, time was associated with the unchanging cycle of the soil. With the circulation of money, the emphasis came to be placed on mobility and men began to believe that 'time is money' and should therefore be used economically. Thus time gradually came to be associated with the ideas of work and of linear progress, and this view was reinforced by religious tendencies in the Renaissance with its emphasis on the individual.†

The invention of a successful pendulum clock in the middle of the seventeenth century had a tremendous influence on the whole concept of time, for at last mankind was provided with an 'accurate' timekeeper that could tick away continuously for years on end. This must have greatly strengthened belief in the homogeneity and continuity of universal time. The linear view of time as continual progression finally prevailed in the nineteenth century, mainly through the influence of the biological evolutionists.

2.4. Temporal perception and judgment(i)

Despite the sophisticated ideas of time in modern science and philosophy, most people still have the feeling that time is something that goes on of its own accord unaffected by anything else and that if all activity were suddenly to cease it would still continue without any interruption whatsoever. For many people even the way in which we measure time by the clock and the calendar is absolute, and by some it has even been thought that to tamper with either is to court disaster. When, in 1916, summer time was first introduced in the United Kingdom by advancing the clock one hour, there were many who objected to interfering with what the popular novelist Marie Corelli called 'God's own time'. Similarly, in 1752, when the British government decided to alter the calendar so as to bring it into line with that previously adopted by the Catholic countries of Western Europe, following the decision of Pope Gregory XIII in 1582, and decreed that the day following the second of September should be styled the fourteenth of September, many people thought that their lives were being shortened thereby. Some workers believed that they were going to lose eleven days' pay. So they rioted and demand 'Give us back our eleven days'!

Even today, although we are all familiar with the idea of altering the clock to suit our convenience, it comes as a shock to some people when

† This attitude was not confined to Protestants. A staunch Catholic like the famous painter Rubens had a somewhat similar ethos of work (Wedgwood 1975).

they travel by air to find that there is, for example, a five hour time difference between London and New York, so that when it is only four o'clock in the afternoon in New York it is already nine o'clock in the evening in London and, except at the height of summer, quite dark. Even stranger is the effect of the International Date Line drawn in a zig-zag fashion down the centre of the Pacific Ocean. When a ship on its way from San Francisco to Hong Kong crosses this line it loses a whole day because of the 24 hour time difference between any position immediately to the east of the line and any position just to the west of it. So, although in this case there is no need to adjust our watches, we have to discard a day from the calendar. Alternatively, if we cross the line in the opposite direction (Hong Kong to San Francisco) we have the unusual experience of an eight-day week. If the crossing is made at midnight we get two Fridays, say, in succession!

Although the time of civil life is measured in a purely conventional way that happens to suit us, have we not some inner feeling for the passage of time that has some deeper significance? In addition to the five senses of sight, hearing, touch, taste, and smell, do we not have some sense of direct temporal awareness? On this question the most diverse views have been held. For example, following a suggestion by J. Czermak (1857), the celebrated Austrian physicist and philosopher of science Ernst Mach (1865) maintained, in opposition to Kant, that time is not an *a priori* condition of mental functioning but is an *a posteriori* sensation. He believed that we have a specific sense of time and that the sense organ involved is the ear. He thought that our temporal awareness depends on the work, or effort, associated with the exercise of our powers of attention and the resulting fatigue of the organ concerned. Guyau pointed out that this sensation of time, if it exists, is vague, irregular, and exceedingly subject to error. Janet went further and categorically rejected the whole idea of a specific sense of time and maintained that Mach was drawing philosophical conclusions on an inadequate factual basis. In particular, he was neglecting the role of the stimulus: if we speak of a sense of duration, Janet argued, we are assuming that things exist outside us in the same way as we conceive them. He concluded that 'La durée, ce phénomène complexe et tardif dans l'évolution, que nous comprenons encore très mal car nos notions sur les temps sont encore très vagues, la durée ne doit pas être interprétée comme une sensation élémentaire' (Janet 1928, p. 47). More recently, another distinguished French psychologist in Paris, Paul Fraisse, has criticized Janet for going too far. In his view, although Janet was correct in his claim that our feelings of duration are reactions to the nature of our actions, he overlooked the fact that some of our actions—

such as the synchronization of our movements to periodic stimuli when we dance—are direct adaptations to time (Fraisse 1964, p. 81).

Although the ear is undoubtedly our most sensitive organ for perceiving change,† the general opinion nowadays is that we do not have any truly primitive experience of time as such. Our direct experience is always of the present, and our idea of time comes from reflecting on that experience. Not only do all animals except man seem to live in a continual present‡, but time only becomes important for most men in an advanced and complex industrial civilization, such as ours, in which their lives are ruled by the clock (and not just by the rising and setting of the sun) because it is necessary to make precise timings. When most people lived in villages and spent their lives close to the soil there was little need for clocks and watches.§ At the other extreme, the remarkable achievements in space travel of recent years would have been quite impossible without the ultra-precise timing of each rocket burn.

So long as our attention is concentrated on the present we tend to be unaware of time. If what we are doing interests us, time seems short. As Diderot said, 'Let us work; work has the advantage of shortening our days and lengthening our lives'. We experience a feeling of duration whenever the present situation causes us to relate it either to our past experiences or to those we desire for the future. In the former case we rely on memory; in the latter, time appears to us as an interval between a need and its fulfilment. When we become conscious of this interval there is often a tendency to regard it as long. A young child is not so capable of waiting as an older child or an adult and only learns to bear waiting as his emotional stability develops. The more attention we pay to time, the longer it seems. Never does a minute seem so long as when we look at the second hand moving round the face of a clock or watch.

† Hearing locates stimuli only vaguely in space, but it locates them with admirable precision in time. Hearing is, *par excellence*, the sense that appreciates time, succession, rhythm, and tempo (Guyau 1890, pp. 74–5).

‡ Carefully analysed experiments on the most intelligent animals, such as chimpanzees, indicate that even actions which might be interpreted as evidence of some idea of the future are purely instinctive (Köhler 1957). Moreover, the fact that dogs appear to display powers of memory when they display wild excitement on seeing their masters after long separation cannot be regarded as positive evidence that they have any image of the past as such.

§ Moreover, very few specific appointments were made. As late as the seventeenth century an important government official, Samuel Pepys, did not possess a watch when he was thirty. He lived by the church bells of London and occasionally a sundial. He moved around from public places to coffee houses and to taverns, hoping to do business. He often went to court to confer with the Lord High Admiral, James Duke of York, and found that the Duke had gone hunting. Pepys never expresses any surprise or resentment. Time had a totally different significance for him from what it has for us (Plumb 1972).

Since our sense of duration depends on the number of stimuli that we have perceived and stored in our minds, it follows that if an interval has many divisions—marked off visually or aurally—it tends to appear longer than an equal interval (by clock time) that has fewer.† On the other hand, when, for some reason, there is a drastic reduction in the amount of external information registered by the brain duration appears to be short. Following the great Messina earthquake of 1908, three brothers were trapped in rubble for 18 days, but when they were freed they thought that only four or five days had elapsed. Similar results have been obtained in experiments by volunteers who have stayed alone, without clocks or watches, in cold dark caves. In 1962 the young speleologist Michel Siffre spent 60 days in the Scarasson cave, in the Alpes-Maritimes department, and emerged thinking that he had been there for only 35‡ (Siffre 1965). In 1966 D. Lafferty spent 127 days in similar conditions in one of the Cheddar caves, and more recently J.-P. Mairetet endured 174 days in a

† If we compare a minute as it appears when a metronome beats 90 times in it and when the metronome beats, say, only 40 times, we find that usually the former minute seems the longer. With the passage of time the effect of the storage in our brains of impressions received in a given interval of time tends to diminish. This can be demonstrated by estimating the respective durations of two successive intervals that are equal by clock time. The tendency is for the first to appear shorter in retrospect than the second. For this reason, Ornstein (1969) rejected the hypothesis (Fraisse 1964, Michon 1967) that the amount of information consciously registered during an interval determines the psychological apparent duration of that interval. He argued that the remembered experience of the earlier interval appears to be shorter because of a greater loss of information from storage, and consequently he concluded that durational experience is a function of the amount of storage required for stimuli perceived during that interval. G. Underwood and R. A. Swain (1973), however, have pointed out that Ornstein's analysis overlooked the effects of attention. Experiments conducted by them on assessments of duration by subjects required to listen to prose passages of equal length, the reading of which was partially masked with varying intensity of white noise, revealed that the passages requiring more attention for analysis were judged to be of longer duration than those requiring less attention.

In a later investigation (Underwood 1975) the solution of the paradox that time seems to pass *either* very slowly *or* very quickly when we are kept busy was shown to depend on whether it is the task or its boring nature that we pay attention to. A watched pot never boils because of 'the enforced selectivity of attention to the pot', but if we are engaged with an interesting task our attention will be diverted away from clock time. Underwood argues that it is 'the organizational aspects of the task which are responsible for the observed quantitative differences in subsidiary-task performance during encoding and retrieval'. Experiments involving 50 seconds of encoding and 50 seconds of subsequent retrieval of lists of nonsense syllables revealed that *retrieval requires more attention than does encoding*, poorly organized material demanding more attention for retrieval (possibly in the search for retrieval clues) than well organized material. (See also Underwood 1976.)

‡ The medical experts who reported on Michel Siffre's physical condition after he left the cave said that very probably he had experienced a state of incipient hibernation with body temperature below 96.8 °F. This would have slowed down his metabolic processes, etc. They thought that this numbing by cold may have been responsible for his accelerated perception of time (Siffre 1965, p. 215).

cavern 70 metres below the earth's surface. Like Siffre, each of them greatly underestimated the length of time that they were incarcerated.

On the other hand, experiments conducted on subjects whose sense of awareness has been enhanced by stimulants such as caffein or psychedelic drugs reveal that duration appears to them to be lengthened relative to normal experience. Thus, de Quincey, in describing the effects of opium, said that it sometimes produced a vast expansion of time, so that he seemed to have lived for 70 or 100 years in one night (de Quincey 1821). A similar effect can be produced by *cannabis indica*, or hashish,† and also by mescaline, which interferes with an enzyme in the brain so that it becomes unable to use glucose properly, although there is usually no interference with the powers of careful observation and reporting.

Similar distortions of our sense of time can occur in sleep. We have all had dreams of apparently long duration when we were in fact asleep for only a few minutes or even seconds. A famous example was the dream of the Marquis de Lavalette who was in prison during the French Revolution. It occupied a few moments while midnight was striking and the guard outside his door was being changed. 'I was in the Rue Saint Honoré. It was dark and the streets were deserted, but soon a diffused dull murmur was heard. Suddenly a troop of horsemen appeared at the end of the street, terrible beings bearing torches. For *five hours* they passed me by riding at full gallop. After them came a vast number of gun carriages loaded with dead bodies ...' (Sturt 1925, p. 110).

One factor that strongly influences our general sense of temporal duration is our age, for as we get older time appears to pass more rapidly. As Guy Pentreath put it in verse:

> 'For when I was a babe and wept and slept, Time crept,
> When I was a boy and laughed and talked, Time walked;
> Then when the years saw me a man, Time ran,
> But as I older grew, Time flew'.

The reason for this apparent speeding up of time is that our organic processes tend to slow down as we grow older, so that, compared with

† According to Walter de la Mare (1930), to whom the poet J. Redwood Anderson described his experiences while under the influence of this drug, 'The first effect—and this remained true for every subsequent occasion—was the alternation of time-values. Time was so immensely lengthened that it practically ceased to exist ... But this slowing-down applied only to physical events; my own movements for instance, and those of other people; it did not apply to the processes of thought. Those, on the contrary appeared to be very greatly accelerated ... I thought with a rapidity comparable to that of dreaming, but with an acuteness and logical sequence very rarely experienced in dreams'. The reverse effect can be induced by drugs, such as pentobarbitol and nitrous oxide, that tend to slow down the vital functions. Similar results can be obtained by living in a rarefied atmosphere (Barach and Kagan 1940) and also in experiments on animals (Grabensberger 1934).

them, civil time, as registered by the clock and the calendar, appears to go ever faster.

As regards temporal *order*, as distinct from temporal *duration*, we generally have difficulty in placing a sensation of one kind between two of another kind if they are in close succession. If, however, we could perceive time directly, the nature of the sensations determining the intervals concerned would be of no particular significance, and the difficulty of placing would be just as great when all three sensations are of the same kind. Fraisse (1952) tested this conclusion and found statistically that the difficulty of placing was *less* in the homogeneous case. He therefore concluded that *we have no specific time sense*. In other words, we have no direct experience of time as such but only of particular sequences and rhythms. Thus it is not time itself, but what goes on in time that produces effects. As we have seen in our discussion of absolute and relational time (pp. 33–9), it is generally believed nowadays that time does not exist in its own right but is produced by the events which we say 'occur in time'. Whereas space is automatically part of our experience and the idea of it results from reflecting on our movements, the idea of time becomes evident to us only after many stages of thinking, and that is why in higher cultures there is more conscious awareness of it than in primitive societies. Our idea of time is a mental construction that we only gradually learn to perform, our awareness of it being based on the number of changes that we observe occurring in a given interval.

From all this evidence we conclude that our conscious sense of time depends on the mechanism of attention and the coding and storage of information in the brain rather than on any specific internal organ of time experience. It is affected not only by our general mental and physical state, including our age, but also on the nature of our surroundings and by the culture in which we live. *Our 'sense' of time is neither a necessary condition of our experience, as Kant thought, nor a simple sensation, as Mach believed, but an intellectual construction.* It depends on processes of mental organization uniting thought and action. It is a late product of human evolution, in all probability closely related to the development of language. Moreover, we have, as yet, no reason to assume that any living creature other than man has the capacity for temporal awareness.

2.5. Temporal perception and judgment (ii)

We have seen that our awareness of time involves factors that we do not associate with the abstract idea of time, notably fixation of attention. Our attention may be kept constantly on the move, so that before it is accommodated to one presentation it is distracted by another, or it may be kept stationary by the repetition of the same presentation. Our conscious awareness of time depends on the fact that our minds operate by *successive acts of attention* and, moreover, is influenced by the tempo of our attention. This depends not only on our physical and mental state, including our memory and expectation, but also on the nature of the subject matter attended to, for it appears that we cannot attend to two simultaneous events and perceive both of them clearly unless they are combined in some way.† For example, we cannot attend to a visual piece of information and an independent aural piece simultaneously presented to us (Mowbray 1954). Attention can, however, switch from one thing to another in about 0.2 seconds. Indeed, attention always wanders, even when only a single stimulus pattern is presented to us, as can be demonstrated by the well-known experiment of looking at ambiguous, or reversible, diagrams such as the Necker cube (see Fig, 7.5). These automatic fluctuations, cause us to see different configurations alternately.

Contrary to the conception of time so eloquently advocated by Bergson, there seems to be a discontinuity in our perception of attention. After a comparatively continuous perception, there comes a break and then a new presentation begins. The duration of these perceptual fluctuations depend on the individual's attitude and conditions of perception. If we lived in a uniform continuous temporal medium, and if we had no discrete inner feelings, we would be unable to perceive time. In other words, we could not be conscious of a *strictly* uniform and continuous time, for as Gaston Bachelard (1950, p. 66) has remarked, 'c'est précisement un temps où il ne se passe rien'. Consequently, he concludes that 'la continuité psychique est non pas une donnée, mais une oeuvre'

† It is true that we can sometimes combine two actions in a single performance, for example skilled playing with both hands on the piano, but when we are told that Julius Caesar was able (lucky man!) to dictate several letters 'at the same time', each to a different copyist, we realise that he must have shifted back and forth—although even that was no mean feat. Two independent actions can be performed if one at least is automatized.

There is, however, some difference of opinion on whether in fact we always have a single unified train of thought. Hebb (1972) maintains that sometimes two independent thought processes are running in parallel, e.g. when a lecturer is speaking one sentence and already thinking about the next. However, Young (1978, p. 216) points out that, although the brain has many parts and different combinations of them may operate at different times, 'the whole system is controlled by one central reticular system and produces in each of us a single stream of consciousness'. This conclusion is important for our recognition of the unidimensional nature of time (see Sections 4.1 and 7.9). It is also supported by the researches of Cherry (1953, 1954) on different speech signals presented simultaneously to both ears: 'the final act of recognition can only be performed upon one *message* at a time.

(1950, p. viii). In particular, rhythms† help to produce our concept of time, instead of themselves being founded on a basis of uniform time, for as previously stated we have no direct experience of pure time but only of particular sequences and rhythms since it is not time itself but what goes on in time that produces effects. As William James (1890, vol. I, p. 620) pointed out, '*we can no more intuit a duration than we can intuit an extension devoid of all sensible content*'.

In his discussion of our experience of time as based primarily upon acts of attention to a succession of distinct presentations, Locke argued—incorrectly, as we now believe—that we have no perception of duration except by reflection on the train of ideas which we observe to succeed one another in our understandings, and he wondered 'whether it be not probable that our ideas do, whilst we are awake, succeed one another in our minds at certain distances, not much unlike the images in the inside of a lantern turned round by the heat of the candle' (Locke 1690, p. 109). He was aware that even when we are awake the 'degree of quickness' of the train of ideas in the mind may be 'sometimes faster and sometimes slower', but he thought that there were 'certain bounds' to their rate of succession 'beyond which they can neither delay nor hasten'. Be that as it may, our estimation of time is also controlled by other mental factors, notably by our sense of judgment. This is particularly evident in the case of dreams, like that of the Marquis de Lavalette (p. 63) where we are the victims of illusion, for it seems improbable that the temporal effects in such dreams are due to an enormous number of mental events occurring during only a few seconds of physical time. On the contrary, it would appear that the illusion of an immense time interval is due to an error of judgment. As Miss Sturt (1925, p. 117), who studied this question in considerable detail, pointed out, 'Dreams occur when we are physically still and they possess hallucinatory vividness. We seem to be acting, and yet there is no movement. In waking life the necessity of physical movement is a continual drag on the speed of our thoughts... In dreams there is a speed of actionless thought combined with the belief that we are acting, and then the judgment of the amount of clock-time occupied by a series of events is at fault'.

Locke's successors, Berkeley and Hume, visualizing mental time simply as a succession of ideas in the mind, concluded that it must be discrete, and so could not correspond to the continuous time variable of Newtonian physics. If we regard our experience of time as dependent upon acts of attention to successive presentations, are we also compelled to accept a

† Bachelard (1950, p. x) makes the stimulating suggestion that 'il y a place, en psychologie, pour une rhythmanalyse dans le style même où l'on parle de psychanalyse'.

similar conclusion and to regard the attribution of continuity as the result of after-reflection? This does not seem to be an adequate description and explanation of what actually happens. James Ward (1918), in his penetrating study of the problem, argued that our perception of a period of time is not strictly comparable to a discrete series of terms any more than to a series of infinitesimals. For, even if the most vivid impressions are discrete, it does not follow that the whole field of consciousness changes discontinuously. Attention does not necessarily move by hops from one definite spot to another but rather 'by alternate diffusion and concentration, like the foot of a snail, which never leaves the surface it is traversing. We have a clear presentation discerned as A or B when attention is gathered up; and when attention spreads out we have only vague and more or less confused presentations. To some extent, such confused presentations are always present, and so serve to bridge over the comparatively empty interval during which attention is unfocused'.

The problem of how time order is constructed by us on the basis of our movements of attention from one presentation to another is beset with difficulties.† We have already stressed the distinction between succession in thought and the thought of succession. Our conscious appreciation of the fact that one event follows another is of a different kind from our awareness of either event separately. If two events are to be represented as occurring in succession, then—paradoxically—they must also be thought of simultaneously. Unfortunately, both memory and the traces in the mind of our movements of attention can be notoriously unreliable guides to the order of events as they actually happened. Piéron (1952, p. 290) has drawn attention to this singular contrast between the unreliability of our capacity for conscious psychological appreciation of time and the physiological exactitude apparent in the establishment of organic rhythms in animal behaviour. More striking still is the fact that patients in a hypnotic trance have been found to possess a far more accurate sense of time than in their normal state, which not only points to the existence in ourselves of permanent organic and mental rhythms but also indicates

†For example, C. Spearman (1923) objected to the idea that our perception of temporal sequence is *due* to shifting of attention on the grounds that, with maximum effort, the latter requires at least 0.2 seconds, whereas the shortest possible time that will yield the perception of succession is much less, for example for hearing successive electric sparks it is only 2 milliseconds (0.002 seconds). He argued that our perception of nowness and sequence 'are just elementary cases respectively of apprehending the characters of experience and educing relations between these characters '—in other words, they are due to direct intuition. Although Spearman was justified in arguing against the *necessity* of 'attention' as a mediating clue for our perception of time, he overlooked the fact that it often greatly influences our conscious awareness of temporal succession.

that in the normal functioning of consciousness all such rhythms are overshadowed by the transience of external events.

In children the development of a conscious 'time sense' comes at a later and more sophisticated stage than the development of a spatial sense, presumably because it requires a greater degree of image representation. At first, the succession of events gives rise to purely motor responses. When it cries with hunger a very young child has its first experience of duration ('waiting time'). Each temporal sequence in the child's experience is isolated and purely egocentric: it begins with desire or effort and ends with success or failure. As Piaget (1969) has stressed, this does not mean that the child has an inborn sense of time; its egocentrism is simply a natural lack of awareness of the distinction between itself and the external world.

Robert Wallis (1966) has pointed out, in his admirable study of human time, that the long delay experience by the young child, as compared with other mammals, in acquiring the ability to walk (about a year) has a profound influence on the general development of the human intellect and, in particular, our sense of time. Initially, distance will be that beyond the child's reach, whereas for a puppy it will be the space it must traverse to get something. His relative immobility forces the young child to construct his knowledge on the basis of the perception of the world seen from his cradle by means of mental images. Wallis suggests that it could well be the child's eagerness to grasp what he cannot reach that infuses in him the first notion of objective time, indissolubly linked to a space that he cannot yet cross. Later, when he can at last walk, to reach is also still to *wait*, and precisely in the 'time' of waiting to traverse a certain space. This feeling of delay gives rise to both our judgment and appreciation of internal time. Moreover, it is crucial for the general development of the human brain because of the connection which it establishes between the time of expectation and the distance from satisfaction.

It has been shown that the gradual acquirement of particular temporal conceptions can be closely correlated with the child's acquisition and use of language (Decroly 1932). The 'construction' of time begins with the correlation of 'velocities', either of human activity or of external motions. The distinction between different durations is not grasped operationally until they can be correlated with a system of successions and simultaneities. At the age of eighteen months the meaning of 'now' can often be grasped and of 'soon' by two. As a rule, by the age of three the child can understand 'not today' and he uses 'tomorrow' and 'yesterday' in that sense (Gesell and Ilg 1943). Gradually temporal sequences begin to be recognized as applying to external events themselves and not merely to the child's own movements and actions, although time still remains an extrapolation of subjective duration inherent in his activity (Piaget 1955). At a much later stage, when time is no longer associated with the child's own activity, it still remains tied to particular objects or movements and subordinate to space. Piaget has found that if, at the age of four or five years, a child sees two moving objects leave the same point and arrive simultaneously at two different terminal points he will acknowledge the simultaneity of the departures but will usually contest that of the arrivals, even when this is clearly perceptible. 'He recognises that one of the objects ceased to move when the other stopped, but he refuses to grant that the movements ceased 'at the same time', because there simply is as yet no time common to different speeds. Similarly, he conceives of 'before' and 'after' in terms of spatial succession and not yet in terms of temporal succession' (Piaget 1950, p. 136). Piaget goes on to make the extremely significant point that, even when these difficulties have been overcome, *there nevertheless still exists a systematic incapacity to combine local times into one single time.* Even when a child of age six or seven years recognizes that two equal quantities of water flowing at the same rate into two bottles of different shape both begin and cease to flow simultaneously, he will deny that water has been flowing for as long into one bottle as into the other. Only at the age of about eight are the relations of temporal order (before and after) co-ordinated with those of duration so as to give rise to the idea of a time common to various movements at different speeds (Piaget 1950, p. 145).

Piaget has been criticized by Fraisse for basing his investigations of the growing child's development of time sense too much on considerations concerning speed. He thinks that Piaget was biased in his work on this subject because he undertook it after he had been asked by Einstein whether our intuitive grasp of time is identical with our intuitive grasp of velocity, and what bearing this has on the genesis and development of the child's conception of time (Piaget 1969, p. ix). Piaget assumed that our basic kinematic intuitions are of distance and speed, whereas Fraisse believes (correctly, in my opinion) that the young child also has a primitive intuition of duration. 'He sees the latter in the elementary form of an interval which stands between him and the fulfilment of his desires' (Fraisse 1964, p. 227). In particular, Fraisse disputes Piaget's contention that rapid action inevitably gives rise to an apparent contraction of time.† Instead, he points out that although it is true that speed sometimes increases the number of changes felt when it entails effort and concentration,

† Piaget's view is associated with the implication of the simple kinematic formula $v = s/t$ that, for a given distance s, the time t varies inversely as the speed v.

as, for example, in running† and in writing quickly, it can in other instances decrease the number when an action can be performed rapidly simply because it is easy. 'In both cases', says Fraisse, 'we judge the duration not in relation to speed but according to the effort made and the changes felt'.

Beyond the age of eight, the development of the idea of time has been studied by E. Michaud (1949), who asked nearly two thousand children between 10 and 15 whether they thought they become older when, in changing over to summer time, all clocks are advanced one hour. He found that the children questioned fell into four categories: (1) those for whom time is a 'real quantity', so that advancing the clock affects time itself and therefore makes them an hour older; (2) those for whom the change of time simply affects their activity, since an hour is lost that could have been useful; (3) those for whom the question asked is essentially a mathematical one, e.g. the 15-year-old who said 'I suddenly grew an hour older because the clock was put forward an hour, but we get the hour back in winter and I get an hour younger then... so my age hasn't changed'; (4) those who have come to realize that clock time is *conventional*, natural uniform time being independent of human action. Michaud found that about three children out of four aged ten did not regard time as an abstraction, and that not until 13 does one child in two realize this. Only at the mental stage described by Piaget as that of logical operations ('adolescence') does the child begin to grasp the abstract concept of time as something distinct from actual clocks. 'Adaptation to time', Fraisse concludes, 'is therefore a function of the development of the intelligence'. He points out that very high correlations have been found‡ between general intelligence and answers to questions concerning temporal orientation, the divisions of time and the ways of dating events, all of this being knowledge which must be learned but which cannot be understood until the processes of the measurement of time have a meaning for the child (Fraisse 1964, p. 280).

As regards adults, it has been found that our judgment of duration is

† In his illuminating discussion of distortions of the time sense in the *Handbook of Clinical Neurology*, Dr. William Gooddy (1969) points out that not all distortions of the time sense can be considered pathological, and in referring to the apparent indefinite prolongation of 'personal time' in conditions of extreme physical and emotional exertion, he draws attention to the experience of Dr. Roger Bannister, who states that when running at the limits of physiological endurance the time sense may be significantly changed. At the moment of achieving the 'four-minute mile', Bannister felt that his body had exhausted all its energy and only willpower kept him going. 'This was the crucial moment when my legs were strong enough to carry me over the last few yards as they could never have done in previous years. With five yeards to go the tape seemed almost to recede. Would I ever reach it? Those last few seconds seemed never-ending...' (Bannister 1955).

‡ In western cultures, of course.

more accurate for short intervals up to 5 or 6 seconds, presumably because we can concentrate on them directly, whereas in the case of longer intervals our minds tend to wander and our judgments are influenced by the number and kinds of events that have filled the time.

Our subjective judgments of duration affect other simultaneous experiences. Thus, the so-called *tau* effect (Helson and King 1931) indicates that judgments of spatial distances depend on the time taken to traverse them. If three points are marked on the skin and the interval of time between stimulating the second and third is greater than the interval between stimulating the first and second, the subject will judge the distance between the second and third as greater than that between the first and second, irrespective of whether it may in fact be equal or less. Similar results have been obtained with visual phenomena. Conversely, temporal judgments of duration have been shown to be affected by associated spatial components—the *kappa* effect (Cohen, Hansel, and Sylvester 1953, 1954, 1955). For example, if one is faced by three sources of light which flash successively and is asked to control the middle flash so that it comes midway between the first and third flash, one will tend to allot a *shorter* time to the interval between the pair of flashes that are at the *greater* distance apart. Similar results have been obtained with auditory phenomena. If the subject listens to two different continuous tones and is asked to assign an equal duration to each, he will tend to allot a shorter duration to the tone of higher pitch. Phenomena such as these indicate that our conscious experience cannot be completely analysed into independent sense data and that its various aspects are interrelated; in particular, the spatial and temporal components are interdependent.†

2.6. Is there a standard unit of perceptual time?

Wilhelm Wundt (1832–1920), who is generally regarded as the man who established psychology as an independent discipline on an experimental basis (although he held a chair of philosophy at the University of Leipzig from 1875), made the time sense one of the main subjects of investigation in his laboratory from its establishment in 1879. He and other psychologists working there, notably J. Kollert (1882) and V. Estel (1884), believed that there exists a definite psychological unit of time available to the mind as a

† This interdependence is also revealed by analysis of *simultaneous* phenomena. According to R. W. Sperry (1952), the perception of simultaneous spatial relations depends upon temporal organization in the brain processes. For example, if we wish to understand the physiological mechanism involved in the visual perception of, say, a triangle, he suggests that we should picture the triangle as if it were being constructed in time out of dots and dashes passed *successively* to the brain so that the perception of the triangle as a whole comes gradually.

standard. They studied intervals between the two beats of a metronome and determined an 'indifference point' (or zone), i.e. a duration for which the average error of estimation is zero, the tendency being for shorter intervals to be overestimated and longer ones underestimated. They concluded that this indifference point exists and is about three-quarters of a second. However, H. Münsterberg (1889) disputed this claim and maintained that there is *no fixed mental unit of duration* and that one should try instead to find *physiological* measures of the time sense. Forty years later the validity of his argument was conclusively demonstrated by H. Woodrow (1930), who showed that differences between individuals cannot be smoothed out to give the desired result, one factor that influences estimates being the subject's 'attitude of observation', a factor that had not been controlled in the original experiments but which tends to change with the length of the duration judged and so influences the location of the 'indifference point'.

Despite there being no psychological manifestation of a precise standard unit of perceptual time in Wundt's sense there is what Fraisse (1964, p. 120) has called 'the development of a *central tendency*' related to the range of durations perceived in a given situation. Generally, we expect a stimulus of any kind (e.g. the apparent weight of a chair when we lift it) to be around some average value and we tend to minimize small differences and overestimate those that are fairly large. Consequently, in a range of stimuli we usually overestimate those below the average and underestimate those above it. This law is particularly well illustrated by temporal intervals. Nevertheless, Fraisse (1964, pp. 121–2) argues that there is a zone around 0.7 seconds that may well correspond to some definite physiological process for it is found in connection with phenomena of various different kinds where no true central tendency is observed.

Some support for this suggestion is provided by researches that enable us to estimate the duration of the complex process involved in a single act of perception. A good example is provided by an ingenious experiment devised by Renata Calebresi (1930). We can usually apprehend a maximum of seven to eight letters at a single glance. If, however, letters are presented to us for a very short interval of time, 10 milliseconds, then the most that we can perceive is 4.1 letters on the average. Calebresi had the idea of presenting two four-letter groups, for 10 milliseconds each, at different temporal intervals between them. She found that the average number of letters retained varied with the length of the interval, ranging from 4.4 letters if the interval were 0.05 seconds to 6.5 letters if the interval were 0.7 seconds, and 7.2 letters if the interval were 1 second or even 1.2 seconds. She concluded that an interval of about 1 second is necessary for the two groups to be totally apprehended together. From this it would appear that the *complete* perceptual process requires about 0.7 to 1 second.†

If we are to perceive that two events occur in succession, there must be a certain minimum separation between them. The reciprocal of this minimum separation can be regarded as a measure of temporal acuity. It depends on the particular sensory mechanism involved and is greatest in our auditory experience. If two brief clicks that are distinguishable (e.g. of unequal intensity) are separated by an interval as short as 2.5 milliseconds, the ear can determine their temporal order (Patterson and Green 1970, 1971). This minimum time in which *conscious*‡ perception of auditory succession is possible is about one-tenth of that for our perception of visual succession, although the temporal interval necessary for discrimination between two successive flashes of light can be shortened by direct stimulation of a particular type of nervous tissue§ in the sub-cortical region of the brain known as the 'reticular formation' (see H. H. Jasper 1958). With visual projection of an object for about 10 milliseconds an observer is conscious of something, but usually cannot say what it is.

J. Cohen (1967) has drawn attention to the penetrating observation by the famous neurologist J. Hughlings Jackson that *time in the form of some minimum duration is required for consciousness*, the loss of consciousness in an epileptic seizure being due to the sheer speed of neural discharge. Cohen points out that our visual system normally acts as a protecting buffer for the brain by regulating the *rate* of information transmitted to it. The minimum duration of a visual experience can be defined in terms of the rate at which the successive presentations of static images is seen as

† Ornstein (1969) has pointed out, however, that although this is the interval we take for our metrical unit of time (whether c.g.s. or SI), this choice is determined by our culture. He refers to an Indian culture that takes as its temporal unit the time required to boil rice, and asks 'how do you suppose that our technological culture could now get along with that sort of basic unit'?

‡ Our *subconscious* powers of auditory discrimination, however, are amazing. Experiments have shown (Klemm 1919, Hornbostel and Wertheimer 1920) that our cues for sound localization are provided by our perception of binaural time differences, differences of binaural intensity being far too feeble. The binaural time difference is greatest when the sound originates close to one ear, but even then is less than a millisecond. When the sound originates 3° to one side of the median plane of the head, the time difference is only about a fortieth of a millisecond. Nevertheless, it still provides an effective cue. The binaural time differences must, of course, be 'decoded' before we can refer them to location. This decoding is primarily a reflex motor process—a quick instinctive turning of the head to the source of the sound.

§ The time unit of serial neurophysiological processes is of the order of a millisecond (see p. 98). Doubtless, this is because if it were nearer one-hundredth of a second these processes would tend to be disrupted by thermal 'noise'.

apparent movement, particularly in the cinema. It is about 16 to 18 a second, each frame lasting about 0.06 seconds. For auditory sensation the minimum frequency is also about 16 to 18 cycles a second, and similarly about 18 tactile signals a second is the rate at which a clearly differentiated series of tactile impressions can be distinguished from an impression of tactile vibration. On this basis the duration of the unit of perceptual time would be of the order of 50 milliseconds. We may therefore take this as the approximate duration of the *mental moment*, i.e. the interval between distinguishable perceptions or decisions.†

2.7. The psychological present

We have seen that in rejecting Kant's theory thinkers like Mach claim that our awareness of time is a sensation, whereas others like Janet argue that it is an intellectual construction. Although we have adhered to the latter view, we must now consider the problem further. First we take note of the fact that direct perception of change, irrespective of whether it is explicitly recognized as succession, requires the simultaneous presence in our awareness of events in distinct phases of presentation. The combination of simultaneity and succession in our perception means that the time of our conscious experience is more like a moving line than a moving point. That this is in fact so was pointed out in 1882, in an anonymous book which we now know to have been written by E. R. Clay. 'The relation of experience to time', he wrote, 'has not been profoundly studied. Its objects are given as being of the present, but the part of time referred to by the datum is a very different thing from the coterminus of the past and future which philosophy denotes by the name Present' (Clay 1882). Thus, all the notes of a bar of music seem to the listener to be contained in the present,‡ just as all the changes of place as a meteor flashes across the sky appear to the observer to be contained in the present. Clay called this finite segment of time which constitutes our immediate experience the *specious present*. William James suggested that, because each stimulus of the nervous system leaves some latent activity which only gradually fades away, we experience at each moment brain

† This agrees with the rate of word-scanning frequency in an easy book, about 20 words a second.

‡ If our perception of the present were truly instantaneous we could not recognize that the notes of a melody make up a tune. In this context we may recall a pertinent remark by Coleridge in *Biographia Literaria*: 'The delicious melodies of Purcell and Cimarosa might be disjointed stammerings to a hearer whose partition of time should be a thousand times subtler than ours'.

processes that overlap each other, the amount of overlapping determining the feeling of the duration occupied (James 1890, p. 635). He adopted Clay's term with enthusiasm on the grounds that the 'true present' must be durationless, a moment of time sharply dividing past from future and utterly distinct from both. This so-called 'true present' will be discussed in Chapter 4; it is a mathematical idealization like the dimensionless point of geometry. Consequently, the term *specious present* is itself somewhat specious, and it would be preferable to use instead a more neutral term such as *psychological present.*

The range and content of the psychological present depends on the focusing of our attention, but it can embrace both primary memory images and immediate expectations or pre-percepts. A classic visual example of pre-perception occurs when a doctor directs his attention to a patient's blood and sees it flow before the lancet penetrates the skin. An equally vivid example of a primary memory image which forms part of the psychological present was suggested by Bertrand Russell, who argued that we sometimes notice that a clock *has been* striking although we did not notice it while it was striking. Mundle (1954) criticized this example on the grounds that our conscious awareness of the sounds in question could be attributed to our hearing a noise which was still going on, caused by the continuing vibration of the striking mechanism, but he too claimed that there are other cases in which one seems able to 'inspect' sounds which have audibly terminated. He described how when falling asleep he had sometimes been disturbed by a short series of sharp blasts from a railway engine in a nearby shunting yard and that he found himself attending to such sounds after, and only after, they had been audibly terminated, the sounds still seeming 'immediately accessible' in the sense that he could count them and discriminate their relative durations.

A visual example that has been frequently discussed is the movement of the second hand of a watch; we seem to see this directly in a way in which we do not see the movement of the minute or the hour hand. C. D. Broad (1923) argued that to see a second hand *moving* is quite different from 'seeing' that an hour hand *has* moved, since in the one case we are concerned with 'something that happens in a single sensible field', whereas in the other we are concerned with a comparison between two different sensible fields. Broad's contention that a movement of definite duration is literally sensed *as a whole* was disputed by H. J. Paton (1951) who maintained that 'if in a moment I can sense several different positions of a second hand, then these different positions would be sensed as being all at the same moment. That is to say, what I should sense would not be a movement, but a stationary fan covering a certain area'. Nevertheless, in the case of a falling star, we *are* aware that one part of the movement is earlier than another, in spite of the whole being comprised within one mental present. Russell (1948) maintained that, if we were not aware of this, we should not know whether the movement had been from A to B or from B to A.

Although James spoke of its permanence of quality, he realized that the specious (or psychological) present is not an interval of fixed duration, but is a variable stretch of time with its content perceived as having one part earlier and another part later. The term 'specious present' as used by psychologists is unfortunately somewhat ambiguous. In its widest sense it can be regarded as signifying a duration of temporal experience compatible with a certain perspective unification. In a more restricted sense it may be confined to an interval of time during which events are not recognized as being earlier or later but are confused in an apparent simultaneity. In the case of events affecting two different receptor systems, such as sight and hearing, two physically simultaneous events can be perceived as successive, and two physically successive events can be perceived as simultaneous—or even in reversed order. These confusions are partly due to the different physiological delays between stimulation of the receptor organ and the resulting perception, and partly to personal factors. Simultaneity of heterogeneous sensations has thus to be estimated from the indefiniteness of their apparent temporal order.

The systematic study of this question, which has an important practical bearing on the confidence we have in our senses as instruments of observation, originated in astronomy, with the discovery of the phenomenon of 'reaction times'. In 1796 the Astronomer Royal Maskelyne dismissed his assistant Kinnebrook because he appeared to be inaccurate in his observation of stellar transits. Nearly twenty years later it occurred to Bessel that the difference between the two astronomers might be personal. It is now universally recognized that even the best observers habitually note the passage of a star across the fixed wires of a transit instrument slightly too early or slightly too late by an amount which varies from one observer to another, called the 'personal equation'.

In analysing the concept of the present, Guyau stressed the connection between time and action. Bergson went further and argued that one must not merely act but must be conscious of acting, that is one must be aware of producing a certain effort. Janet (1928, p. 309) claimed that this too is insufficient and that the present must be viewed as an intellectual act uniting narration with action: it should be regarded as 'un récit de l'action que nous nous faisons à nous-même pendant que nous sommes en train d'agir'. He cited convincing evidence for this conclusion from his detailed analysis of a bizarre form of amnesia (often due to chronic alcoholism) known as the Korsakoff syndrome.† In this illness the patient appears perfectly normal except that he never speaks about the present and suffers from retarded memory. He has no powers of conscious reflection on the present but only on the not too recent past. Thus, when Charcot's patient, Mme D., who had developed this syndrome following severe shock and hysteria, was severely bitten by a dog some time later, she said that she had a pain in her leg but had no idea what it was. Whenever Charcot questioned her she replied intelligently, but as soon as his back was turned he ceased to exist for her. Nevertheless, her recent memory must have existed in her dreams, for in them she talked about events of the same day and cried out against the dog that had bitten her. Janet discovered that Mme D., whom he studied carefully for several years, required at least eight days to organize an observation. He concluded that 'organization of the present' depends on our immediate memory.

Thus, so far from being a simple matter of direct sensation, our psychological present must be regarded as the product of an elaborate construction. It is intimately related to our past, since it depends on our immediate memory, but it also determines our attitude to the immediate future. The vagueness which seems to characterize its temporal extension is inherent in its nature. 'La construction du présent', wrote Janet (1928, p. 313), 'empêche une détermination précise de sa durée'. For, in the words of Whitehead (1920), 'the temporal breadths of the immediate durations of sense-awareness are very indeterminate and dependent on the individual percipient... What we perceive as present is the vivid fringe of memory tinged with anticipation'.

2.8. Memory and the concept of the past

Although awareness of the present is the most fundamental temporal experience, its relation to the past can be complicated, as has been shown by Janet (1928, pp. 321–4) in his discussion of the curious psychological

† Korsakoff (1889 a, b); for recent literature on this syndrome see Talland (1965) and Barbizet (1970). A useful short historical survey of the literature since 1889 was given by Rapaport (1961, pp. 226–31).

No less remarkable for its disorientation effect on the patient's sense of time is the post-encephalitic syndrome described so brilliantly by Oliver Sachs (1973). Some patients who survived the *encephalitis lethargica* (sleepy sickness) pandemic of 1917–27 passed into a timeless state which deprived them of all sense of history and happening. For example, Rose R. in 1969 was 64 years old but she felt as if she were still in 1926, the year of the onset of her illness. Dr. Sachs believes that she felt her 'past' as present. 'Is it possible', he asks, 'that Miss R. has never, in fact, moved in from the "past"? Could she still be "in" 1926 forty three years later'? (Sachs 1973, p. 69). In the case of another of his patients (Mrs. Y), treated with the drug L-dopa, Sachs reports the occurrence of curious distortions in the time ordering of her perceptions so that they sometimes appeared to occur in the wrong sequence.

phenomenon of *déjà vu.* This is the sense of false familiarity which occasionally characterizes the entire present situation so that we have the uncanny feeling that we have experienced long ago what is happening now—and also that which is going to follow—despite the fact that we know from causal reconstruction that this is impossible.† This false sense of pastness is due to the feeling of non-presence ('*c'est le sentiment de l'absence du présent qui caractérise ces malades*'). In *déjà vu* it seems that whatever occurs is regarded as familiar. It may be that we unconsciously apprehend before we are consciously aware, so that the ensuing conscious perception is, in fact, a memory of what we subconsciously registered a moment before (Cassirer 1957). Be that as it may, the phenomenon clearly reveals both the semantic correlation of the 'familiar' with the 'past' and the crucial role of memory in our construction of the latter.

The connection between familiarity and memory was stressed by Bertrand Russell in his reformulation of a theory due to Hume. In his search for the essential characteristic which distinguishes memory from imagination, Hume (1738) came to the conclusion that remembering consists in having mental images of 'superior force and vivacity', but this solution has long been recognized as quite inadequate, and often false. According to Russell (1921), a memory image is distinguished from other images in the mind by the feeling of *familiarity*, and it is this feeling which provides the sense of pastness. Although this solution is clearly superior to Hume's, it has been questioned on the grounds of circularity, since it might be argued that familiarity itself presupposes memory. However, Russell's point was that the two are synonymous. Hence the objection fails. Nevertheless, Russell's theory leaves open two important questions:

(i) What guarantee have we for the reliability of memory?
(ii) Is memory necessary and sufficient for knowledge of the past?

The difficulty in answering the first question is to avoid a *petitio principii*, for any test of the reliability of memory would appear to involve

† A good description of the *déjà vu* phenomenon is given by Dickens in *David Copperfield* Chapter XXX IX (p. 530 in the Everyman edition): 'We all have some experience of a feeling, that comes over us occasionally, of what we are saying and doing having been said and done before in a remote time—of our having been surrounded, dim ages ago, by the same faces, objects and circumstances—of our knowing perfectly well what will be said next, as if we suddenly remembered it!'

The *déjà vu* phenomenon may have been one of the psychological sources that gave rise to the doctrine of metempsychosis taught by the Pythagoreans and others. It is probable that the hippocampus is involved in any temporal lobe fit which causes *déjà vu* (Ritchie Russell 1959). The hippocampal system (two elongated prominences on the floor of each ventricle of the brain) appears to play an essential role in our recognition of the familiar. Patients who have had the hippocampus removed on both sides forget the incidents of daily life as quickly as they occur, although they clearly remember details of their childhood.

an appeal to memory. A subtle analysis of the validation of self-authenticating memory was given by Harrod (1942), who maintained that the contention that memory is informative is a hypothesis 'which must take its chance in the rough and tumble of experience along with other hypotheses'. He believed, however, that the hypothesis could be 'verified', i.e. justified without appeal to intuition, if we agreed to accept *a priori* a principle of induction in the form of the conditional proposition that, *if* certain things have been found to remain stable for some time, then they are *likely* to continue to do so for a little longer. He admitted that it was 'with reluctance' that he introduced this *a priori* principle,† but he argued that, if it were accepted, then memory could be validated by the 'fulfilment of predictions'. In order to justify belief in the latter *without invoking memory*, he appealed to predictions which are made and fulfilled within a so-called specious present.

Harrod's ingenious argument was criticized by Furlong (1951, pp. 58 *et seq.*) who objected to invoking an inductive principle and also to assigning so crucial a role to the specious present (although he himself fell back on this concept in his own theory!). Instead, he believed that we cannot *prove* that memory is infallible. All we can, and all we need, do is to account for our belief in its general reliability by appeal to experience. We are, in point of fact, continually finding that our memories are dependable informants, although we often must appeal to the confirming evidence of other men's memories as well as our own. The distinction between remembering and imagining is, therefore, *logical* rather than psychological. This point has been clearly expressed by Peters: 'The test of whether a person remembers or imagines is not the subjective criterion of pastness accompanying the imagery, but the evidence which confirms or refutes what is asserted about the relationship between the situation thought about and the thinker's participation in actual events. And to establish whether or not such a relationship holds—i.e. whether it *is* a case of remembering rather than imagining—a person's private conviction is a good guide but an unreliable test' (Peters 1956).

Turning to the second question, we find that philosophers have often tried to answer that in memory we are *directly acquainted with the past.*‡ Apart from its implausibility (which alone would not be sufficient reason for rejecting it), this hypothesis fails to explain the 'pastness' of past events, the fact that they once *were* but now are *not.* Indeed, the

† He was however, careful to state (p. 62) that he was *not* invoking a *general* principle of uniformity of nature but only 'uniformity limited in space and time and scope of application' and that he was only claiming probability, not certainty.

‡ For example, Samuel Alexander (1920): 'The object is compresent with me as *past*'.

hypothesis makes it difficult for us to avoid self-contradiction on this score, for we cannot occupy two different positions in time simultaneously. On the other hand, philosophers who reject the hypothesis and maintain that in memory we merely *image* events are faced with the problem that this imagery tells us little about temporal context. Instead they usually appeal to our conscious awareness of the *succession* of events in the mental present. They argue that this gives rise to our initial notion of pastness which we then gradually learn to extend beyond that range. Indeed, Furlong claimed that children develop their powers of memory in just this way. 'At first, they have very little ability to recollect. Even at the age of two they may be quite unable to remember where they hid a toy a few hours before. By three they can tell what happened yesterday, but other past events are remembered as having happened 'a long time ago'. The learning of the days of the week and the months of the year is the work of a still more mature age' (Furlong 1951, p. 96).

Eva Cassirer suggested that the notions of temporal succession and of the past-as-remembered should be regarded as alternative descriptions of the same thing. She argued that, if we focus attention on A during a short interval of time at the end of which we experience B, we can say that we are experiencing B while remembering A, in the sense of having A as a lingering sensation (*not* an after-image), or a thought on which our attention was fixed, *retained* in the mind. At the moment of experiencing B our attention is ready to shift to it from A; and the slight effort that it takes to retain A in our attention up to and through B (during the time of experiencing B our attention is actually divided) constitutes what she calls the 'memory-effort'. 'We judge that part of our quasi-compresent experience which is associated with this effort as the earlier of the two events. The difference which we recognize as such by a shift of attention is experienced as the succession of two events within, or at the fringes of, one span of attention i.e. within the 'specious present'. The notions of succession and of the past-as-remembered arise in one situation and are alternative descriptions of the same thing' (Cassirer 1957, pp. 39 *et seq.*).

The association of perception with the 'past' was stressed by Bergson in his famous book *Matter and Memory*. He contended that the so-called 'present' consists mainly of the immediate past and that every perception involves memory: '*Practically we perceive only the past*, the pure present being the invisible progress of the past gnawing into the future'. (Bergson 1911, p. 194). Russell has rightly objected that Bergson's definition of our past as 'that which acts no longer' (Bergson 1911, p. 320) is circular. He argued that Bergson's theory of time was based on a confusion of past occurrences with our present memory of them (Russell 1946) and, in fact, *omits time altogether*! It is, in his view, an account of the difference between perception and recollection—both *present* facts—and not, as Bergson believed, an account of the difference between the present and the past.

Nevertheless, Russell too fell into error in his own view of the relation between memory and time when he maintained that it is merely an 'accident' that memory works backwards and not forwards, in other words that it reveals the past and not the future (Russell 1917). This view implies that our relations to past and future would be symmetrical were it not for some fortuitous quirk of mind. It overlooks the fact that, when we remember the past, just as much as when we attempt to foresee the future, our thinking tends to run forward in time (we think of events in the order in which they happened), and that to reverse this natural sequence in remembering requires considerable effort. This is not fortuitous, for the very nature of mental activity is to reach out towards the future in order to anticipate the event which is about to happen.†

A similar mistake underlies Ayer's contention that 'There is no *a priori* reason why people should not succeed in making true statements about the future in the same spontaneous way as they succeed, in what is called the exercise of memory, in making true statements about the past. In neither case is their state of mind important; all that matters is that they get the answers right without having to work them out' (Ayer 1956, p. 166). In an essay-review, H. H. Price (1958, p. 457) pointed out that this 'cheerful view of the matter' was a natural consequence of Ayer's whole-hearted acceptance of the idea expressed in the epigram *Le temps ne s'en va pas, mais nous nous en allons* (Ayer 1956, p. 153). For this idea of our temporal position continually changing has a close affinity with the theory of 'the block universe',‡ despite the fact that this theory implies that past (and future) events coexist with those that are present—a view which Ayer categorically rejected. Consequently, it is not surprising that Ayer was puzzled by the fact that we normally stipulate that cause cannot succeed effect. His solution was to appeal to the *fact* that we know very little about the future compared with the past: that is why 'our reliance on memory is an important factor in forming our idea of the

† For example, by providing information of distant events our eyes function as biological 'early warning' systems enabling us to probe the future, e.g. by anticipating danger. Inadvertent acts of anticipation include spoonerisms and the inversion of letters in typing; according to Lashley (1951), it is as if the aggregate of word units are 'in a state of partial excitation held in check by the requirements of grammatical structure but ready to activate the common path if the effectiveness of the check is in any way interfered with.'

‡ According to this theory external events permanently exist and we merely come across them (see p. 274).

causal direction of events' (Ayer 1956, p. 175). However, if it ever were to happen that we came to rely equally on precognition, we should presumably take a 'spectator attitude' and regard causality as reversible.

Despite the lucidity of their analyses, neither Ayer nor Price succeeded in resolving the point at issue. Indeed, Price even remarked on the difficulty of talking sense about time (Price 1958, p. 454). Curiously enough, neither referred to the following elementary, but essential, difference between past and future events. Consider, for the sake of illustration, two machines, one of which automatically records a particular sequence of events, for example the barometric pressure at a certain meteorological station, whereas the other forecasts a similar sequence. Apart from the fact that the second machine would presumably be far more complex than the first, there is a fundamental difference between their respective modes of functioning: each reading recorded by the moving pen of the barometer as the clock-controlled, paper-covered cylinder revolves is made *simultaneously* with the event to which it refers, whereas those printed on the tape emerging from the pressure-predictor would not be produced simultaneously with the events to which they refer. This illustrates the essential difference between memory and prediction and reflects the asymmetry between past and future. It explains why, on the whole, we place far more credence in our memories than in our forecasts.†

Of course, we can have no logically irrefutable proof of the absolute reliability of any memory, whether human or mechanized, because the hypothesis that everything, including simulated memories, came into existence a moment ago, although in conflict with our generally accepted explanations of phenomena, is not formally inconsistent with our present experience. All statements about the past must in the end be based on our readiness to accept as axiomatic some statement concerning the past, for example, that the wavy line on a barometric chart refers to a genuine temporal sequence, i.e. to events that actually happened one after another and not simultaneously.

The great bulk of our knowledge of the past is based on historical record and theoretical inference from archaeological, geological, and other data. Nevertheless, although our own personal memories do not go very far back, they are of vital importance. Most historical records are based on personal memories of events experienced by the writer or his

† Even our faith in the predictions of, for example, Her Majesty's Nautical Almanac Office is based on our *memory* of the past reliability of the *Nautical Almanac*. Of course, we can also make *retrodictions* which are similar to predictions in so far as they are not made simultaneously with the events to which they refer. The difference between retrodictions and predictions turns on the extent to which the former depend on events which were recorded when they occurred and the latter on other predictions.

contemporaries. Moreover, although memory as such must be distinguished from the perusal of recorded memories, and the past—which is constructed by the critical co-operative effort of mankind—kept distinct from our individual past-as-remembered, personal memory is an essential factor in our knowledge of the *near* past.

2.9. Time and the psychological aspects of memory (i)

The term 'memory', like so many words in common use, has a number of distinct meanings. We use it to denote both the *retention* and the *recall* of our perception of specific past events (and of our past thoughts) in their time setting. We also use it to denote 'immediate memory', remembering proper names and much that we read and hear, and also the recognition of familiar afferent stimuli (associated with previously encountered people, objects, places, etc.). We share our capacity for this last type of undated conscious memory with some animals. Also, like them, we remember how to perform certain routine and skilled operations (habit memory etc.). At a lower level, we possess *unconscious* spinal cord memory which controls, for example, much of our lower limb activity. However, since our concern in this book is with time, we shall concentrate attention mainly on the highest type of memory, our memory of past events.

In this, as Aristotle repeatedly remarked, the idea of time is an essential feature. In his *De memoria et reminiscentia* (449b24) he wrote: 'only those animals which perceive time remember, and the power by which they perceive time is also that by which they remember', for 'when someone is actively engaged in memory, he perceives in addition that he saw this, or heard it, or learned it earlier; and earlier and later are in time' (Sorabji 1972, p. 49). As Marjorie Grene (1963) has pointed out, Aristotle's short treatise deserves more attention than it usually gets from philosophers because of its contrast not only with Plato's standpoint but also with modern associationist and behaviourist theories of mind like that of Hume.

According to Hume, in whose philosophy there is no perception of time but only a succession of disconnected atoms of sensation, the difference between a memory (an 'idea' in his terminology) and a percept (or 'impression') is only explicable by what Miss Grene calls 'the limping criterion' of vividness, a criterion which Hume himself did not regard as entirely satisfactory. Nevertheless, even Aristotle has been criticized (Spearman 1937) for ultimately succumbing to *the widespread fallacy that memory can be defined without reference to time* by appealing to the notion of copying when he wrote: 'memory and remembering are . . . the having of an image regarded as a copy of that of which it is an image' (Sorabji 1972, p. 52). To this Spearman retorted 'Shall we say that a footprint remembers the foot which made it'? Spearman did not confine his criticism to Aristotle. He drew attention to the fact that in his famous address delivered before the Imperial Academy of Science at Vienna, *Das Gedächtnis als allgemeine Funktion der organisierten Substanz*,† the physiologist E. Hering left awareness of time wholly out of account. Hering's views attained widespread influence through their detailed presentation by the zoologist R. Semon (1904), who postulated that *any* stimulus acting upon 'irritable matter' leaves behind it a *permanent* imprint, or physical trace, which he called an *engram*. When studying the psychological aspects of memory we must therefore be on our guard against neglecting the temporal dimension, and this means that we must be careful to distinguish between 'retention' and 'recall', although the latter (whether voluntary or induced) is the only test so far available for the former.

It has often been observed that we remember, that is recall, best those thoughts that are connected with our special interests. For, as William James remarked about Darwin's phenomenal memory for biological facts: 'Let a man early in life set himself the task of verifying such a theory as that of evolution, and facts will soon cluster and cling to him like grapes to their stem' (James 1890, Vol. I, p. 662). On a far humbler level, it is a common fact of experience that an athlete who is innocent of other intellectual accomplishments will often have a phenomenal knowledge of statistics relating to games and sports. This is because he is constantly going over these things in his mind, so that they form for him not so many odd facts but a concept system full of interrelations, every fact being retained by the combined suggestive power of the whole mass.

The importance of the associations and 'setting' of our individual memory elements can hardly be overestimated. If we recall a past event without any associates or definite setting, we find it extremely difficult to decide whether it is an act of memory or imagination. On the other hand, it is well known that, if we entertain an imaginary proposition for a long time and are continually referring to it, we may ultimately come to believe that it is a true memory—witness George IV's conviction in later life that he was present at the battle of Waterloo and led a cavalry charge! Moreover, in old age the superior tenacity of the mental associations formed by events remembered in childhood often contrasts painfully with

† 'On memory as a universal function of organized matter' (for an English translation, see Hering (1905)).

the inability to recall what happened even only a few minutes previously. Whatever our 'memory-organ' may be, in this state of presbyophrenia it is almost impossible to make any new record in it.†

The systematic experimental investigation of memory was inaugurated‡ by Ebbinghaus who reported his results in 1885 in his famous monograph *Über das Gedächtnis.*§ In order to study the subject objectively, he devised experiments (on himself) involving nonsense syllables. He investigated forgetting quantitatively, in particular by determining the number of repetitions needed for relearning given material after varying intervals of time. He found that the curve of retention falls rapidly at first but later flattens out asymptotically, indicating—in his view—that associations once formed are never entirely lost. He also discovered that with a given number of repetitions more was learnt when they were separated by intervals of time—and the more numerous the intervals, the better the results. Other experiments showed that, in learning a series of syllables, associations are formed not only between neighbouring syllables but also between more distant members of the series. These associations occur in both temporal directions, namely, as learnt and retroactively.

The importance of forgetting for the useful functioning of memory had already been stressed a few years previously by the French psychologist Ribot. He pointed out that all recollected items undergo 'foreshortening' owing to the omission of an enormous number of facts that originally filled them. This foreshortening is, however, a tremendous advantage for us, for, if in order to recall an event in our distant past it were necessary to traverse the entire series of intervening items, memory would be impossible because of the length of time required for the operation. We therefore arrive at the paradoxical conclusion that an important condition of remembering is that we should be able to forget! 'Forgetfulness, except in certain cases, is not a disease of memory, but a condition of health and life' (Ribot 1885). Indeed, idiots with rote memories cannot recall a

† The disappearance of memories from our powers of recall begins with the most recent and hence unstable and proceeds to the older and more stable, ending with learned skills, habits, and instinctive memories. This was designated by Ribot (1885, p. 122) as 'the law of regression or reversion'. The disappearance of words is governed by a rule of grammatical order that has come to be known as Ribot's Law: proper names go first, then common nouns, and finally verbs. This order of progressive forgetfulness, which to some degree we all suffer from as we grow older, has been explained by Bergson (1911, pp. 151–2): forgetfulness proceeds from words to gestures, and because verbs approach more closely actions that can be mimicked by us they are the last to go (see also Delay 1961).

‡ Apart from slightly earlier pioneer studies by Francis Galton.

§ For an English translation, see Ebbinghaus (1913).

particular memory without recounting a whole train of events, however unimportant and irrelevant, in serial order.†

It is well known that pre-literate peoples tend to have phenomenally good memories judged by our standards.‡ Many of the oldest epics were originally handed down orally. Plato commented on the fact that the art of writing was detrimental to the cultivation of memory. Children often display remarkable spontaneous memory for detail shortly after the material is originally presented to them. Presumably, they retain their initial impressions with greater facility because they do so with less discernment.§.

The two most famous modern theories of memory, Bergson's and Freud's although in other respects very different, agree in postulating that *all* forgetting is due to failure of recall and not of retention. In other words, forgetting is assumed to be a reversible process and memory, in the sense of 'unconscious retention', is irreversible. This hypothesis cannot be refuted by purely psychological tests, since these concern recall rather than unconscious retention. There is some striking, although inconclusive, evidence in its favour, for it is well known that, either as the result of illness or an accident or under hypnosis, patients often recall in meticulous detail events which previously they had appeared to have completely forgotten.‖ A remarkable example has been cited by R. W.

† Warrington and Weiskrantz (1970) have suggested that the primary disability of amnesic patients is an inability to suppress unwanted information. They believe that normally some mechanism of selective forgetfulness prevents the brain from being overloaded with information and thereby rendering the recall of an appropriate item difficult.

‡ D. Kay (1888) tells how the celebrated missionary Dr. Moffatt was surprised to find shortly after preaching a long sermon to a group of African natives that one of them, a simple-looking young man, was repeating it in full to an attentive crowd with uncommon precision, imitating as nearly as he could the manner and gestures of the original.

§ Some individuals, however, are naturally endowed with phenomenal powers of memorizing which they retain in adult life. A celebrated case was the journalist William Woodfall (1746–1803). Although newspapermen in his time were forbidden to take note of speeches in Parliament, he founded in 1789 a daily paper in which for the first time parliamentary debates were published the morning after they had taken place. Woodfall used to sit in the gallery with his elbows on his knees and his head in his hands memorizing the speeches.

Calculating prodigies tend to have remarkable memories for numbers, but otherwise they are often of mediocre intellect, although there have been some notable exceptions. A similar facility for languages requires even more phenomenal powers of memorizing. A famous example was Cardinal Mezzofanti (1774–1849) who spoke over fifty languages fluently, besides having a less perfect acquaintance with many more – but in other respects his intellectual abilities were modest (Russell 1858).

‖ Some interesting cases of the recall of supposedly forgotten memories can be found in Forbes Winslow's classic *On obscure diseases of the brain and disorders of the mind* (1861). For example, he cites the case of a French countess who, having left her native land during the Revolution and settled England, suffered a severe attack of fever and in her delirium

Gerard (1953) of a bricklayer who, under hypnosis, is alleged to have 'described correctly every bump and grain on the top surface of a brick he had laid in a wall twenty years before', but doubts have been thrown on the validity of this claim.

Freud's other hypothesis was that all forgetting, even minor slips of the tongue and pen which we normally attribute to 'chance', is in fact *motivated*, i.e. due to emotional inhibition of recall (Freud 1914). Unquestionably, this theory has thrown light on many curious phenomena of memory that had never previously been explained.† Nevertheless, it is not an easy theory to test experimentally, for, as has been shown by A. F. Zeller (1950), it is difficult to distinguish unambiguously between the effects of forgetting and of poor learning. He maintained that even the apparent repression in everyday life of the circumstances of a shameful act may be due to the latter, the humiliated individual tending to withdraw into himself and to become temporarily unobservant of his surroundings.

The general hypothesis (which includes Freud's as a special case) that forgetting is due to retroactive interference by immediately subsequent impressions and feelings was tested experimentally in an ingenious way by Steinberg and Summerfield (1957). First, they found that the administration of the central depressant drug nitrous oxide impaired the formation of associations. Then they showed that the administration of this drug

---

cried out in an unintelligible jargon. A Welsh servant, however, declared that she understood the countess who, she said, was speaking in Welsh. On recovery, the lady spoke to her friends intelligibly and, when they related what had happened, she informed them that during her infancy she had been taught the dialect of Lower Brittany (which resembles Welsh) by a nurse who was a native of that part. She had totally forgotten the dialect many years before the attack of fever 'which in so curious a manner revived the impressions that had been so long obliterated'.

† Freud concentrated on the forgetting of particular memories, but emotional factors can inhibit general recall. H. Syz (1937) found this in the case of a patient who had suffered complete amnesia for all events occurring in the previous three years until hypnotic and analytic treatment exposed the emotional cause of this wholesale repression. Thus the function of recall may be abolished for years and yet the capacity for it remain intact. For therapeutical applications of hypnosis, and of 'time distortion' under hypnosis, to enable patients to recall emotionally distressing repressed experiences such as rape, see Cooper and Erickson (1959).

Rapaport (1961) gives a valuable general study of the influence of emotion on memory. Striking examples have been provided by patients suffering from *allomnesia*. This is a form of memory disturbance in which 'memories', although not altogether false, are pathologically distorted. A particularly dramatic case mentioned by Rapaport was that originally described by Pick (1905). While talking to his physician, a patient suddenly jumped on a passing attendant and assulted him severely. He then fell backwards clutching his head and crying 'Jesus Maria, this guy killed me, he broke my skull'! Afterwards he had no recollection of having struck the attendant but declared that he would stab and kill the man who had assaulted him!

immediately after learning reduced forgetting, presumably by reducing interference learning.

Bergson (1911) believed, like Freud, that time lapse has no effect upon memory retention, but whereas Freud's is a theory of why we forget Bergson's is a theory of why we remember. He propounded a 'motor' hypothesis of recall, maintaining that the process of remembering is an inclination to react to, or to act in, a given situation by adopting a particular bodily attitude. By a recognition of resemblance, present perception evokes an appropriate image from the unconscious store of memories. Unfortunately, this theory fails to account for any memory which is not intimately related to perception. For example, it cannot explain our recalling the *date* of a past event—an essential feature in our use of memory to reconstruct the past. Moreover, despite his characteristic emphasis on the active nature of remembering, Bergson visualized memory retention *statically*, the mind storing alongside of each other (quasi-spatially) all our states in the order in which they occur.

A far more satisfying analysis of memory was given by Bartlett (1932) in his book *Remembering*. He produced concrete evidence that retention, as well as recall, depends on *dynamic* factors. His investigations were based on a constructive criticism of the pioneer researches of Ebbinghaus to which we have already referred.

In devising his experiments, Ebbinghaus was influenced by the 'associationist' viewpoint, then widely held, that all mental activity could be regarded as the automatic organization of sensory impressions due to stimulation of the various sense organs. Complex ideas were thought to be generated by association of the simple ideas derived from these impressions. Memory was explained as the effect of the more or less stable association of one impression or idea with another, so that the occurrence of one called forth the other. The inadequacy of this point of view came to be realized by Ebbinghaus's contemporary G. E. Müller (1911–17) who improved his experimental techniques. Müller discovered that *interaction* among associations could not be neglected and that in memorizing the mind plays a *creative*, and not a purely mechanical, role, for it does not receive the material passively, but groups it, hears it rhythmically, establishes meanings, and so forth.

Attention was thus drawn to the significance of 'setting' in connection with memory. Bartlett not only fully appreciated this but broke new ground in his criticism of Ebbinghaus's use of nonsense syllables. Ebbinghaus believed that it was essential for the proper functioning of the experimental method to use material which had the *same* significance for everybody and argued that this condition was satisfied by material which signifies *nothing*. However, as Bartlett (1932, p. 3) pointed out this is only true initially, for uniformity and simplicity of stimuli do not necessarily provoke uniformity and simplicity of response, particularly at the human level. Convinced that the experimental constraints introduced by Ebbinghaus interfered with, rather than assisted, the study of the most characteristic features of remembering, Bartlett used instead subtly selected meaningful pictures and prose. In particular, he investigated the serial reproduction over a period of years of carefully chosen legends. He showed conclusively that 'long-distance' remembering is not the re-excitation of innumerable fixed, lifeless, and fragmentary traces, but instead is 'an imaginative reconstruction', dependent upon one's 'attitude' at the time of recall and using only a few striking details which are actually remembered, the active settings which control human remembering being determined by our 'interests'.

In place of the traditional idea of the passive 'trace' as the basic element of memory mechanism, Bartlett introduced the concept of the 'schema'. By this he meant 'an active organization of past reactions, or past experiences', i.e. a schematic form of the past. It was an ingenious adaptation of an idea originally formulated by the neurologist Sir Henry Head in order to account for the temporal regulation of co-ordinated bodily movements.† Head's 'schemata' were built up in chronological order. Similarly, as Bartlett pointed out, all relatively low-level remembering tends, in fact, to be rote memory, i.e. the repetition of a series of reactions in the order in which they originally occurred. Indeed, even on a high level of behaviour we often tend to respond by serial reactions whenever our critical faculties are relaxed, as when we are tired, delirious, or intoxicated. However, although this process is the most natural way of maintaining a completed 'schema' as far as possible undisturbed, it has obvious drawbacks. Higher mental activity would be impossible if we could not break up this chronological order and rove over the events

† For example, in every skilled bodily performance involving a sequence of movements, each is made as if the position reached by the limbs at the end of the preceding stage were somehow recorded and still functioning, although that stage is already past. Head found the notion of individual images, or traces, inadequate to explain the way in which past movements still retain their regulative function. Instead, he introduced the notion which he called the 'schema'. 'By means of perpetual alterations in position we are always building up a postural model of ourselves which constantly changes. Every new posture of movement is recorded on this plastic scheme, and the activity of the cortex brings every fresh group of sensations evoked by altered posture into relation with it. Immediate postural recognition follows as soon as the relation is complete ... The sensory cortex is the storehouse of past impressions. They may rise into consciousness as images, but more often, as in the case of spacial impressions, remain outside central consciousness. Here they form organised models of ourselves which may be called schemata. Such schemata modify the impressions produced by incoming sensory impulses in such a way that the final sensations of position or of locality rise into consciousness charged with a relation to something that has gone before' (Head 1920).

which have produced our present 'schemata'. On the other hand, because their sensory equipments and their movements are limited, lower animals are subject to habit behaviour, i.e. the constant repetition of reactions in fixed chronological sequence. However, with the increase in the number and variety of reactions in the higher forms of life, rote recapitulation and habit behaviour are increasingly wasteful and it becomes more and more necessary to go direct to that portion of the organized setting of past responses most relevant to the needs of the moment. To do this the organism must acquire the capacity to turn round upon its own 'schemata' and construct them anew. *This is where and why consciousness comes in.* It may be that what then emerges is an *attitude* towards the integrated effects of a series of past reactions. But how do we pick out a specific past event—other than the last to occur? According to Bartlett, it is our *interests* (and, at a lower level, our appetites and instincts) which predispose us to particularize in this way. The traces which function in memory recall must be regarded as interest dominated and interest carried and they must change as our interests change (Bartlett 1932, p. 212).

The main conclusion of Bartlett's acute psychological analysis of memory was that the memory 'trace' elicited by normal recall, that is by conscious remembering without hypnotic or other abnormal aid, is not a static engram but is influenced dynamically by the changing framework of associations determined by the evolution of our interests and by our powers of reason and imagination. In other words, *recall is a constructive process* which never literally repeats a past experience or activity. As Hunter (1964, p. 203) has pointed out, there seems to be no biologically sound reason why it should. 'Memory seems to have been evolved to deal only incidentally with those rare situations where we are required to give a flawlessly accurate account of the past. Its primary function is not to conserve the past but to make possible adjustment to the requirements of the present'.

2.10. Time and the psychological aspects of memory (ii)

Since our awareness of time seems to be closely linked to our sense of hearing, it might be expected that our ability to remember sounds would depend significantly on auditory imagery. Surprisingly, this is not so. For example, auditory images are seldom found to be sufficiently precise and detailed to be useful for memorizing music. Instead, we tend to rely far more on kinaesthetic, or motor, imagery. To recall a tune or a rhythm we usually hum it or tap it out. In other words, to recapture the sounds we have heard we do not listen internally but instead we repeat a series of physical actions. Behaviourists go so far as to regard all forms of recall as purely kinaesthetic, but I do not believe that this extreme view can be sustained, for even in the case of music the memorizing of a complex instrumental work involves, in addition to kinaesthetic experience, what has been called 'rational remembering', the process of rational analysis of the work facilitating the process of memorizing (Mainwaring 1954).

Despite the inadequacies of *auditory* imagery, it is possible that some insight into the mechanism of human memory can be obtained from the study of *visual* imagery, for of all mnemonic techniques the most famous depended on such imagery. This was the 'place system' originally devised, according to Cicero, by the Greek lyric poet Simonides of Ceos in the 6th century B.C. Aristotle refers to it four times, but the only extant descriptions of it are Roman, notably those by Cicero and Quintilian. Mnemotechnics, or 'the art of memory', although unfamiliar in modern times, played a vital role in ancient and medieval civilization since, before printing or the use of paper for note taking, a trained memory was essential for many avocations. (Mnemosyne was regarded by the Greeks as the mother of the Muses.) The 'art of memory' was based on the idea that vision is the keenest of the senses, so that perceptions received by the ears or by reflection can be most easily retained if they are also conveyed to our minds by the mediation of the eyes (Cicero, *De oratore*, II, 1xxxvii, 357).

The clearest description of the architectural type of image used is given by Quintilian, the leading teacher of rhetoric in Rome in the first century A.D. The images by which, for example, a speech is to be remembered are placed, in imagination, in the different places which have been memorized in a building as spacious and varied as possible. As soon as the memory of the various items is required, the mind visits in turn the memorized places. The method ensures that the points are remembered in the right order, since the order is fixed by the sequence of places in the building. Moreover, the background places can supply a connection when the items to be remembered have no memorable connection themselves. One can also run through the things memorized in the reverse as well as in the original order, and one can even skip over certain places and visit, say, every third one. For skilful practitioners remarkable achievements have been claimed. In modern times a technique similar to the ancient 'place system' has been practised by the Russian mnemonist Shereshevskii (Luria 1969) who can not only repeat immensely long lists of items after a single hearing but can recall them after many years.†

† As regards similar claims made in antiquity, Hippias in Plato's dialogue *Hippias Major* (285E) boasted that he had a technique for repeating fifty names after a single hearing and Seneca the elder (*Controversarium Libri*, Lib 1, Praef. 2) could repeat two thousand names in the order in which they had been given.

The psychological study of visual imagery was initiated in the last century by Francis Galton (1822–1911). To his surprise, he found that many men of science claimed never to have experienced such imagery, and he concluded that 'an over-ready perception of sharp mental pictures is antagonistic to the acquirement of habits of highly generalized and abstract thought' (Galton 1883). Moreover, he came to the conclusion that poor imagery could not be associated with poor memory. In 1957, however, psychologists at the University of Pennsylvania (Wallace, Turner, and Perkins 1957) obtained remarkably successful results from a memory experiment with paired words in which the subjects were instructed to try and form mental images linking each pair. For example, an elephant and ice could be linked by visualizing an elephant carrying a block of ice on its back. Subjects tested after seeing 500 pairs once only scored as high as 99 per cent correct responses. After seeing 700 pairs once only their success rate was still 95 per cent.

Nevertheless, we must not jump to the conclusion that mental images are the internal representations of memory. All that we can say is that we do not yet know whether mental imagery is part of the mechanism of memory or merely an epiphenomenon.† However useful they may be for some people, images may not play any essential role in our psychological processes.‡ In other words, it may be that what we retain mentally as the result of, say, a visual experience does not resemble a picture but rather *a coded description or abstract representation* of the original perception. This would explain why we can recall an object better than we are able to draw it and also why, when we remember something imperfectly, the memory we have of it is not like a photograph with some parts missing but instead lacks some important general feature, e.g. there were steps up to the door of the house but we cannot recall how many.

In line with this idea of coded description, many psychologists engaged

† There are no grounds for believing that some people possess what is often loosely called 'photographic memory'. The nearest approach to this is the remarkable form of visual imaging known as 'eidetic' found in some children, which was first studied in detail by the German psychologist E. R. Jaensch and named by him after a Greek word signifying 'pertaining to images'. Although these sharp visual images of perception-like clearness which last for many seconds and sometimes minutes strongly resemble the original stimuli they are not completely photographic (Haber 1969). There is no reason to suppose that they depend on memory.

‡ As we have seen, imagery can undoubtedly often perform a useful mediating function. In tests of subjects asked to produce intervals ranging from a second to a minute some used an imaginary clock with a seconds hand, but others used an imaginary ticking, listening to a sequence of numbers, foot tapping, etc. (Doob 1971). It has been suggested that one reason why visual imagery may be particularly effective is because of its *simultaneous* nature: the relationship between two or more stimuli may be represented within *one* image (Herriot 1975).

in the study of memory in recent years have been guided by the growing influence in our society of the 'computer'. As a result, they no longer appeal primarily to the sensory concepts of stimulus and response but rather to those of information science, particularly the ideas of *encoding*, *storage*, and *retrieval*. As Baddeley (1976) explains, this is because in the last two decades the dominant view of memory has become increasingly functional. Whereas the characteristic view of memory was formerly that of a store that merely held information which might or might not be used in other information-processing tasks, the tendency now is to regard memory as an integral part of other information-processing tasks such as perception, pattern recognition, comprehension, and reasoning. This has led to a growing awareness of the importance of coding, since this involves the processing of information in different ways. The currently popular analogy with the 'computer' should, however, be regarded as a purely heuristic device and not as being necessarily any more realistic than the now discredited analogy between the central nervous system and a telephone exchange.

A distinction has long been drawn between *short-term*, or immediate, memory and *long-term* memory. Immediately after hearing or reading an unfamiliar seven-digit telephone number most of us can dial it correctly, although a few minutes later we have forgotten it. Indeed, this transitory type of memory, with its rapid decay, is vital for many types of skilled performance, such as typing from shorthand notes. The decay of immediate memory depends on how many fresh stimuli are dealt with. If one does nothing else after hearing a telephone number, one can recall it after a much longer interval than if one continues paying attention to other stimuli. The average span of immediate memory (the number of items, such as digits, which can be correctly repeated after one presentation) has been shown to be about seven items for young adults. It is less for young children and shows a decline with age after about thirty. In a classic paper G. A. Miller (1956) showed that this span is invariant in terms of independent units, or 'chunks', irrespective of the amount of information (number of 'bits') in each chunk. For example, if subjects are trained to encode binary symbols into decimal digits they can reproduce after one presentation a long string of binary digits corresponding to about the same number of decimal digits as their usual span. These and other results indicate that, although by organizing the stimulus input hierarchically we can break the informational bottleneck, so that the apparent size of the immediate memory span depends on the complexity of the encoding processes that the individual has learned, the limitation to about seven units that can be handled at any one time is fundamental. Immediate memory would therefore seem to be a process of strictly limited capacity. The biological usefulness of a short-term store like this, which can handle material required for immediate action without placing it in a much larger long-term store, is evident.

Immediate memory appears to be particularly influenced by acoustic similarity. It has been found that if we memorize a string of consonants we make more errors when they are presented visually instead of aurally (Conrad 1964). This suggests that some form of acoustic coding is involved. Moreover, it is the order of the consonants that is disrupted; acoustic similarity actually helps retention of the letters. Long-term memory, however, depends more on semantic similarity.

Immediate memory seems to conform more closely to something like an engram, or memory-trace, linking reception and recall in a one-one relation, than does long-term memory, which usually involves reorganization of the material. Broadbent (1958) has advocated a filter theory of immediate memory based on the idea that there are two separate short-term storage systems, S and P, with different characteristics. The storage system S holds information that arrives simultaneously and so works in parallel. It acts as a kind of buffer to allow time for later selective operations in the perceptual system P, which processes information sequentially and so is a parallel-to-serial converter. Since the filter mechanism controls this conversion, it is similar to the concept of attention (Murdock 1974, p. 153). The short-term memory system S holds a fixed number of items, and if a fresh item arrives it can only be held in S by knocking out some item already there.†

Most psychologists now believe that long-term memory is a separate storage system‡ of indefinite capacity that can hold items displaced from S. This has been confirmed by testing an experimental subject for recall with the aid of hints of a number of words that he has been given. For each word he has forgotten he is given an associated word, e.g. if the word is 'sky' he is given the hint 'blue'. It has been found (McLeod, Williams, and Broadbent 1971) that, if he still does not remember the word and is given a second hint, this has more effect than if the first had not been given. Both hints work together, and it is possible to remember words which one would not recall given either hint alone. This indicates that in the long-term storage system a particular memory is stored under more than

† This theory, as presented by Broadbent (1958), was later elaborated by him (Broadbent 1971) by interpolating between the short- and long-term systems an 'address register' which holds information about items (rather than the items themselves) that can be used as retrieval clues indicating actions that should be taken. If the short- and long-term storage systems are like in-trays and filing cabinets, respectively, Broadbent's 'address system' is like a system of notes pointing to items of high priority for retrieval.

‡ The distinction between short-term and long-term memory *stores* is not universally accepted. Craik and Lockhart (1972) have advocated the concept of different depths of processing instead of different processes. See also Baddeley (1976, pp. 165–9).

one heading and is 'cross indexed'. There is nothing corresponding to this in our short-term memory, in which retrieval is based soley on 'recency'.

Although we still know very little about the nature of long-term memory, it is generally believed, as I have already mentioned, that most forgetting (failure to recall) is due to associative interference between different memories, rather than to the decay with time of some engram or trace. Confirmation of this conclusion came from an important experiment by B. J. Underwood (1957). Most previous experiments on forgetting over long periods of time had been on subjects who had learned many lists. Underwood showed that if an experimental subject is given a list of nonsense syllables to learn without any previous experience of such a list he will forget very little of the list once he has learned it. If, however, he learns another list subsequently, his forgetting of that list will be faster.† Consequently, it is doubtful if any forgetting would occur at all in the absence of some previously or subsequently learned material.

The concept of interference by associated learning can explain the fact, to which I have already alluded, that our ability to recall in detail what we have heard tends to vary inversely with our cultural level and to be greatest in pre-literate societies. Moreover, we can see why it is that a bright person will often forget items more rapidly than a slow-witted one, since he has more on his mind and consequently his memory storage system may be expected to contain more interfering associations.‡

It is clear that long-term memory should not be considered as a single system operating on a common code. The words learned by an actor or a singer are coded phonologically, whereas when a scientist recalls the contents of a scientific paper it is usually in a far from literal form. Moreover, a distinction must be drawn between remembering one's past actions and one's past trains of thought.

Recently, Tulving (1972) has emphasized the division of long-term memory into two parts, corresponding to memory with record, or *episodic memory*, and memory without record, for which he has introduced the useful term *semantic memory*. The former contains information about temporally dated episodes or events and the spatio-temporal relations between these events. This information is stored in terms of its references

† In a later investigation (Underwood 1964) a subject learnt a list of verbal items until he was able to repeat it perfectly. He was then tested 24 hours later, when it was found that he could recall only 80 per cent of the items correctly. Subsequently he was made to learn a second list and was tested for that after a further 24 hours, when it was found that the percentage of items forgotten had increased. Repetition of this process yielded a continuously increasing rate of forgetting until, when tested with the twentieth list 24 hours after he had learnt it the subject was found to have forgotten eighty per cent.

‡ Of course, this is not incompatible with the fact that men of great learning often have remarkable memories for the things that interest them.

to our autobiographical experience. The act of retrieval of information from the episodic memory store is itself a special type of input and so changes the content of this store. On the other hand, semantic memory is the memory required for the use of language. It does not register perceptions but their cognitive referents, i.e. symbols, concepts and their relations. It permits the retrieval of information not directly stored in it, and retrieval leaves its contents unchanged. It is concerned with cognitive rather than autobiographical information. Forgetting not only occurs more readily in the episodic store, but we know much more about how it occurs in that store than in the semantic store.

The characteristic difference between these two kinds of long-term memory concerns time: episodic memory is time related, whereas semantic memory is largely time independent. The importance of time in relation to the study of memory is not confined, however, to long-term memory. A significant experiment that reveals its significance for short-term memory has been devised by Graham Hitch (Broadbent 1973, pp. 95–6). Letters of the alphabet are presented to a subject in different spatial positions around him. If there is no correlation between the spatial positions and the temporal order in which they are selected for presentation, the errors that tend to appear in recalling the particular item at a given place are not usually the items that have been placed on the right or left of it but those that come before or after it in *time*. 'It seems likely therefore', Broadbent comments, 'that the various spatial positions are remembered in a sequence in time, and also that the particular letters are recalled as a sequence in time, so that retrieval proceeds from a spatial position to the corresponding time in the sequence, and then from time in the sequence to a particular item'. In other words, it does not proceed directly from spatial position to item. Since patterns of error are, in general, a good guide to the system of filing that is being used, Broadbent concludes that time may well be a more fundamental category than space in human short-term memory.

Future study of the psychological aspects of memory is likely to involve detailed experimental investigations of retrieval processes from semantic as well as from episodic storage and of the respective roles of automatic and conscious processing. Particular attention will continue to be paid to problems of coding, since memory is coming to be regarded as a part of man's general mental activity as the coder of his environment. Although physiological correlates of the memory processes postulated by psychologists will continue to be sought, it can safely be predicted that the psychological aspects of memory will never be entirely subsumed under the physiological, since the difference between the two is analogous to the basic distinction drawn by the computer analyst between 'software' and 'hardware'. This distinction can be made irrespective of whether one is a reductionist or not, for it is now generally recognized that the functional analysis of a physical system can be separated from its material structure. As Broadbent (1970) has so pertinently remarked, 'It is highly likely that people in different cultures, or indeed different individuals within our own society, organize their memories in different ways. In due course, these are the problems which psychologists will have to study without help from physiology'.

2.11. Time and the neurophysiological basis of memory

In psychology we study behaviour and speculate on the physiological structure which may give rise to it. In neurophysiology, on the other hand, we investigate the functioning of the central nervous system and try to discover how it produces, or influences, behaviour. Only since the end of the last century, indeed mainly in the last few decades owing to the invention of new microtechniques, has significant progress been made in our understanding of brain physiology, and our knowledge still lags behind that in the much older, albeit more controversial, field of psychology. Nevertheless, we must now try to discover whether the main conclusions drawn from our discussion of the psychology of memory are compatible with recent advances in neurophysiology.

It was first shown, about 1890, by the Spanish physiologist Ramon y Cajal that the nervous system is composed of discrete nerve cells (afterwards known as *neurons*) all of similar general structure, the functional contacts between them being effected by close contact of free ends and not by syncitial continuity as in the rival reticular theory of Gerlach and Golgi. In 1897, Sherrington gave the name *synapse* to these functional connections. His great achievement was to show how the reactions of the nervous system could be explained by the integrated behaviour of independent neurons, each functioning as a unit and exerting graded excitatory or inhibitory synaptic actions on other neurons. The number of these cells in the human brain is of the order of 20 thousand million (in a bee's a million). Encephalogram studies suggest that they are all in a state of almost constant activity. This has an important bearing on the problem of the physiology of memory, for it is clear that any neurons which retain any kind of special memory trace of an experience must also be involved in many other activities.

It was shown many years ago, by Helmholtz, du Bois-Reymond, and others, that the nervous impulse is electrical in nature, although it is not a simple electric current. Later it was found that nerve, like muscle, is refractory, so that a second electrical impulse does not occur if one stimulus succeeds another too rapidly, the recovery time being of the order of one-hundredth of a second. About 1912 Lucas and Adrian showed that a neuron 'fires' on what Lucas called the all-or-none principle, i.e. it transmits an impulse down its axon (or output† fibre)—which then affects other cells—only when the stimulus is strong enough, the neuron afterwards reverting to its inactive state. Just over a decade later, Adrian made the first electrical recordings of individual nerve fibres and discovered that the action potentials of electrical pulses in neurons do not vary in amplitude but only in frequency—up to about a thousand pulses a second. He thus established the fundamental law: *neural information is frequency modulated* (Adrian 1928). The chemist Ostwald, and others, discovered that the axon is covered with a very thin polarized semi-permeable membrane (negatively charged on the inside and positively on the outside) which separates the internal fluid from an external fluid of very different composition. The potential difference across this membrane (a few tenths of a micron wide) is of the order of 60 millivolts, and the electrical pulse is generated by its temporary localized breakdown. The resulting electromagnetic surge field induces a similar breakdown in the adjacent 'down-stream' region of the axon, while the original potential difference in the first region is being restored. This process continues down the axon and is what we mean by saying that the neuron 'fires'. The speed of travel is uniform. It varies as the square root of the fibre diameter and is of the order of 5 centimetres per millisecond.

By inserting microelectrodes into the giant axon of a squid, which has a diameter of about 400–800 microns, Hodgkin and Huxley (1939, 1945) were able to measure directly the potential difference between the inside and outside of the membrane. They discovered that, contrary to what had previously been thought, when the neuron fires the potential difference does not disappear but is reversed, there being a transitory action potential, about twice as great as the original 'resting' potential, that lasts in any one place for about a millisecond. Small local currents flowing out from it depolarize the adjacent area until it too attains the 'action' potential. In this way the nervous impulse traverses the axon without loss of amplitude. The actual mechanism responsible for this step-by-step process is provided by the fluids inside and outside the axon. Normally,

† Each neuron has a large number of input, or receiving, fibres, known as *dendrites*, but only one axon. At its far end the axon splits up into a number of smaller branches. The synapses are the small regions where these make contact with the dendrites of other neurons. According to Mary Brazier (1977, p. 75), 'It is unlikely that the simple case of a single fibre synapsing with a single secondary neuron ever occurs in the human nervous system. Each cell lies in a forest of arborizing fibre endings'. Except for its many dendrites, a typical neuron is somewhat like a kite. Some neurons are more than a metre in length! A useful textbook with simple quantitative discussions is that by Katz (1966).

the fluid outside contains about ten times the concentration of sodium ions to be found inside, whereas the concentration of potassium ions is about thirty to forty times as great internally as externally. About 1950, Hodgkin and Huxley discovered that when the critical potential is reached as an electric current is passing through an axon the local permeability of the membrane to sodium ions suddenly increases. For about 0.2 milliseconds there is an inflow of these ions from the external fluid causing the inside of the membrane to become positively charged, followed by a small loss of potassium ions that restores the local membrane potential to its 'resting' value.† Subsequently, the original ionic concentration is also restored by a cellular metabolic process that causes sodium to be pumped out and potassium to be pumped in.

The one-way action of the nervous impulse (away from the cell nucleus) is believed to be entirely due to the synapses, for it has been found by experiment that, were it not for the synapses, impulses could travel either way along the axons. For a long time it was generally assumed that in the central nervous system the transmission of the impulse across the synapse, or 'synaptic cleft' as it is sometimes called, to the next neuron was electrical in nature because it is so rapid. However, with the development of intracellular recording techniques it has been accepted by most neurophysiologists since the early nineteen-fifties that at most synapses in the vertebrate (and, as far as is known, in the invertebrate) central nervous system the transmission process is chemical.‡ When the impulse reaches the end of the nerve fibre it

† The brilliant experimental researches of Hodgkin and Huxley were supplemented by a no less successful mathematical analysis, for they formulated differential equations with which they were able to predict with considerable accuracy many factors relating to the electrical behaviour of the squid's giant axon such as the form, duration, and amplitude of the 'action' potential, the conduction velocity, and the associated ionic movements (Hodgkin and Huxley 1952).

‡ A number of synapses with electrical transmission have, however, been detected in the lower vertebrates and invertebrates (Gray, 1974, p. 388). We still have much to learn about the way in which both synapses and axon membranes function. Recently, W. D. Branton (1978) and his colleagues at the University of California, San Francisco, have discovered, in the abdomen of a mollusc known as the sea hare, a new mode of transmission of neural signals by chemicals, probably peptides, by applying an electrical stimulus to a knot of nerve cells called the bag cells that have nerve endings that are not close enough together to form synapses. Other recent evidence suggests that this new mode of transmission, the responses to which are markedly different from those associated with synapses, may be limited neither to sea hares nor to peptides as messengers. A similar explanation has been applied to those nerves in the brains of mammals that use amines as messengers. It may well be that further research will lead to some considerable revision of current ideas concerning nerve and brain action.

For an historical account of the controversies associated with the hypothesis of the chemical transmission of nerve impulses (the first empirical evidence was provided by Otto Loewe of Graz in 1921) see Bacq (1975).

liberates a minute jet of a chemical substance which passes across the gap and stimulates the next nerve cell in the chain of conduction.† Chemical synapses have a synaptic delay time of about a millisecond.

The unidirectional transmission of signals through the synaptic cleft appears to be the result of an important difference between the structure of the pre-synaptic and post-synaptic endings of nerve cells. With the aid of the electron microscope the former have been found to contain many tightly packed small vesicles of uniform size close to the synaptic membrane, and it is presumably from them that the transmitter substance is released in specific amounts ('quantal release'). On the other hand, the post-synaptic cellular region does not normally contain any vesicles and consequently no transmitter substance.

Each synaptic ending influences the conductivity of only a small part of the post-synaptic cell membrane, and the effect of the transmitter action may be insufficient to produce an impulse in the whole post-synaptic cell. The probability of this occurring is increased when two or more impulses arrive together over separate axons. The number of synaptic endings that may be found on any one nerve cell can be very large, estimates for a cell in the cortex of a mammal ranging up to 50 000, although they are not necessarily all from different neurons (Mark 1974, p. 11). Since the initiation of an impulse in a nerve cell may depend on the co-operation of many other neurons, *timing is an important factor in neural transmission.*

The all-or-none principle of the nerve impulse may be the reason why we feel instinctively compelled to base our 'laws of thought' on two-valued logic. Be that as it may, neural action has come to be widely regarded as similar to the coding of numbers in terms of the binary digits 1 and 0. These digits are also isomorphic with the short and long signals of the Morse code. If a message is to be coded as a particular, and in general lengthy, sequence of the two symbols 1 and 0, it is natural to ask whether it can be recoded more concisely by storage *en bloc* of certain chosen standard sequences, i.e. specific sequential patterns of the two symbols, for example, 00, 01, 10, 001, 010, 100, etc. This idea was used by Oldfield (1954) with the object of determining a possible physiological basis for Bartlett's schemata. He suggested that a cerebral Morse code signal corresponding to a specific memory might be analyzed by the brain into a number of recognized standard patterns. Oldfield argued that not only would this be an economical process but also that the corpus of

† Several different substances, in particular acetylcholine, are thought to act as transmitters at different synapses in the central nervous system, but each neuron utilizes only one (Dale's principle). After crossing the synaptic cleft acetylcholine is eventually destroyed by a chemical reaction with an enzyme (choline-esterase). Poisonous 'nerve gases' act by preventing the destruction of nerve transmitter substance, thereby causing its action to persist. They are the most toxic compounds known.

standard patterns, with the various possible linkages between them, might provide a mechanism for the schemata. As we have already remarked, long-term memory usually seems to involve reorganization of the material and what is stored may well be like a précis.† It was pointed out by Gerard (1953) that patterned memory could easily change with time, if occasionally certain neurons or synapses dropped out, and that this might account for the variations of a specific memory at successive recalls.

Oldfield's hypothesis illustrates the modern tendency, to which we have referred in § 2.10, to regard the operations of electronic computers as useful clues to the general functioning of the brain when engaged in higher mental processes. The limitations of this analogy‡ were pointed out by one of the most brilliant mathematicians of the century, John von Neumann (1958), who was himself a leading authority on computing machines. Although there are far more neurons in the brain than there are electronic components in even the largest of computers, the most significant difference between the brain and the computer concerns their respective times of operation. The reaction time of a neuron, between possible successive stimulations, is about 10^{-2} seconds, whereas that of the vacum tube or transistor is about 10^{-6}–10^{-7} seconds. Hence, the brain has more and slower components and the computer fewer and faster. Because of its slower speed and far larger number of working units, von Neumann argued that the brain will tend to pick up and process as many informational (or logical) items as it can in parallel, i.e. *simultaneously*, whereas the machine will be more likely to do things in series, i.e. *successively*. This *temporal* difference has far-reaching consequences, for not every serial set of operations can be replaced by a set in parallel, since in the former certain operations can only be performed *after* certain others and not simultaneously with them. The transition to a serial

† This could explain our ability to recognize a melody first heard in another key.

‡ J. Z. Young (1964) has argued, I believe convincingly, that the neural memory is probably more like an *analogue* computer, made by selecting parts from a code set, than like a *digital* computer. 'Moreover', he writes (Young 1964, p. 268), 'we cannot assume that the effects of past experiences are stored in the brain in the same individual particulate manner in which this is done in a computer. In animals operations are performed actually in the memory and by the actions of the code elements themselves'. On the basis of his celebrated experiments on the octopus, Young (1964, 1966, see diagram on p. 40) has concluded that the record made in its memory during learning is the result of a specific choice between alternatives. In the octopus the situation has the experimental advantage of simplicity: attack or retreat when some object moves into the visual field, and the taking or rejecting of an object by the arms. Young believes that the essential feature is the 'switching off' of neural pathways that are not required, the gradual process of learning consisting in the accumulation of cells that are suitably connected; but, if the process is irreversible once achieved, how is forgetting possible? This may be due to interference between representations (even in an octopus) or the onset of other conditions that make it impossible for the creature concerned to 'read out' properly from the memory (Young 1964, p. 181).

scheme from one in parallel may be impossible, or possible only if the logical procedure is changed. 'Specifically', von Neumann wrote, 'it will almost always create new memory requirements, since the results of the operations that are performed first must be stored while the operations that come after are performed. Hence the logical approach and structure in natural automata may be expected to differ widely from those in artificial automata. Also, it is likely that the memory requirements of the latter will turn out to be systematically more severe than those of the former' (von Neumann 1958, pp. 51–2).

From von Neumann's penetrating comments it is clear that we ought to pay special attention to the temporal aspects of memory and of the general functioning of the central nervous system. This accords with the fact that the basic method by which neurons transmit information is chronometric, for, to use the language of the engineer, it is a frequency-modulated pulse code, since the all-or-none principle of nervous impulse results in the intensity of a stimulus being transformed into frequency of pulses, i.e. the number transmitted in unit time, the rule being that the greater the stimulus the higher the frequency. Since weaker signals are transmitted with longer intervals between them than those between stronger signals, the system can be regarded as an ingenious natural device for circumventing the deleterious effects of 'noise' (due to random effects of molecular disorder, thermal vibration, etc.) in contrast to amplitude modulation, where the weaker the signal the more it is affected by noise.

There is also, of course, an important spatial aspect of neural functioning based on anatomical interconnections between cells. For example, each muscle in the body is controlled by a group of neurons in the central nervous system that have only this function. Similarly, specific neurons exist in the cortex that respond only to particular features of the visual field observed by the eyes. Nevertheless, the function of a retinal receptor is to stimulate repetitive nerve activity as a function of the intensity, or change of intensity, of incident light. In so far as nerve activity is either all or none, there is only one method of transmitting information concerning the pattern of intensities on an array of optical cones and rods: the spatial pattern from the environment must be converted into a temporal pattern of nerve activity at the cortical termination of the optic nerve if the pattern is to be perceived (Myers 1965).

The importance of the temporal aspect of neural activity in general was emphasized long ago by Descartes. In his *Traité de l'Homme* he compared nervous function with the harmonic structure of organ music. At the Wellcome Symposium in 1957 on 'The History of Philosophy of Knowledge of the Brain and its Function', Walther Riese (1958) drew attention to this analogy and described it as 'particularly fortunate', since it anticipated by nearly three hundred years the 'kinetic melodies' of Christian von Monakow, whose great treatise *Die Lokalisation im Grosshirn* appeared in 1914. In this massive tome, Monakow demolished the idea of the (verbal) engram as a static trace, or imprint, and replaced it by the concept of what he called 'chronogenetic localization'. Thus, instead of picturing the engram as having a precise location, he regarded it as having a definite history, during the course of which it may undergo profound changes of content and meaning, involving ever more widely diffused cortical structures.

Monakow's hypothesis that memory (and other intellectual functions) cannot be localized in specific cortical areas is supported by the extensive researches of the American physiological psychologist K. S. Lashley (1929). He conducted many experiments on the effect of major cortical excisions in animals, notably rats and apes. He found that, in general, these excisions made remarkably little difference. For example, memory of specific visual forms was retained when all but some 20 000 of the million or so cells in the visual cortex of the rat were removed, despite the evidence that no parts of the rat's cerebral cortex except the visual areas are essential for visual perception and memory. Indeed, the rat can retain so many memories (acquired by learning) after the destruction of so many parts of its cortex that it is clear that no one specific part can be regarded as essential for these memories. Nevertheless, the rat cannot dispense with its entire cortex. Lashley therefore concluded that memories do not depend on localized engrams but on factors affecting the cortex, or a particular region, as a whole. Some years later he suggested that memories might perhaps be more or less stable resonance or interference patterns of neuronal activity which are reduplicated throughout the entire cortex or a particular region (Lashley 1950).

Commenting on Lashley's discoveries and conclusions, the famous brain surgeon Wilder Penfield (1954, p. 472) argued that as we pass up the evolutionary scale we find evidence of increasing specialization and decreasing replaceability of different parts of the cortex. He discovered that in the case of patients suffering from focal epilepsy the application of a stimulating electrode† to the cortex of the dominant temporal lobe could automatically induce them to recall specific memories of earlier experiences. He claimed that this pointed to a more or less precise localization of the memory trace, but his argument was criticized on the

† The electric current employed is usually generated by a potential of a few volts, the pulse rate being from forty to a hundred per second, and each pulse lasting from two to five milliseconds. Unless he is told, the patient does not know when the electrode is being applied, since no pain is ever felt when the cortex is being interfered with by the surgeon.

grounds that it does not automatically follow that memories must be stored in those areas of the brain from which they can be elicited. Moreover, as Penfield found, when a large part of the cortex of the dominant lobe is excised, although the patient may be freed from the recurrent epileptic hallucination associated with a particular unpleasant memory (involuntarily recalled in this way), he can still *voluntarily* summon this memory. He therefore concluded that there must be an identical memory trace stored in the non-dominant temporal lobe, and that, since the memory is not merely of the past event but of the individual's thinking and feeling about it, the highest integrating processes do not occur in the cortex. Believing that they must take place somewhere, he argued that we must look to that part of the brain which has symmetrical functional relations with both cerebral hemispheres, and he therefore suggested the higher brain stem, which includes the thalamus, or old brain, found even in the most primitive animal species. In support of this hypothesis concerning the ultimate seat of consciousness, Penfield observed that pressure on the thalamus produces unconsciousness. However, striking as this evidence is, particularly in view of the fact that the cerebral cortex can apparently successfully resist violent manipulation, it is not necessarily a proof of Penfield's hypothesis, for it merely confirms that the thalamus is necessary for consciousness, just as oxygen and blood-sugar are. Indeed, it is difficult to believe that man's highest mental processes actually take place in the most primitive part of his brain. Generally speaking, we should not assume automatically that any region must necessarily be credited with those functions which cannot be efficiently performed when it is disturbed or disorganized.

Penfield's investigations have, however, been of considerable importance in support of the hypothesis that the brain, or mind, retains a complete record of the stream of consciousness, i.e. of all detail which was recorded mentally† at the time of occurrence, although later most of it is entirely lost to voluntary recall. The particular experiences elicited by electrical stimulation of the cortex presumably depend on chance, but once produced they tend to recur on subsequent stimulation. In other words, the same 'strip of time' tends to be re-activated.‡ According to Penfield (1958), 'This is not memory, as we usually use the word, although it may have some relation to it. No man can recall by voluntary effort such a wealth of detail. A man may learn a song so that he can sing

† Including 'infra-conscious awareness' as revealed, for example, by hypnosis.

‡ Two distinct 'strips of time' are never activated together. It is as if, when one is activated, some all-or-nothing mechanism inhibits the recall of others. In some cases, however, different 'flashbacks' have been produced from successive stimulations of the same point (Penfield 1975, p. 22).

it perfectly, but he probably cannot recall in detail any of the many times he heard it'. Most things that we are able to recall to memory are generalizations and summaries. Patients say that the experience brought back by the electrode is 'much more real than remembering', and is like living through the past once again. These 'flashbacks' are usually of utterly unimportant incidents which the patient would never recall voluntarily. The electrode activates all those things to which he happened to pay attention in the interval of time concerned. But, despite this doubling of consciousness, the patient remains fully aware of the present situation. Indeed, he often cries out in astonishment that he is hearing and seeing friends whom he knows are in fact, far away, or even no longer alive. His recall of the past is accompanied by the same thoughts and feelings as he experienced then.

Besides these remarkably vivid examples of 'flash-back', Penfield discovered another type of response which he described as 'interpretive'. For, when he stimulated a part of the cortex at the posterior limit of the (right) temporal lobe to which no previous function had been assigned by anatomists, he was surprised to discover that the patient responded by saying that he had an interpretive 'feeling' about the present situation, namely that it was 'familiar', 'strange', etc. Penfield therefore called this region of the brain the 'comparative-interpretive cortex' and advanced the following hypothesis concerning its normal functioning.

When we meet an acquaintance of years gone by, whom we have not thought of since, we may be startled first by a sudden signal of familiarity, because of the sound of his voice, his smile, his way of talking. Almost instantaneously some strange mechanism in the brain is providing us with a standard of comparison. We see how this present man differs from the acquaintance in the past—the man we have not thought of for many years. A moment earlier we could not have pictured him. Now we can compare the past with the present in great detail. We detect the slightest change in the face and hair. We note that his movements are slowed, the hair thinned, and the shoulder stooped. But his laugh, perhaps, has not changed.

Penfield assumed that in situations like this the comparative-interpretive cortex of the temporal lobes manages to select and activate the short strips of past conscious experience in which this man was once the focus of our attention. It makes possible the scanning process by which past experiences are selected and made available now for the purpose of comparative interpretation.

On this view, hidden in the interpretive cortex there is a mechanism that unlocks the past and scans it for the automatic interpretation of the present. Moreover, it probably also serves us as we try to make a conscious comparison of present experience with past. More recently, Penfield has drawn attention to the hippocampal zone, which lies below the cortex on the undersurface of each temporal lobe, as a region that appears to play some role, together with the interpretive cortex, in scanning the record of past experience and in memory recall. For, although the surgical removal of one hippocampus appears to have little effect on memory, the removal of both abolishes the ability to recall fairly recent past experiences, although other forms of memory, such as short-term memory and memory of language, skills and non-verbal concepts, as well as of events in the *remote* past, are retained (Penfield and Mathieson 1974).

On the other hand, Brenda Milner (1966, 1972), as a result of lengthy investigations of the patient H.M. who suffered removal of the hippocampus on both sides together with some of the overlying cortex and thereby lost the ability to remember new experiences for more than a few seconds,† has suggested that the hippocampus serves to facilitate the entry of information into a long-term memory store located elsewhere in the cortex. (There is no reason to suppose that any memories are actually stored in the hippocampus.) After bilateral temporal lobectomy, however, it is possible that the resulting defect is not a failure of storage, as Milner has always supposed, but a failure to inhibit inappropriate responses at the time of retrieval, as suggested by Warrington and Weiskrantz (1970).

The apparent clash between Penfield's results and Lashley's has not been easy to resolve, for, although the former concerned the long-term memories of human beings for specific events and the latter the learned behaviour patterns of rats trained to run mazes, Lashley believed that his hypothesis of the dispersion of memory was not confined to animals. Not only was it in general accord with Bartlett's theory of the role of 'schemata' in the recall of meaningful material, but it also appeared to throw light on Ebbinghaus's results relating to the memorizing of nonsense syllables: in particular, on Ebbinghaus's discovery that associations are formed not only between adjacent, but also between remote, syllables. Lashley regarded this as an elementary illustration of the general principle that every memory becomes part of a more or less extensive organization. When discussing Penfield's contention that there is greater cortical differentiation in man and the higher apes than in lower animals, Lashley (1950, p. 486) drew attention to the fact that the surgical removal of particular parts of the human frontal lobes does not produce such pronounced defects as usually result from widespread traumatism.

† For example, he would read the same magazines over and over again without finding the contents at all familiar!

Nevertheless, even if we agree with Lashley's general conclusion that memory traces cannot be precisely localized in the cortex and that Penfield's results do not, in fact, indicate where specific memories actually reside, it does not follow that we must accept Lashley's hypothesis that these engrams are more or less stable resonance patterns of neuronal vibration over comparatively large areas.

As just mentioned, Lashley referred to the effects of traumatism, but his resonance hypothesis is difficult to reconcile with the well-established fact that long-term memories can survive great changes in the overall activity of the brain. After the deepest anaesthesia or severe electric shocks many times repeated, which one would expect to extinguish all vibrating patterns in the cortex, one's memories usually return intact. Similarly, although needles thrust into the brain of a hibernating animal (a hamster) which had been artificially chilled failed to pick up any electrical activity, no loss of memory was observed when the animal was tested, after recovery, for retention of learning of a simple maze learned beforehand (Gerard 1953).

On the other hand, although no direct relationship between synapses and memory has yet been established, it is possible that the engram of a short-term memory of a few seconds' duration may be an electrochemical pulse circulating in a closed loop of neurons. It has been estimated that the period of such a feedback loop would be of the order of one-hundredth of a second. It is therefore difficult to imagine how interference between different memory patterns could be avoided and a reverberating circuit be kept in phase for periods longer than a few seconds. The fact that post-traumatic amnesia usually abolishes the memory of all events immediately preceding a severe shock (while leaving older memories unimpaired) is compatible with the hypothesis that short-term memories are maintained by neuronal circuit vibrations. Moreover, the conclusion that this hypothetical mechanism, if applicable at all, can apply only to short-term memory is in accord with the current belief that short-term and long-term memory are maintained by quite different processes.

Lashley's investigations have been interpreted by Mark (1974, p. 71) as signifying not that each part of the cortex somehow contains a fragment of each memory but rather that long-term memories, at least of behaviour, are made at a level of organization in the brain where integration of all the information provided by the different senses can contribute to the selection of the most useful patterns of action. Mark argues that a certain amount of redundancy in the neuronal processes collecting this information is to be expected. Among the various theories that have been suggested to explain how such a distributed memory could work without involving the neuronal circuitry resonance mechanism suggested by Lashley, one of the most intriguing, advocated by K. H. Pribram, H. C. Longuet-Higgins, and others in the late 1960's, was based on the physical analogy of the hologram (Pribram 1969). A hologram† is a photographic plate in which a scene is recorded in the form of a complex interference or diffraction pattern instead of as a normal photographic image. When the pattern is illuminated by a coherent laser beam the original image is reconstructed. The essential point of the analogy is that every element in the original image is distributed over the entire plate. If even a small part of a hologram is appropriately illuminated, the entire scene reappears although, depending on the size of the part, it may be somewhat blurred. Pribram suggested that memories may be produced in a broadly similar way in the brain by wavefronts travelling through neural networks that interact with each other to produce interference effects, the function of the cortex being to organize the processes of assimilation and reconstruction. However, despite its initial appeal as an explanation of Lashley's results, the hologram model has come to be regarded as being no more than a physical analogy, the only experiment so far devised for distinguishing between the consequences of a hologram-type model of memory and any other type having proved negative (Valentine 1968).

The difficulties of accounting for long-term memory by means of electrical connections between cells have led some investigators to concentrate on chemical processes within cells and in their connections with other cells. Two general classes of chemical theory of memory have been studied: transmitter theories and those involving cellular protein and ribonucleic acid (RNA). According to the former, memories depend on chemical transmissions at synapses. Unfortunately, the only chemical transmitter substance about which much is known is acetylcholine, and so far very little has been discovered about the role of acetylcholine synapses in learning and memory. More attention has been paid to theories based on molecular processes occurring within nerve cells. The most likely molecules to be involved are proteins and RNA, neurons being rich in both. Protein molecules are continually being formed and disappearing, none lasting more than about a month. Consequently, if learning involved the formation of particular protein molecules, memories would not be retained for longer than that. This means that the formation of such molecules cannot be regarded as the physiological basis of memory. More generally, as has been made clear by the Swedish neurochemist Holger Hydén (1968), there is no evidence that brain cells contain any special

† Holography was invented by D. Gabor (1948) and gained him the 1971 Nobel Prize for Physics.

'memory molecules' that store information in a linear way, i.e. record and reproduce it tape-recorder fashion.

Instead, Hydén has investigated the RNA content of neurons to discover whether any changes occurred in it during the process of learning. He experimented with rats. These animals normally have a preferred paw when reaching for food, but they can learn to use the other paw if that is the only way to reach it. Hydén trained right-pawed rats to take food with the left paw, and he then examined biochemically cells taken from the right sensori-motor cortex (corresponding to the left paw) and compared them with cells taken from the left (corresponding to the right paw). He found that there was a significant increase (up to 40 per cent) in the amount of RNA present in the neurons on the side of the cortex involved in the learning. Hydén also discovered that RNA taken from that side differed in the ratio of its nucleotide bases from RNA taken from the other side. He inferred that the learning task had stimulated the formation of new RNA in the particular side of the brain concerned, and he concluded that generally protein synthesis in the brain during learning is necessary for the formation of long-term memory.

Despite the ingenuity of his experimental techniques, Hydén's conclusions have been much criticized, for it is by no means certain that the molecular changes occurring in neurons during learning are specifically due to the learning itself instead of being merely a by-product of enhanced neural activity. Also, there is no evidence that within the RNA concerned there is any specific storage of information derived from external stimuli (Barbizet 1970).

Other evidence for a chemical theory of memory that attracted considerable attention for a time concerned 'transfer' experiments with planaria, or flatworms, carried out by J. V. McConnell (1962, 1964) of Ann Arbor, Michigan. These primitive creatures, not more than a centimetre or so in length, are capable of learning a conditioned response of the type studied by Pavlov. Normally, they reach towards a bright light but curl up when they receive an electric shock. Given both light and shock simultaneously, they respond to the latter and can be trained to curl up even when only the light is turned on. They have learned to associate light and shock. McConnell claimed that if planaria trained in this way are chopped up and then fed to untrained ones the latter learn to respond to light by curling up more readily than if they had been fed on untrained flatworms. In other words, he claimed that learned behaviour could be transferred to untrained animals in their food! Although McConnell's experiments at first caused quite a stir, they later came under severe criticism from other workers. Similar experiments were later performed on rats. RNA was extracted from the brains of rats that had been trained to approach a food dispenser whenever a light flashed or a clicking noise was made. This RNA was then injected into untrained rats, and it was claimed that the learned behaviour was transferred to them (Jacobson *et al.* 1965). In none of these experiments, however, was there any firm evidence that *memory* had actually been transferred from one animal to another and not just some substance that speeds up the learning process.

Whatever the ultimate explanation of such experiments may prove to be, it is a far cry from the simple type of conditioned-response learning in planaria and rats to the long-term retention and recall of specific events that characterizes human memory. Indeed, in view of the irreversible destructive effects of 'noise', it is difficult to see how any structural modification of a particular set of molecules in the human brain could function as the trace of an isolated memory of a unique event fleetingly observed many years ago.

Biochemical coding theories of memory are no longer regarded as acceptable even for simple learning processes. The current tendency is to regard chemicals that are synthesized when learning occurs as containing no special information. It is now thought by some investigators (e.g. Blundell 1975) that they serve to direct impulses along specific circuits. These circuits become functional when learning occurs and are reactivated when recall of information is needed, the chemicals concerned acting merely as 'traffic signals'. According to Young (1978, p. 83), our present knowledge concerning the brain suggests that its coding system depends on the use of numerous distinct channels, each carrying a slightly different feature of the information relating to a particular memory. This hypothesis of *multichannel coding* is in accord with the suggestion, based on psychological studies of memory of pictures and sentences, that what we learn are fragments made up of clusters of features, any fragment acting as a clue for the whole cluster (Baddeley 1976, p. 346). As Young has pointed out, this 'fragmentation hypothesis' also agrees with ideas based on physiological studies of the function of cells in the primary visual cortex. This region consists of hundreds of thousands of columns of cells at right angles to the cortical surface. Each column comprises about a hundred cells and responds to a particular visual contour, or boundary between light and dark, the action of all the columns of cells together determining what one sees (Young 1978, p. 51). Although Young thinks that, in view of the evidence in favour of the 'fragmentation hypothesis', we may after all be not 'so very far from understanding of the coding system and this may help us to find how alterations in the network produce records in the memory' (Young 1978, p. 96), he is careful to point out that there is still no conclusive evidence for the belief, now widespread among neurophysiologists, that the basis of learning and memory is the formation of new neural connections.

Whether or not this belief is justified, or whether some synaptic changes are associated with the storage of memories in animals and man, we have hardly any evidence on the reading of this information, that is to say on the process of recall and how a specific memory can be recalled at a specific time. Even if it is the case, as it may well be, that short-term memory depends on processes involving neuronal circuits, we are forced to conclude that the physiological mechanism of human long-term memory is still a complete mystery. In particular, as Young (1978, p. 94) has remarked, despite the attractiveness of the schema hypothesis of Head and Bartlett, 'no one has been able to provide any clear notions about the neuronal organization that constitutes the building of a schema'.†

2.12. Time, memory, and personal identity

Memory has long been recognized as the concomitant of personal identity. It is the means by which the record of our vanished past survives within us, and this is the basis of our consciousness of self-identity. We have reason to believe that far more of our past survives in us unconsciously than the small part that is consciously recalled. Why then do we suffer from *total* amnesia concerning the events of the first three or four years of early childhood? Freud, who was the first to consider the problem, postulated a subconscious 'censor' who suppressed from consciousness all recollection of 'infantile sexuality'. A more general and convincing answer has since been given by E. G. Schachtel (1963), who claims that childhood amnesia results from a time lag in the development of the conceptual and conventional memory schemata that are, as was shown by Bartlett, necessary for conscious recall.‡

Memory, as we have seen, depends not only on schemata but also on coding. It is essentially related to what Suzanne Langer has called man's *symbolic transformation of experience.* As she has suggested, the origin of the concept of 'self'—which is usually thought to mark the beginning of conscious memory—may well depend on the process of epitomizing our feelings symbolically. 'To project our feelings into outer objects is the first way of symbolizing and thus of *conceiving* those feelings. This activity belongs to about the earliest period of childhood that memory can

† His own hypothesis is that it involves the introduction of limitations into an initially redundant network, since memory is 'a selective process'.

‡ Moreover, in the first few months of life the neural connections in the brain are largely undeveloped, and the cerebral rhythms revealed in electro-encephalograms suggest that the cortical processes of consciousness are not present. Memory is an important part of consciousness.

recover' (Langer 1957). Whatever physiological mechanism may be involved, the mnemonic trace functions for us as a symbol.† Moreover, the brain itself when studied scientifically is also transformed into a symbol, for its analytical description depends on what Lord Brain called 'the abstract and symbolical terms of neurophysiology' (Brain 1952). 'One thing is certain', wrote Descartes, 'I know myself as a thought and I positively do not know myself as a brain'.

The relationship between consciousness and brain is still very little understood. Not all brain events are mental events, and we do not know the neural correlate of every mental process.‡ Consciousness is associated with some nerve fibres but not with all, for if an incoming stimulus is blocked in its path before it reaches the cerebral cortex we never become aware of it. On the other hand, much of our mental activity is independent of sensory inputs. It is often argued that consciousness is merely an epiphenomenon of brain because a machine could, in principle, be devised to perform actions like ours. However, such a machine would have to be programmed, and if this were done by another machine a third machine would be required to programme that, and so there would be an infinite regress. As Sir Cyril Hinshelwood said in his Presidential Address at the Anniversary Meeting of the Royal Society in 1959, 'In its higher functions the human brain is programmed, not by other mechanisms, but by aesthetic and moral elements which somehow have their seat in consciousness, by elements, that is to say, which belong to the half of reality concerned with the observer rather than the observed' (Hinshelwood 1959).

An important difference between mind and brain concerns spatial localization. Even from the point of view of those who believe that mental processes can be entirely reduced to neurophysiology, mind is regarded as the complex interaction of various parts of the brain not to be localized in any one of them. It has even been suggested that mind may be located in a space of more than three dimensions. J. R. Smythies (1956) has postulated that brain and mind occupy different three-dimensional subspaces of this higher space, although possibly they may share the same time dimension. Smythies regards mind as spatially

† This interpretation throws light on the otherwise puzzling fact, to which W. Riese (1958, p. 133) has drawn attention, that recognition of the symbolic nature of thought and of its linguistic expression has been found 'to promise greater insight into the dynamics of speech defects resulting from brain lesions than their description in merely physiological terms, those of purely motor or sensory type'.

‡ Two phenomena in particular which have been cited as illustrations of the distinction between mind and brain are our experience of pain and the fact that, to quote Darwin, 'thinking what others think of us excites our blushes' (Beadnell 1934). For without awareness there could be no pain, and blushing is possible only because *attention* can influence capillary circulation.

extended. Indeed, if he had not, there would have been no point in his introducing additional spatial dimensions. I believe this to be an unnecessary hypothesis. Instead, we should concentrate on the fact that whereas brain, because it is a material entity, exists both in three-dimensional space and in time, mind, as manifested in consciousness, is a phenomenon pre-eminently associated with time alone. In other words, mind is essentially temporal in nature, like a tune. Consequently its 'interaction' with brain can occur only in time, and therefore mentally. It is like the interaction between a tune (in the mind) and the corresponding musical score (on paper). The root difficulty in discussing the problem at all is that mind is a self-referring concept, so that when we try to analyze what we mean by it we are like a man who is trying to pull himself up by his own shoe-laces!

If a mind is to be regarded as a memory-based process of integration, conservation and modification of personal identity, with temporal extension and temporal position but no spatial extension and no precise spatial location, must it be associated uniquely with the region of space occupied by a particular brain? Normally, we assume that this is so, except possibly in the case of paranormal phenomena, such as telepathy. However, even if we leave that contentious problem on one side, there are other complicating issues to be considered. The Pythagorean and Buddhist belief in metempsychosis, or the transmigration of souls, implied that the same mind can be associated at different times with different bodies (and not only human bodies), but this hypothesis is not open to scientific testing. On the other hand, apparently irrefutable cases are known of the converse association of different minds with the same body, as in the remarkable cases of multiple personality studied by Morton Prince and others.† In such cases a particular personality is associated with a particular body at different intervals of time separated by gaps when other

† The term 'multiple personality' designates the manifestation by a single person of two or more relatively distinct and different 'personal identities', alternating or co-existing 'co-consciously', the activity in the period ruled by one 'personality' usually, but not invariably, not being remembered in the period ruled by the other 'personalities' (see, for example, Rapaport 1961, 206–14).

This concept of multiple personality must be distinguished from that of the 'split personality' revealed by the ingenious tests devised by Roger Sperry and his collaborators at the California Institute of Technology to explore the effects of severing the corpus callosum, the band of some 200 million nerve fibres that joins one side of the cortex to the other. According to Sperry (1968), this operation, performed only on patients suffering from severe epilepsy, leaves them with two separate minds, or spheres of consciousness, that which is experienced by the right hemisphere being entirely outside the realm of awareness of the left, although only one-usually the left-does all the talking. (This lateral specialization seems to be unique to man and may have been related to the evolution of language, see Ornstein 1972). On the other hand, the distinguished neurophysiologist Sir John Eccles believes that the outstanding discovery of these investigations is the *uniqueness* of the dominant hemisphere in respect of conscious experience (Eccles 1977).

personalities take over, although it can happen that the different personalities overlap, so that as Prince (1920, p. 98) reported in the celebrated case of Miss Beauchamp, 'there is a doubling of consciousness without any true division of the normal self'.

Although cases of genuine multiple personality are very rare, the condition manifests itself in such a variety of forms that no two cases are quite the same. Nevertheless, a good idea of the phenomenon is provided in the earliest well-documented case. It concerns a young woman named Mary Reynolds who was born in England in 1794 and taken to Pennsylvania by her family when she was four years old. She grew up melancholy and shy and was given to solitary religious meditations and devotions. When she was about nineteen she had a hysterical attack which left her blind and deaf for several weeks. Three months later, she slept for nearly twenty hours and awoke seeming to have forgotten everything she had learned. She soon became familiar again with her surroundings but it took her a few weeks to learn to read, calculate, and write, although her handwriting was much cruder than it had been. She was buoyant, witty, and fond of company. After about five weeks of this new life, she again had a long sleep and awoke as her original self with no recollection of what had happened meanwhile. Thereafter, her two personalities alternated, but gradually the second tended to predominate and finally took over completely when she was thirty six and lasted until her death in 1854 (Hunter 1964, p. 243).

Since our concept of personal identity is so intimately related to time, it is not surprising that self-consciousness only develops as the growing child becomes aware of memory and thereby ceases to live in a continual present. The relation between man's personal time and the universal time of the world around him depends on his cultural environment, but as is evident from the phenomenon of multiple personality it is often far from simple. However, to explore further the relations between the time of the individual and universal time we now turn to biology.

References

Adrian, E. D. (1928). *The basis of sensation.* Christopher's, London.

Alexander, S. (1920). *Space, time and deity,* Vol. I, p. 113. Macmillan, London.

Ayer, A. J. (1956). *The problem of knowledge.* Penguin Books, Harmondsworth.

Bachelard, G. (1950). *La dialectique de la durée.* Gallimard, Paris.

Bacq, Z. M. (1975). *Chemical transmission of nerve impulses: a historical sketch.* Pergamon Press, Oxford.

Baddeley, A. D. (1976). *The psychology of memory.* Harper and Row, New York.

Bannister, R. (1955). *First four minutes*, p. 192, Putnam, London.

Barach, A. L. and Kagan, J. (1940). Disorders of mental functioning produced by varying the oxygen tension of the atmosphere. *Psychosom. Med.* **2,** 53–67.

Barbizet, J. (1970). *Human memory and its pathology* (transl. D. K. Jardine), p. 165. W. H. Freeman, San Francisco.

Bartlett, F. C. (1932). *Remembering*. Cambridge University Press, Cambridge.

Beadnell, C. M. (1934). *The expression of the emotions in man and animals* (by C. Darwin). Chap. XIII. Watts, London.

Bedini, S. A. (1963). The scent of time: a study of the use of fire and incense for time measurement in oriental countries. *Trans. Am. phil. Soc.* (*New Ser.*), **53** (5), 1–51.

Bell, P. M. (1975). Sense of time. *New Sci.* 15 May, p. 406.

Bergson, H. (1911). *Matter and memory* (transl. N. M. Paul and W. Scott Palmer). Allen and Unwin, London.

Berkeley, G. (1713). The first dialogue between Hylas and Philonaus. In *A new theory of vision and other writings*, p. 225. Everyman Library Edition (1910), Dent, London.

Blundell, J. (1975). *Physiological psychology*, pp. 117–8. Methuen, London.

Brain, W. R. (1952). *The contribution of medicine to our idea of the mind* (Rede Lecture), p. 22. Cambridge University Press, Cambridge.

Brandon, S. G. F. (1959). The origin of religion. *Hibbert J.* **57,** 349–55.

Branton, W. D., Mayerl, E., Brownell, P. and Simon, S. B. (1978). Evidence for local hormonal communication between neurones in *Aplysia*. *Nature* (*Lond.*) **274,** 70–2.

Brazier, M. A. B. (1977). *The electrical activity of the nervous system*. Pitman Medical Publishing Co., Tunbridge Wells.

Broad, C. D. (1923). *Scientific thought*, p. 351. Routledge and Kegal Paul, London.

Broadbent, D. E. (1958). *Perception and communication*. Pergamon Press, London.

—— (1970). Review lecture: psychological aspects of short-term and long-term memory. *Proc. Roy. Soc. B.* **175,** 333–50.

—— (1971). *Decision and stress*, Chap. VIII. Academic Press, London.

—— (1973). *In defence of empirical psychology*. Methuen, London.

Calebresi, R. (1930). *La determinazione del presente psichico*, p. 188, R. Bempored, Florence.

Callahan, J. F. (1948). *Four views of time in ancient philosophy*, pp. 180–4. Harvard University Press, Cambridge, Mass.

Cassirer, Ernst (1945). *Rousseau, Kant, Goethe*, p. 56. Princeton University Press, Princeton, N.J.

Cassirer, Eva (1957). *The concept of time; an investigation into the time of psychology with special reference to memory and a comparison with the time of physics*. Ph.D. Thesis, University of London.

Cherry, E. C. (1953). *J. acoust. Soc. Am.*, **25,** 975–9.

Cherry, E. C. and Taylor, W. K. (1954). *ibid.*, **26,** 554–9.

Clay, E. R. (1882). *The alternative: a study in psychology*, p. 167–8. London.

COHEN, J. (1967). *Psychological time in health and disease*, pp. 61–3. Charles C. Thomas, Springfield, Ill.

COHEN, J., HANSEL, C. E., and SYLVESTER, J. D. (1953). A new phenomenon in time judgment. *Nature* (*Lond.*) **172,** 901.

—— —— —— (1954). Interdependence of temporal and auditory judgments. *Nature* (*Lond.*) **174,** 642.

—— —— —— (1955). Interdependence in judgments of space, time and movement. *Acta Psychol.* **11,** 360–72.

CONRAD, R. (1964). Acoustic confusions in immediate memory. *Brit. J. Psychol.* **55,** 75–84.

CONTENAU, G. (1954). *Everyday life in Babylon and Assyria* (transl. K. R. and A. R. Maxwell-Hyslop), p. 213. Arnold, London.

COOPER, L. F. and ERICKSON, M. D. (1959). *Time distortion in hypnosis.* Williams and Wilkins, Baltimore.

CRAIK, F. I. M. and LOCKHART, R. S. (1972). Levels of processing: a framework of memory research, *J. verb. learn. and verb. behav.*, **11,** 671–84.

CZERMAK, J. (1857). Ideen zu einer Lehre vom Zeitsinn. *Sitzungsber. Akad. Wiss. Wien, Math.-Naturwiss. Kl.* **24,** 231–6.

DECROLY, O. (1932). *Etudes de psychogenèse*, Chapter IV. Lamertin, Brussels.

DE LA MARE, W. (1930). *Desert islands*, pp. 95–6. Methuen, London.

DELAY, J. (1961). *Les dissolutions de la mémoire.* Presses Universitaires de France, Paris.

DE QUINCEY, T. (1821). *The confessions of an English opium-eater*, p. 114. Everyman Library Edition (1907), J. M. Dent, London.

DIELS, H. (1882). *Simplicius*: *In Aristotelis physicorum libros commentaria*, 790. Reimer, Berlin.

DOOB, L. W. (1971). *Patterning of time*, pp. 200–1. Yale University Press, Newhaven, Conn.

EBBINGHAUS, H. (1913). *Memory*: *a contribution to experimental psychology* (transl. H. A. Ruger and C. E. Bussenius, with a new introduction by E. R. Hilgard). Dover Publications, (1964), New York.

ECCLES, J. C. (1977). In *The self and its brain* (by K. R. Popper and J. C. Eccles), p. 315. Springer International, Berlin.

EIBL-EIBESFELDT, I. (1970). *Ethology*, pp. 422 *et seg.* Holt, Rinehart and Winston, New York.

ELIADE, M. (1954). *The myth of the eternal return* (transl. W. R. Trask). Pantheon Books, London.

ESTEL, V. (1884). Neue Versuche über den Zeitsinn. *Phil. Stud.* **2,** 37–65.

EVANS-PRITCHARD, E. E. (1937). *Witchcraft, oracles and magic among the Azande*, p. 347. Clarendon Press, Oxford.

FRAISSE, P. (1952). La perception de la durée. *Année Psychol.* **52,** 39–46.

—— (1964). *The psychology of time* (transl. J. Leach). Eyre and Spottiswoode, London.

FRANKFORT, H. (1956). *The birth of civilisation in the near east*, p. 9. Doubleday, New York.

FREUD, S. (1914). *Psychopathology of everyday life* (transl. A. A. Brill). Allen and Unwin, London.

FURLONG, E. J. (1951). *A study in memory.* Nelson, London and Edinburgh.

GABOR, D. (1948). *Nature* (*London*), **161,** 777–8.

GALTON, F. (1883). *Inquiries into human faculty and its development.* p. 60. Everyman Library Edition (1907), Dent, London.

GERARD, R. W. (1953). *Sci. Am.* **189** (3), 118–26.

GESELL, A. and ILG, F. L. (1943). *Infant and child in the culture of today*, p. 24. Harpers, New York.

GONSETH, F. (1972). *Time and method* (transl. E. H. Guggenheimer), p. 153. Charles C. Thomas, Springfield, Ill.

GOODDY, W. (1969). Disorders of the time sense. In *Handbook of clinical neurology*, Vol. 3, p. 236. North-Holland, Amsterdam.

GRABENSBERGER, W. (1934). Experimentelle Untersuchungen über das Zeitgedächtnis von Bienen und Wespen nach Verfütterung von Euchinin und Iodothyreoglobulin. *Z. Vgl. Physiol.* **20,** 338–42.

GRANET, M. (1934). *La pensée chinoise*, p. 330. La Renaissance du Livre, Paris.

GRAY, E. G. (1974). The synapse. In *The cell in medical science*, Vol. 2: *Cellular genetics, development and cellular specialisation* (ed. F. Beck and J. B. Lloyds), pp. 385–416. Academic Press, London.

GRENE, M. (1963). *A portrait of Aristotle*, pp. 161 *et seq.* Faber and Faber, London.

GUYAU, M. J. (1890). *La genèse de l'idée de temps.* Alcan, Paris.

HABER, R. N. (1969). Eidetic images. *Sci. Am.* **220,** (4), 36–44.

HARDIE, R. P. and GAYE, R. K. (1930). In *The Works of Aristotle* (ed. W. D. Ross), Vol. II, *Physica*, Book IV, 14. Clarendon Press, Oxford.

HARROD, H. R. F. (1942). *Mind* **51,** 47–68.

HEAD, H. (1920). *Studies in neurology, Vol. II, pp.* 605–6. Clarendon Press, Oxford.

HEBB, D. O. (1949). *The organization of behaviour; a neuropsychogical theory.* Wiley, New York.

—— (1958). *Textbook of psychology* (1st edn.), p. 84. Saunders, Philadelphia.

—— (1972). *Textbook of psychology* (3rd edn.), p. 91. Saunders, Philadelphia.

HELSON, H., and KING, S. M. (1931). An example of psychological relativity. *J. exp. Psychol.* **14,** 202.

HERING, E. (1905). *On memory* (4th edn.). Open Court, Chicago.

HERRIOT, P. (1975). *Attributes of memory*, p. 79. Methuen, London.

HINSHELWOOD, C. N. (1959). Presidential Address. *Proc. Roy. Soc. A* **253,** 447.

HODGKIN, A. L. and HUXLEY, A. F. (1939). Action potentials recorded from inside a nerve fibre. *Nature* (*Lond.*) **144,** 710.

—— —— (1945). Resisting and action potentials in single nerve fibres. *J. Physiol.* **104,** 176–95.

—— —— (1952). A quantitative description of membrane current and its application to conduction and excitation in nerve. *J. Physiol.* **117,** 500–44.

von Hornbostel, E. M. and Wertheimer, M. (1920). *Sitzungsber. K. Preuss. Akad. Wiss.* 388–96.

Hume, D. (1738). *A treatise of human nature*, p. 87. Everyman Library Edition (1911), Dent, London.

Hunter, I. M. L. (1964). *Memory: facts and fallacies.* Penguin Books, Harmondsworth.

Hydén, H. (1968). Biochemical approaches to learning and memory. In *Beyond reductionism: new perspectives in the life sciences* (ed. A. Koestler and J. R. Smythies), pp. 85–117. Hutchinson, London.

Jacobson, A. L., Babich, F. R., Bubash, S., and Jacobson, A. (1965). Differential approach tendencies produced by injection of RNA from trained rats. *Science* **150,** 636–7.

James, W. (1890). *The principles of psychology* (reprinted 1950). Dover Publications, New York.

Janet, P. (1928). *L'évolution de la mémoire et de la notion du temps.* Chahine, Paris.

Jasper, H. H. (1958). *Reticular formation of the brain.* Churchill, London.

Katz, B. (1966). *Nerve, muscle and synapse.* McGraw-Hill, New York.

Kay, D. (1888). *Memory: what it is and how to improve it*, p. 18 (footnote). Kegan Paul and Trench London.

Kemp Smith, N. (1934). *Immanuel Kant's Critique of pure reason*, p. 48. Macmillan, London.

Klemm, O. (1919). *Arch. Gesamte Psychol.* **38,** 71–114.

Köhler, W. (1957). *The mentality of apes*, p. 234. Penguin Books, Harmondsworth.

Kollert, J. (1882). Untersuchungen über den Zeitsinn. *Phil. Stud.* **1,** 78–89.

Korsakoff, S. S. (1889a). Etude medico—psychologique sur une forme des malades de la mémoire. *Rev. Phil.* **28,** 501–30.

—— (1889b). Über eine besondere Form psychischer Störung, combiniert mit multipler Neuritis. *Arch. Psychiatr.* **21,** 669–704 (transl. M. Victor and P. I. Yakovlev (1955). *Neurology* **5,** 394 *et seq.*)

Langer, S. (1957). *Philosophy in a new key* (3rd edn.), p. 124. Harvard University Press, Cambridge, Mass.

Lashley, K. S. (1929). *Brain mechanisms and intelligence.* University of Chicago Press, Chicago.

—— (1950). In search of the engram. *Symp. Society Experimental Biology* **4,** 479.

—— (1951). The problem of serial order in behaviour. In *Cerebral mechanisms in behaviour: the Hixon Symposium*, (ed. L. A. Jeffress), pp. 112–36. Wiley, New York.

Leach, E. R. (1954). Primitive time-reckoning. *In A History of Technology.* (ed. C. Singer *et al.*), Vol. I, p. 126. Clarendon Press, Oxford.

Locke, J. (1690). *An essay concerning human understanding* (ed. A. S. Pringle-Pattison, 1924). Clarendon Press, Oxford.

Loehlin, J. C. (1959). The influence of different activities on the apparent rate of time. *Psychol. Monogr.* **73,** no. 474.

LURIA, A. R. (1969). *The mind of a mnemonist.* Cape, London.

MACH, E. (1865). Untersuchungen über den Zeitsinn des Ohres. *Sitzungsber. Akad. Wiss. Wien, Math.-Naturwiss. Kl.* **51,** (2), 135–50.

MCCONNELL, J. V. (1962). Memory transfer through cannibalism in planarium. *J. Neuropsychiatr.* **3,** (Suppl. 1), 542–8.

—— (1964). Cannibalism and memory in flatworms. *New Sci.* **21,** 465–8.

MCLEOD, P., WILLIAMS, C. E., and BROADBENT, D. E. (1971). Free recall with assistance from one and two retrieval clues. *Brit. J. Psychol.* **62,** 59–65.

MAINWARING, J. (1954). Memorizing. In *Groves Dictionary of Music and Musicians* (ed. E. Blom), (5th edn.), Vol. V, pp. 669–73. Macmillan, London.

MARK, R. (1974). *Memory and nerve cell connections.* Clarendon Press, Oxford.

MICHAUD, E. (1949). *Essai sur l'organisation de la connaissance entre* 10 *et* 14 *ans.* Vrin, Paris.

MICHON, J. A. (1967). *Timing in temporal tracking.* Van Gorcum, Utrecht.

MILLER, G. A. (1956). The magical number seven, plus or minus two: some limits on our capacity for processing information. *Psychol. Rev.* **63,** 81–97.

MILNER, B. (1966). Amnesia following operation on the temporal lobe. In *Amnesia* (eds. C. M. Whitty and O. L. Zangwill), pp. 112–5. Butterworths, London.

—— (1972). Disorders of learning and memory after temporal lobe lesions in man. *Clin. Neurosurg.* **19,** 421–46.

MORLEY, S. G. (1947). *The ancient Maya* (2nd edn.), p. 305. Stanford University Press, Stanford, Calif.

MOWBRAY, G. H. (1954). *Q. J. Exp. Psychol.* **6,** 86.

MÜLLER, G. E. (1911–17). *Zur Analyse der Gedächtnistätigkeit und des Vorstellungsverlaufes* (3 vols). Teubner, Leipzig.

MUNDLE, C. W. K. (1954). *Mind.* **63,** 42.

MÜNSTERBERG, H. (1889). *Beiträge zur experimentellen Psychologie.* **11,** 1–68.

MURDOCK, B. B. (1974). *Human memory: theory and data.* Lawrence Erlbaum Associates, Potomac, Maryland.

MYERS, O. B. (1965). Conversion from spectral to temporal pattern. *Nature (Lond.)* **206,** 918–9.

NEEDHAM, J. (1956). *Science and civilization in China,* Vol. 2, pp. 288–9. Cambridge University Press, Cambridge.

—— (1965). *Time and eastern man.* Royal Anthropological Institute Occasional Paper, No. 21.

VON NEUMANN, J. (1958). *The computer and the brain.* Yale University Press, New Haven, Conn.

OLDFIELD, R. C. (1954). *Br. J. Psychol.* **45,** 14–23.

ORNSTEIN, R. E. (1969). *On the experience of time,* p. 23. Penguin Books, Harmondsworth.

—— (1972). *The psychology of consciousness,* p. 63. Freeman, San Francisco.

PATON, H. J. (1951). *In defence of reason,* p. 107. Hutchinson, London.

PATTERSON, J. H. and GREEN, D. M. (1970), Discrimination of transient signals having identical energy spectra. *J. acoust. Soc. Am.,* **48,** 894–905.

—— —— (1971). Marking of transient signals having identical energy spectra. *Audiology* **10,** 85–96.

Penfield, W. G. (1954). *Epilepsy and the functional anatomy of the human brain.* Little, Brown, Boston.

—— (1958). Some mechanisms of consciousness discovered during electrical stimulation of the brain. *Proc. Natl. Acad. Sci.* U.S.A. **44,** 51–66.

—— (1975). *The mystery of the mind.* Princeton University Press, Princeton, N.J.

Penfield, W. G. and Mathieson, G. (1974). Memory. An autopsy and a discussion of the role of the hippocampus in experimental recall. *J.A.M.A. Arch. Neurol.* **31,** 145–54.

Peters, R. (1956). *Hobbes*, p. 133. Penguin Books, Harmondsworth.

Piaget, J. (1950). *The psychology of intelligence,* (transl. M. Piercy and D. E. Berlyne). Routledge and Kegan Paul, London.

—— (1955). *The child's construction of reality* (transl. M. Cook), Chap. IV. Routledge and Kegan Paul, London.

—— (1969). *The child's conception of time* (transl. A. J. Pomerans). Routledge and Kegan Paul, London.

Pick, A. (1905). Zur Psychologie des Vergessens bei Geistes— und Nervenkranken. *Arch. Krim. Anthrop. Kriminal* **18,** 256–7.

Piéron, H. (1952). *The sensations, their functions, processes and mechanisms.* (transl. M. H. Pirenne and B. C. Abbott). Frederick Muller, London.

Pines, S. (1955). *Proc. Am. Acad. Jewish Research,* **24,** 11.

Plumb, J. H. (1972). *In the light of history,* p. 231. Allen Lane, The Penguin Press, London.

Pribham, K. H. (1969). The neurophysiology of remembering. *Sci. Am.* **220,** 73–86.

Price, H. H. (1958). *Mind,* **67,** 457.

Prince, M. (1920). Miss Beauchamp: the theory of the psychogenesis of multiple personality. *J. abnormal Psychol.,* **15,** 67–135.

Pusey, E. B. (1907). *The confessions of Saint Augustine* (transl. E. B. Pusey), Book XI, paragraph 31, p. 270. Everyman Library, Dent, London.

Ralling, C. (1959). A vanishing race. *Listener* **62,** 87 (16 July, 1959).

Randall, J. H. (1960). *Aristotle,* pp. 61 *et seq.* Columbia University Press, New York.

Rapaport, D. (1961). *Emotions and memory.* The Menninger Clinic Monograph Series No. 2, Science Editions, New York.

Reid, T. (1785). *Essays on the intellectual powers of man* (ed. A. D. Woozley, 1941), p. 210. Macmillan, London.

Ribot, Th. (1885). *Diseases of memory* (3rd edn.), p. 61. Kegan Paul, Trench, London.

Riese, W. (1958). Descartes's ideas of brain function. In *The history and philosophy of knowledge of the brain and its functions* (ed. F. N. L. Poynter), p. 118. Blackwell, Oxford.

Ritchie Russell, W. (1959). *Brain, memory and learning,* p. 37. Clarendon Press, Oxford.

RUSSELL, B. (1917). *Mysticism and logic*, p. 202. Allen and Unwin, London.
—— (1921). *The analysis of mind*, pp. 157 *et seq.* Allen and Unwin, London.
—— (1946). *History of western philosophy*, p. 835. Allen and Unwin, London.
—— (1948). *Human knowledge*, p. 226. Allen and Unwin, London.
RUSSELL, C. W. (1858). *The life of Cardinal Mezzofanti.* Longman, Brown, London.
SACHS, O. (1973). *Awakenings.* Duckworth, London.
SCHACHTEL, E. G. (1963). *Metamorphosis; on the development of affect, perception, attention and memory*, Chap. 12. On memory and childhood amnesia. Routledge and Kegan Paul, London.
SEMON, R. (1904). *Die Mneme.* Engelmann, Leipzig.
SIFFRE, M. (1965). *Beyond time* (transl. and ed. H. Briffault). Chatto and Windus, London.
SMYTHIES, J. R. (1956). *Analysis of perception.* Routledge and Kegan Paul, London.
SORABJI, R. (1972). *Aristotle on memory* (*De memoria et reminiscentia*). Duckworth, London.
SPEARMAN, C. (1923). *The nature of 'intelligence' and the principles of cognition*, p. 318. Macmillan, London.
—— (1937). *Psychology down the ages*, Vol. I, p. 283. Macmillan, London.
SPERRY, R. W. (1952). *Amer. Sci.* **40,** 305.
—— (1968). *The Harvey Lecture Series*, **62,** 293–323.
STEINBERG, H., and SUMMERFIELD, A. (1957). *Q. J. exp. Psychol.* **9,** 138–45, 146–54.
STERN, W. (1930). *Psychology of early childhood* (transl. A. Barwell), p. 112. Macmillan, New York.
STURT, M. (1925). *The psychology of time.* Routledge and Kegan Paul, London.
SYZ, H. (1937). *J. gen. Psychol.* **17,** 355–87.
TALLAND, G. A. (1965). *Deranged memory: a psychonomic study of the amnesic syndrome.* Academic Press, New York, London.
THOMAS, K. (1971). *Religion and the decline of magic*, p. 617. Weidenfeld and Nicolson, London.
THOMPSON, J. E. S. (1956). *The rise and fall of Maya civilization*, p. 149. Gollancz, London.
TULVING, E. (1972). Episodic and semantic memory. In *Organisation of memory* (eds. E. Tulving and W. Donaldson), Chap. 10. Academic Press, New York.
UNDERWOOD, B. J. (1957). Interference and forgetting. *Psychol. Rev.* **64,** 49–60.
—— (1964). Forgetting. *Sci. Am.* **210** (3), 91–9.
UNDERWOOD, G. (1975). Attention and the perception of duration during encoding and retrieval. *Perception* **4,** 291–6.
—— (1976). *Attention and memory.* Pergamon Press, Oxford.
UNDERWOOD, G. and SWAIN, R. A. (1973). Selectivity of attention and the perception of duration. *Perception* **2,** 101–5.
VALENTINE, J. D. (1968). Is visual memory holographic? *Nature* (*Lond.*) **220,** 474–5.

VAN GENNEP, A. (1909). *The rites of passage,* (transl. M. B. Vizedom and G. L. Caffee (1960)). Routledge and Kegan Paul, London.

WALLACE, W. H., TURNER, S. H., and PERKINS, C. C. (1957). *Preliminary studies of human information storage.* Signals Corps Project 132C, Institute for Cooperative Research, University of Pennsylvania, Philadelphia, Pa.

WALLIS, R. (1966). *Le temps,* quatrième *dimension de l'esprit,* pp. 51 *et seq.* Flammarion, Paris.

WARD, J. (1918). *Psychological principles,* p. 220. Cambridge University Press, Cambridge.

WARRINGTON, E. K., and WEISKRANTZ, L. (1970). Amnesic syndrome: consolidation or retrieval? *Nature (Lond.)* **228,** 628–30.

WEDGWOOD, C. V. (1975). *The political career of Peter Paul Rubens (Seventh Walter Neurath Memorial Lecture),* p. 12. Thames and Hudson, London.

WHITEHEAD, A. N. (1920). *The concept of nature,* pp. 72–3. Cambridge University Press, Cambridge.

WHORF, B. L. (1956). An American Indian model of the universe. In *Language, thought and reality* (ed. J. B. Carroll), p. 57. M.I.T. Press, Cambridge, Mass.

WINSLOW, F. (1861). *On obsure diseases of the brain and disorders of the mind.* p. 361. John W. Davies, London.

YOUNG, J. Z. (1964). *A model of the brain.* Clarendon Press, Oxford.

—— (1966). *The memory system of the brain.* Clarendon Press, Oxford.

—— (1978). *Programs of the brain.* Oxford University Press, Oxford.

ZAWIRSKI, Z. (1936). L'évolution *de la notion du temps,* p. 71. Gebethner and Wolff, Cracow.

ZELLER, A. F. (1950). *J. exp. Psychol.* **40,** 411–23.

3

BIOLOGICAL TIME

3.1. Physiological processes and man's sense of time

We have seen that man's time-sense† is the product of many factors, sociological as well as psychological. Moreover, when we are subjected to abnormal environmental conditions, our estimation of time can be greatly distorted, as we have seen in the case of Michel Siffre who stayed two months in a dark cave with no external time cues. On the other hand, there are frequent cases in which human beings have been able to estimate with a high degree of accuracy periods of several hours in the absence of external indicators. This phenomenon is familiar to those who are able to wake up at a given time without the aid of an alarm clock. This 'head clock', as it is sometimes called, works more precisely under hypnosis. An order to perform a particular action after a prescribed time interval will usually be obeyed with considerable accuracy. Even without hypnosis, the 'head clock' can often function remarkably precisely over fairly long periods. In a classic experiment, Macleod and Roff (1936) found that two subjects shut up in a sound-proof room for 48 hours and 86 hours respectively estimated the time with such accuracy that their respective errors amounted to less than one per cent, 26 min in the former case and 40 min in the latter. John Locke's belief that men have no perception of time 'but by reflection on the train of ideas they observe to succeed one another in their understanding' (Locke 1690) is obviously quite inadequate to explain the high degree of accuracy achieved in this experiment. Although the need for some external cues in order to estimate the time of day correctly when the subject is confined in a dark cave may be partly attributable to the peculiar psychological effect of this highly abnormal environment, not only was the time Siffre spent in the cave far longer than the times spent by Macleod and Roff in their experiment but also his body temperature was significantly reduced from its normal level.

The hypothesis of a temperature-dependent internal chemical clock as the physiological basis for man's estimation of time was put forward by Hudson Hoagland in 1933. Ten years previously, Henri Piéron (1923) had suggested that, if our bodily processes were artificially speeded up

† Use of the term 'time-sense' does not imply that a special time-sense organ exists (see p. 61).

or slowed down, our estimation of clock-time would be correspondingly affected. Four years later, Marcel François (1927) confirmed this hypothesis on subjects asked to tap a Morse key at the rate of three taps a second. He found that the rate of tapping increased when their body temperature was raised by diathermy. Hoagland's interest in the problem was aroused when his wife, who had developed a high temperature due to influenza, insisted that his absence for 20 minutes had lasted much longer. In experiments on her ability to count at an estimated rate of a digit a second, he found that as her temperature rose so her counting became more rapid. When her temperature was 98.4 °F she took 52 seconds to count 'a minute', but when it rose to 103 °F her 'minute' lasted only 37.5 seconds. She herself was unaware that her counting-rate had altered. Hoagland proposed that these findings were due to the operation of an internal chemical clock, its rate increasing with temperature in accordance with the well-known reaction-rate equation of Arrhenius† (Hoagland 1933). By comparing his results with those of François's study of Morse-key tapping, he found an average value for the energy of activation μ of about 24 000 cal. He argued that 'These results definitely imply the existence of a unitary chemical process serving as a basis for the subjective time scale, a process quite possibly irreversible in nature and perhaps catalysed in a specific way'. He concluded that 'our sense of time depends on the speed of a chemical step in some group of cells, presumably in the brain, that acts as a chemical clock'.

Hoagland's hypothesis was subjected to further experimental testing by C. R. Bell (1965). He obtained some correlation between increase in body temperature and speed of counting (at a rate estimated by the subject to be one digit a second) and speed of tapping (at an estimated rate of three taps a second). One subject failed to demonstrate a change in his counting speed in the predicted direction. When the energy of activation μ was calculated for the other subjects (whose counting speed had increased), a mean value of 74 000 cal was obtained within a range of 9000–139 000 calories. Bell concluded that in view of this variability in the Arrhenius-equation energy of activation value, 'the constancy of which is essential to the hypothesis of a potent biochemical influence on time estimation performance in humans', Hoagland's hypothesis that man's control of his time estimation of short intervals is by means of a chemical clock 'would seem to be untenable' (Bell 1966). More recently, Bell (1975) has developed a new cooling technique which eliminates

† According to this equation, the reaction-rate k is proportional to $\exp(-\mu/RT)$, where μ is the activation energy per mol, R is the gas-constant and T the absolute temperature. The value of μ denotes the amount of kinetic energy per mole in excess of the average energy which the molecules must acquire for the reaction to take place.

many of the potentially fatal dangers formerly associated with body cooling and has used it on subjects who were asked to estimate a 60-second interval in both warm and cold conditions. Although most of the cold subjects clocked a 'long' minute, the results were not in accordance with the proportional changes expected on Hoagland's hypothesis, and some showed a speeding-up. A more complex mechanism is required to explain this variability. In Bell's view, even if Hoagland's clock exists, the mind's reading of it is subject to vagaries.

Whatever the dependence of our sense of time awareness and powers of time estimation on our bodily metabolism may be, it is highly probable that man's homeothermy is the crucial factor linking our individual physiological time in the short term (i.e. from day to day) with universal physical time and preventing the relationship between them from becoming too erratic.† From the point of view of time in the long term, physiological processes in man fall into two groups: those, such as the pulsations of the heart, which are periodic or rhythmic and do not change much with advancing years, and those, such as the sclerosis of tissues and arteries, which are progressive. Some of the repetitive processes, however, also gradually undergo progressive change. A pioneer in the study of this phenomenon of aging in man was Lecomte du Noüy. In 1916 he investigated the healing of superficial wounds of a certain size in adults of various ages by measuring the diameter of the wound as a function of time after the wound was inflicted and thereby calculating an index of the rate of healing. Since, in his opinion, the time needed for a given unit of physiological work of repair appeared to be, on the average, about four times longer at fifty years of age than at ten, he claimed that 'Everything, therefore, occurs as if sidereal time flowed four times faster for a man of fifty than for a child of ten' (du Noüy 1936). In fact, however, his sample was so small and the variance of the individual 'cicatrization indices' that he obtained so great that no precise age-dependence relationship of this type could be derived. Years later, Howes and Harvey (1932), on studying the rate of growth of the cells of fibrous tissue in healing wounds, found that young and old differ not so much in the rate itself but in the time of delay before regrowth begins, there being a greater time lag with greater age. Moreover, although du Noüy believed in the existence of a basic physiological time associated with cell growth and regeneration,

† Animals that have no internal temperature control (reptiles, amphibia, fishes and invertebrates) are greatly influenced by the changes in the ambient temperature, being more sluggish when it is lower. For these animals time would appear to pass more slowly on warm days and more rapidly on cold days. Moreover with ambient temperature variations time would not appear to pass steadily, i.e. in terms of universal physical time. Aspects of the behaviour of animals with no internal temperature control are consistent with these conclusions. (Hoagland 1966)

not all physiological processes are slowed down at the same rate by advancing age. Physiological time, as defined by du Noüy, is not only non-uniform in the same individual at different stages in his life but it also tends to vary from one individual to another.

Physiological time differs from universal physical time because it is essentially an inner time associated with a region of space inhabited by living cells which are relatively isolated from the rest of the universe. Physiological time is controlled by the response of cells to changes occurring inside this region, such as the rate of accumulation of waste products. The biological basis of aging is, however, not yet understood. In particular, it is not even known whether it is genetically programmed. However, irrespective of whether aging is the predetermined consequence of the succession of developments associated with cellular differentiation in the embryo or whether it is due to a random series of accidents ('chemical noise'), it is generally agreed that cellular information-loss is its manifestation and 'clock'.† In a living active organism energy must be continually supplied for the maintenance of its complex structure, and its stability is due to the equilibrium of the processes of degradation and regeneration that are continually going on within it. Its stability is therefore of a dynamic nature. The length of life is controlled by slow progressive modifications of serum and tissue that cannot be arrested and lead to a gradual slowing down of our physiological processes. As previously mentioned (p. 63), this is the main reason why as we grow older, we feel that time tends to race by even more rapidly, for although this illusion is also fostered psychologically by other factors (our lives usually tend to become fuller and each new unit of physical time is a smaller and smaller fraction of our past life), even for those whose lives are comparatively empty physical time normally seems to pass by more rapidly as they grow older. Compared with external physical time, our physiological time is slowing down.

The death of an organism is obviously a sufficient condition for the complete cessation of its physiological time. Surprisingly, it is not a necessary condition. So long as the structure of the organism can be kept intact while no energy is being supplied for the continuance of its normally essential physiological processes, it can be maintained indefinitely in a state of suspended animation. For certain organisms in the

† 'If human aging as observed in the intact subject is not timed by genetic information loss, the nature of its leading "clock" and the accessibility of this to interference remain wide open' (Comfort 1968). According to Strehler (1977), the multicellular state of organisms may be unstable, either because of the failure of the communication system to transmit adequate signals to provide for co-ordination or because the ancestors of organisms are single cells that have evolved in competition with others for space, energy, etc. 'Relics of this competitive ancestry may well become dominant as the coordinative forces, whatever they may be, gradually lose control'.

appropriate circumstances this state of anabiosis, or latent life, can be brought about by dehydration, but generally the most effective agency is very low temperature (Keilin 1959). The stability of an organism in such a state is of a static nature and while therein the organism does not age. In other words, *its physiological time stands still.*† The artificial prolongation of human life by means of refrigeration is, however, unlikely to have any practical application that can be foreseen at present, except possibly for travel in interstellar space.

Although it is now generally believed that man's time sense is not associated with any bodily organ such as the ear, it may be based on central nervous processes such as cerebral rhythms. This hypothesis has been supported by various physiologists and others, notably the neurologist William Gooddy (1958), the mathematician and founder of cybernetics Norbert Wiener (1958), and more recently the Czech electrophysiologist Josef Holubář (1969) who made an important contribution to the experimental investigation of the problem.

Since the pioneer researches of the English physician R. Caton, in 1875, it has been known that the brain generates detectable electric currents. With later improvements in recording apparatus, Hans Berger discovered, in 1924, incessant rhythmical electrical activity in the brain. However, it was not until 1934, following a convincing demonstration by Adrian and Matthews, that electro-encephalograms (EEG) obtained from the potential differences between pairs of electrodes attached to the outside of the skull were generally accepted as valid records of cerebral activity. The voltages concerned are very feeble, being of the order of 10 μV, but the frequencies of the oscillations are more significant than the amplitudes. The harmonic analysis of the records is complicated, but four main types of rhythm, each characterized by a particular frequency range, have been recognized. Of these the most prominent in normal adults (especially at the back of the skull) is the so-called *alpha rhythm*, which ranges from about 8 to about 12 Hz with a central frequency of approximately 10 Hz.

Of all bodily rhythms the alpha rhythm is thought by many to be the most closely associated with our sense of time. This hypothesis has, however, been challenged because the alpha rhythm tends to disappear when the brain is most active. As a rule it is most clearly evident when the eyes are shut and the subject is relaxed. If he opens his eyes or begins to think intensively about a problem, this rhythm becomes increasingly difficult to detect. Of course, his sense of time continues! It may be, however, that the predominance of the alpha rhythm when the brain is resting is due to the synchronized fluctuations of large groups of cells,

† Strictly speaking, this does not happen for animals that are merely in a state of hibernation.

whereas the lower-voltage pattern of electrical activity revealed by the alert brain corresponds to the highly diversified activities of its various parts.

Norbert Wiener (1958) pointed out that, because the alpha rhythm can be generated artificially by submitting the eye to a visual flicker of external impulses at the rate of about ten per second, it is reasonable to suppose that the natural rhythm is the response of the brain to a flicker caused by its own internal oscillations. He argued that careful analysis of the records reveals that around a central frequency close of 10 Hz there is a rather empty range with a sharp peak of great intensity and small width in frequency in the centre, and he concluded that this narrow band of frequencies constitutes a 'clock' in the brain. In support of this conclusion he drew attention to the analysis of reaction times, and he claimed that the various processes involved in perception impose on reaction times irreducible minima of about 0.1 seconds, the period of one cycle of the alpha rhythm.

The hypothesis that the alpha rhythm is a pacemaker for our sense of time has also been strongly supported by the Czech physiologist Josef Holubář (1969). He devised an ingenious experiment to test it. A peculiarity of brain rhythms is that they can be altered significantly by optical stimulation of the eyes with interrupted light, or flicker, of a suitable frequency which produces a synchronous rhythm of electrical activity that spreads rapidly over most of the brain (Adrian 1947). As an appropriate test of the sense of time Holubář used optical flicker in conjunction with two indicators that could not be intentionally affected by the subject and so could be regarded as objective, namely the electro-encephalogram and the galvanic skin reflex (change of electrical resistance of the skin in response to an external stimulus). Holubář found that the intervals of temporally conditioned galvanic skin responses could be altered by visual flicker, depending on the relation between the rate of flicker and the frequency of the alpha rhythm. When the frequency of flicker was significantly different from that of the alpha rhythm, e.g. 7 or 14–15 per second, the intervals were only about half as long as when the subject was not subjected to flicker. When, however, the frequency of flicker coincided with that of the alpha rhythm, i.e. 10 per second, the intervals remained the same as in the absence of flicker. They also remained effectively unchanged for frequencies of one-half and double the alpha frequency. These results led Holubář to conclude that for intervals of the order of minutes the alpha rhythm serves the human organism as a fundamental reference rhythm for the measurement of time.

In support of his hypothesis that the alpha rhythm is a time-measuring mechanism, Holubář drew attention to the role of the thalamus. According to Dempsey and Morrison (1943), the thalamus and its connections with the cortex are involved in the generation of the alpha rhythm. Furthermore, since Spiegel *et al.* (1955) found that after operations on this part of the brain transient disorders of temporal orientation and time estimation occurred in 23 out of 30 patients, it would appear that the thalamus is important for our awareness of time. As for the alpha rhythm being most easily detected when the brain is least active, so far from regarding this as a difficulty for his hypothesis, Holubář argued that the very circumstance that favours the manifestation of the alpha rhythm, undisturbed quiet, is also the most conducive for the accurate maintenance of our sense of time.

Nevertheless, it is not yet possible to conclude that the alpha rhythm is definitely associated with our sense of time, for, although since 1934 it has been accepted by many physiologists that the alpha rhythm is generated by the brain, a contrary hypothesis that has attracted much attention has been advocated by O. Lippold (1970, 1973). He claims that the alpha rhythm is the electrical concomitant of physiological tremor in the muscles that move the eyeballs. He points out that the alpha waves are of greatest amplitude in the occipital regions near the back of the skull and are particularly evident in those subjects with some defect of vision. Moreover, although about one person in twenty has no detectable alpha rhythm, there is no evidence that they continually miss their trains or their appointments! On the other hand, as Lippold points out, it has been reported that those subjects with slower brain rhythms tend to have slower mental processes. 'Such results', writes Lippold (1973, p. 28), 'of course, provided they are true, seem to argue for the alpha rhythm being of cortical origin; I cannot explain them on my hypothesis'. At present, the issue is undecided and Lippold's hypothesis presents a serious challenge to those who believe that the alpha rhythm functions for us as a cortical master clock. As Holubář (1969, p. 85) himself concluded, 'The problem of the sense of time is still far from being resolved'.

3.2. Time and animal navigation

Despite the difficulties and complications that confront us in our attempts to understand the physiogical basis of our sense of time, we now have abundant experimental evidence for the existence of reliable biological 'clocks' in animals and plants. This evidence has come mainly from three different fields: the study of animal navigation, the study of photoperiodism (the responses of living organisms to seasonal changes in the lengths of day and night), and the study of daily and other periodic rhythms in the behaviour and activity of living organisms.

Some of the most exciting research concerning biological time-keeping processes has resulted from the study of bird navigation. It has long been known that migrant birds can fly great distances to specific destinations and that even quite young birds can make their way independently of adults of the same species. The first important advance in the scientific study of the directional orientation of these birds was made by Gustav Kramer in 1949. It was well known that when migrant songbirds are kept in captivity they become very active at night during the periods of their normal spring and autumn migration. Kramer observed that starlings placed in circular cages out of doors at migration time indicated by their behaviour the direction in which they wished to travel. They tended to head in a particular direction, but Kramer noticed that they did not do this when the skies were overcast. On shielding the birds from the direct rays of the sun and using mirrors reflecting the sun's light at right angles, so that the direction of the sun would appear to have been rotated through 90 °, Kramer found that the birds changed their direction by the same amount. Consequently, Kramer concluded that the birds' sense of migrational direction depended on taking their bearings from the apparent position of the sun (Kramer 1950). He also discovered that if they were kept in enclosures that were illuminated by an artificial sun kept in a fixed position, they systematically changed their orientation during the day at the appropriate rate corresponding to the earth's rotation. He inferred that the birds had some form of internal clock enabling them in effect to measure the passage of time (Kramer 1952). He trained starlings to feed in a given compass direction at a particular time of day and then tested the birds at another time. He found that they still took up the training position. The birds obtained their compass direction by sighting on the sun and making due allowance for its regular daily motion. Kramer also found that if a starling trained in this way was confined in a cage with an artifical day and night six hours out of phase with the natural day and night and was later placed in natural sunlight, it would seek food in a direction at about 90° from the true direction. This indicated that the internal clock used by these birds to obtain direction from the sun was maintained by the daily light–dark cycle of the locality, and so in this case was six hours out of phase with local time.

Starlings migrate during daylight, but some birds migrate mainly at night and yet travel long distances. Among these species are the warblers. Similar experiments to those devised by Kramer were made on these birds by E. G. F. Sauer, who showed that they could use the stars to obtain direction. He confined some in a sound-proof cage within a planetarium (Sauer 1957). Although they were given no external indication of the time of year, when autumn came they began to flit about restlessly as if informed by an internal clock that it was time to take wing. So long as the dome was illuminated with diffuse light they showed no particular orientation, but when the star patterns appropriate to the time and place were projected they indicated the direction of their normal migration. No particular stars or constellations seemed to be involved in their direction-finding ability but only the general pattern of the night sky. Since they made due allowance for the apparent rotation of the sky during the course of the night, Sauer concluded that these birds navigate with the aid of an internal clock that allows them to relate the appearance of the heavens at each season to terrestrial geography. The results obtained by Sauer, assisted by his wife, were presented in a series of reports including one at the international Symposium on Biological Clocks held at Cold Harbor in 1960 (Sauer and Sauer 1960). Those experiments in which the planetarium sky was shifted so that the bird's apparent position was changed in latitude gave consistent results for the bird's direction of flight, although the bird had been raised from the egg and so had had no previous navigational experience. On the other hand, experiments simulating changes in longitude gave results that were not easy to interpret. Differences of latitude correspond to differences in the altitude of the Pole Star, but differences of longitude are more difficult to determine because they can only be detected with the aid of a chronometer that is set independently of local time. As Matthews (1968) has remarked, the ease with which the clock controlling the sun–compass reaction can be shifted gives rise to the question of whether a bird has a chronometer that is sufficiently 'rigid', i.e. uninfluenced by outside events, to measure displacement in longitude. He has suggested that the bird might have two clocks, 'one, easily malleable, the clock for sun–compass orientation, the other, more rigid, the chronometer for longitude measurement'.†

A biological clock seems also to be involved in the homing instinct of pigeons. Experiments have shown that, if pigeons are exposed to an artificial day and night out of phase with local time and then released some way from home, they will usually head in the wrong direction, but nevertheless they often reach home eventually. There has been considerable divergence of opinion on how they manage this. There is some

† On the other hand, as the result of studying with the aid of a planetarium the night-flying migratory bird, the North American indigo bunting, Emlen (1975) maintains that these birds rely on the geometrical pattern of the stars for navigation and not on an internal time sense. He claims that there is considerable adaptive redundancy in the buntings' recognition of constellations in the northern sky, thereby making it possible for them to migrate on nights when there is variable cloud cover.

indication that they make use of more than one compass system for determining direction, including sensitivity to the earth's magnetic field (Bliss and Heppner 1976). It is now evident that recognition of landmarks plays a very small role in the homing process (Keeton 1974). Since pigeons can correctly orient themselves homeward under overcast skies, they cannot always rely on the sun–compass method, and it is probable that an internal timing mechanism is involved.†

Migrating birds and homing pigeons are not the only animals that use an internal clock for direction finding. The sand hopper (*Talitrus saltator*) does also. This animal inhabits the wet sand of beaches, and if the beach becomes too dry it travels towards the sea in a direction at right angles to the shoreline. It determines the required direction by means of the sun's position (Hamner 1966). To use the sun's position for this purpose it must rely on an internal clock. Experiments have shown that this clock can be shifted by altering the light–dark cycle.

A remarkably sophisticated application of an innate time sense to direction finding is made by honey bees. Their ability, when laden with food, to fly straight back to the hive has long been known and has given rise to the term 'bee-line' for a straight line joining two places. On the other hand, their time-keeping ability has only been recognized this century. The pioneer in this field was a Swiss doctor, August Forel (1906). He was in the habit of breakfasting on the terrace of his house in the mountains. Bees from a local hive used to come and sample small portions of the jam left over at the end of the meal. Forel discovered that they continued to come at the same time each morning, even after he had started breakfasting indoors and there was no longer any food on the terrace to attract them. Since their searching the terrace for food occurred at the same time each day and at no other time, Forel attributed to the bees a memory for time.‡

A few years later the German zoologist von Buttel-Reepen (1915) introduced the word *Zeitsinn*, or 'time sense', to describe the precisely timed daily visits of bees to fields of buckwheat, which secretes its nectar at the same hour each day. Since the existence of a time clock was not the only possible explanation for this regularity, von Frisch and his colleagues

† There is increasing evidence that terrestrial magnetism is important in pigeon homing. Recently it has been announced that Charles Walcott of New York State University has discovered a minute piece of magnetic tissue (probably magnetite) between the eye and brain of the pigeon. Similar cells have been found in bees.

‡ Von Frisch (1967, p. 253) says 'I know of no other living creature that learns so easily as the bee when, according to its 'internal clock', to come to the table'.

in the University of Munich studied the sense of time of bees under controlled conditions of light and temperature. In 1937 von Frisch claimed that all the evidence indicated that bees have an internal clock that is independent of recurrent changes in the external environment, for although bees could be trained to come to a particular feeding station at the same time each day, they could not be trained to come to the same place at different times. On the other hand, they could be trained to feed at two or more different places at two or more different times of day. Frisch concluded that the time sense of bees is not based on the learning of intervals but depends on an internal clock with a period of twenty four hours.

This hypothesis was later confirmed by a decisive experiment (Renner 1955). Bees that had been trained to collect nectar between 10 a.m. and noon in Paris were flown to New York. It was found that they continued their nectar-collecting activity in phase with the time in Paris and not with the time in New York, which is five hours later. The experiment was then repeated by retraining the bees in New York and flying them back to Paris, where they continued to act on New York time.

The ability of foraging bees to communicate with one another, so brilliantly investigated by von Frisch, also depends on the use of a biological clock. When a successful forager returns to recruit other bees to visit a source of nectar or pollen that he has discovered he performs a dance indicating its distance and direction. If the flowers are nearby (not more than 50–100 metres from the hive), the scout bee performs what von Frisch has called a 'round' dance, turning around once to the right and once to the left and repeating these circles with great vigour for half a minute or more. If the flowers are a long way off, the bee performs a 'waggle' dance by running a short distance and waggling its abdomen from side to side. It then makes a complete turn leftwards, runs forward again in the same direction as before still waggling its abdomen, makes a complete turn to the right, and repeats the whole sequence of movements again and again. This dance provides information about the location of the source. Its distance is indicated by the number of turns made in a given time, the smaller the number the greater the distance. Strictly speaking, however, the rate of turn is not an indication of distance alone, for it also depends on the direction of the wind. A head wind on the way to the source has the same effect as an increased distance and slows the dances, whereas a tail wind has the opposite effect.† The basis of the

† Bees that fly against a head wind to the feeding place have a tail-wind on their return flight, but only the outward flight seems to influence the 'distance' indicated by the dance.

bee's estimate of 'distance' seems, therefore, to be the effort needed to reach the flowers.†

The direction of the flowers is indicated by the straight part of the dance. The dance is performed on the vertically arranged combs inside the hive. The angle between the forward waggling part of the dance and the upward direction, i.e. the direction directly opposite to gravity, indicates the angle between the sun's azimuth and the line of flight to the source of nectar.‡ The sun is thus used as a kind of navigating compass, for it has been found that the direction of the dance changes during the course of the day by approximately the same angle as the sun. Von Frisch and his colleagues were surprised to discover that, even when part of the sky is cloudy and the sun is obscured, bees can still indicate the correct direction of a feeding place relative to that of the sun. It was found that they utilize the relationship between the sun's position and the polarization of the blue sky. More remarkably, if bees are induced to dance during the night, they indicate the direction of the place at which their daily feeding time is closest to the time of the dance (Lindauer 1960). In one experiment a foraging bee announced at 9.31 p.m. a feeding place in the east were it had been fed every day at 6 p.m., but at 3.54 a.m. it indicated a feeding place in the south where it had been fed every day at 8 a.m. Further experiments revealed that not only can bees store in their memories both feeding times and feeding sites but also the azimuth of the sun at any time of day even when they have not seen it for several weeks. Sometimes a scout bee will dance all night when communicating the location of a possible new home to a swarm (Lindauer 1971). In such performances it is as if the bee can calculate the course of the sun, for one can observe the angle between the dance-line and gravity change at approximately 15° per hour. According to von Frisch (1967, p. 353), this

† According to von Frisch (1967, p. 113) in view of the well-developed time sense of bees, one might expect that the distance would be estimated by the time required for the (outward) flight. However, various observations contradict this conclusion. For example, in experiments on a steep slope bees that flew uphill to the feeding station danced more slowly than those whose place of feeding was downhill, although the time of flight was the same in both cases. Consequently, 'we come to the conclusion that their estimation of distance must be based on the expenditure of energy for the flight', and he comments, 'That is a yardstick that truly is strange to man'. However, Ribbands has suggested that it is possible that certain delicate hairs forming parts of special sense organs in a bee's antennae may become bent when a bee is in flight and a stream of air is quickly passing by them, and that the bee senses the length of time during which these hairs have been displaced from their resting position, thereby 'measuring' the distance travelled.

‡ The ability to transpose the angle of the sun from the realm of visual perception to that of gravitational perception is not confined to bees. Even when this has no biological significance, many arthropods, in the absence of light, can transfer automatically to gravity an angle of orientation to a light source (von Frisch 1967, p. 137).

is a clear indication that 'bees read the time of day not from external factors but from an "internal clock"; in other words, their sense of time is not exogenous but endogenous'.

Lindauer (1960) has tried to discover which components of the bees' time-compensated solar orientation are inherited and which learned. He hatched bees in an incubator and raised them for several weeks in a cellar without any view of the sky in an artificial daily cycle of 12 hours light and 12 hours dark until they reached the age of foragers. On the first day of free flight in the open air they were trained to a given compass direction. On the next day they were incapable of finding this direction again. However, if such inexperienced bees were trained in a certain direction for at least five days, they were then able to find the correct direction even in an unknown territory and at a time of day different from the training time. By then they had learned to calculate correctly the sun's motion. It appears that although the use of the sun as a clue to orientation seems to be innate, the way to use it, based on knowing how it moves, has to be learned. A surprising capacity that seems to be innate is the ability to be able to extrapolate from a fraction of the sun's path its whole diurnal course.

The remarkable phenomena associated with the navigational achievements of animals cannot be understood unless the creatures concerned possess some kind of internal time-keeping 'mechanism'. Although it has not been possible so far to discover any of the physiological processes involved, it seems that in these animals there must be some rhythmic processes occurring which can serve as reliable clocks. Although these rhythms can often continue in the absence of external changes, their pace is maintained by external rhythmic events, the light–dark succession of day and night being the most important.

3.3 Photoperiodism

The first clear account of the measurement of time by any living organism was published in 1920 by W. W. Garner and H. A. Allard. They introduced the term *photoperiodism* to designate the response of organisms to the relative length of day and night. Previously it had been generally assumed that annual changes in temperature, light intensity, rainfall, etc. were responsible for organisms developing in accordance with seasonal changes. However, these factors are often unreliable guides to the course of the seasons, and when this is so photoperiodism is much more effective for in photoperiodic responses organisms measure the period of daylight (or darkness) without being influenced by any naturally occurring changes in the intensity of light. As early as 1852, in a book on

The Vegetation of Europe, A. Henfrey suggested that the length of day is a factor in the distribution of plants. M. J. Tournois (1912) regarded it as a factor in flower initiation. He induced earlier flowering in hop and hemp by restricting exposure to daylight to six hours daily. However, in a later paper (Tournois 1914) he attributed precocious flowering to impoverishment of reserve materials in the plant. Meanwhile, G. Klebs (1913) had discovered that, although he could not induce winter flowering in *Sempervivum funkii* by changes of nutrition and temperature, he could do so by a few days of electric illumination. He concluded that in nature the time of flowering is determined by the period of daylight exceeding a certain critical amount, light functioning as a kind of catalyst.

Garner and Allard (1920) were much more thorough. They were concerned with the problem of why, outside the tropics, marked regularity in the time of flowering and fruiting is the rule in plants. Instinctively we think of temperature as the decisive factor causing one season to differ from another in its effect on plants, but they pointed out that this does not explain why a plant, for example the common iris which flowers in May or June, will not flower in winter when grown in a greenhouse under the same temperature conditions as early summer. They systematically investigated one factor after another which might conceivably initiate flowering and fruiting, such as nutrition, humidity, and light-intensity, and eliminated each of them. Finally, after much deliberation, they concluded that the only remaining seasonal phenomenon that could be responsible was change in the relative length of day and night. The importance of this conclusion was that they clearly dissociated length of day from the amount of solar radiation received. By means of simple but carefully controlled experiments they found, for example, that a variety of soya beans which germinate in May and ordinarily flower in September could be forced into blossom in June simply by artificially shortening the period that they were exposed to daylight, while another lot of plants of the same species that were kept under precisely similar conditions, except for being exposed throughout to the normal amount of daylight, did not flower until the usual time of year. They discovered that other typical autumn-flowering plants, such as chrysanthemums, could also be made to flower in summer by shortening the length of daily light exposure. Thus a large group of plants, including most of the so-called summer annuals which regularly flower after midsummer, do so as a result of decrease in the length of day. While relatively short days favour flowering and fruiting in these plants, long days are more favourable to rapid and extensive vegetative development. Some of these plants, therefore, if they receive the full benefit of the long days of summer, may reach giant proportions before being brought into the flowering condition. 'Thus we can understand why it is that when the farmer plants some crops too early there is a tendency towards excessive development of leaf and stem with little fruit or flowering. Late planting, on the other hand, may lead to dwarfing in growth but abundant flowering and fruiting'.

Whereas some plants flower in the autumn, others regularly flower in the spring and early summer. Garner and Allard showed that these plants will only flower when the days are longer than a certain minimum. Subsequently, on the basis of flowering response to the duration of daily exposure to light, they divided plants into 'short-day' plants, which include those investigated by Tournois, 'long-day' plants, which include the *Sempervivum* investigated by Klebs, and 'day-neutral' plants which flower under virtually any daylength condition.

The first conclusive proof that the length of daylight could influence the reproductive behaviour of animals was obtained by the Canadian zoologist William Rowan (1926). Schäfer (1907) had suggested nearly twenty years previously that daylength might be the controlling factor in the migration of birds as well as in their annual gonad cycles. Rowan studied a bird, known as the greater yellowleg, which migrates to Patagonia in the autumn and returns to its breeding grounds in Canada every spring. Although the round trip is about 16 000 miles, the precision of the bird's timing is such that its eggs are always hatched between 26 May and 29 May. Rowan studied this bird for fourteen years and considered all possible factors that might be involved. He concluded that the only one that was sufficiently regular and precise to act as the required synchronizer was the variation in the length of daylight. To test this conclusion he took birds of another species that winters in Canada, the slate–coloured junco, and subjected them to daylengths which were artificially lengthened gradually until after a few weeks the birds were experiencing daylight conditions that normally do not occur until late spring. He then found that they were already in a a condition to breed, whereas control specimens kept under natural winter daylength conditions were not. He subsequently released birds at various states of their reproductive development and found that they migrated when they were between the inactive and the full breeding state.

Rowan's investigation was a landmark in the history of the discovery of photoperiodism. From his researches and others it became evident that photoperiodism is a crucial factor in the reproductive cycle of many species of animal, invertebrate as well as vertebrate, although as a rule it is probably not the only factor. In some animals there appears to be an internal reproductive rhythm which enables them to develop the full breeding state even if they are kept in continual darkness, but in general environmental factors such as photoperiodism serve to 'set' the internal 'clock'.

An important advance in our understanding of time measurement in photoperiodism was the discovery by K. C. Hamner and J. Bonner (1938) that the effects of a long period of daylight could be obtained after a short day if the dark period were interrupted by a comparatively short spell of light. This result led many biologists to believe that it is not the length of the daylight period that is decisive but rather the length of the night. The time when the strongest effect is obtained is not as a rule in the middle of the dark period but often occurs a certain number of hours after it begins. In some cases, however, the beginning of the light period has more influence on the timing of the critical point of highest sensitivity in darkness than the beginning of the dark period itself.

Similar effects of light interruption in a long dark period have been observed in many species, animal as well as plant. For example, Hart (1951) investigated the initiation of the breeding period in ferrets. Breeding can be initiated in two months if they are subjected to 18 hours of continual light daily, but 12 hours of light each day can be just as effective if 1 hour of light is given at the appropriate time in the dark period. Even 6 hours of light a day will suffice, so long as the daily cycle consists of 4 hours of continual light and the 20-hour dark period is interrupted by light from 17 to 19 hours after the beginning of the light period (Hammond 1953). Another important discovery was that the short-day plant *Xanthium pennsylvanicum*, which when maintained on long days (18 or more hours of light) does not form any flower buds at all, will flower abundantly if exposed to a single short day and hence a single long dark period (Hamner 1960). This discovery was additional evidence that the crucial timing process for flowering occurs in the dark period. A variety of organisms have been tested for changes of sensitivity to interruption during this period. The variability of sensitivity with time suggests that these organisms can 'measure' duration. In other words, it appears that the physiological processes concerned are controlled by some internal 'clock', and that this clock is but little affected by changes of temperature.

In the study of plant photoperiodism it has been possible to determine not only when in the dark period light interruption is most effective but also which wavelengths of light are involved, for when relative response was plotted as a function of wavelength to obtain an 'action spectrum' it was found that the resulting curve has a marked peak in the red region. Moreover, it closely resembled the corresponding curves already obtained for other light-controlled processes, such as the germination of some light-sensitive seeds. This indicated that the pigment system acting in photoperiodism is the same as that acting in other photoresponses.

In 1952 it was discovered that the effects of red light in initiating lettuce seed germination could be completely reversed by subsequently exposing the seeds to light of wavelengths in the far red (Borthwick and Parker 1952). A similar reversal was also found to occur with photoperiodism, and this suggested that the pigment involved was converted into a different form by absorption of red light and then back to the original form by absorption of far-red light. Since a short-day plant is exposed to sunlight which is relatively rich in red before being subjected to the long dark period and since this same light is effective in interruption of the dark period, it seemed that the pigment must be automatically converted in the dark from the far-red absorbing form to the red absorbing form. It was soon realized that the far-red absorbing form is the biologically active one. The pigment was finally isolated (from seedlings of different plants grown in darkness) by Butler *et al.* (1959). This was a great achievement, because this pigment, which its discoverers called *phytochrome*, is present only in extremely small quantities and the measurement of its optical absorption is beset with difficulties.†

The basic properties of phytochrome may be briefly summarized. It exists, probably in all plants, in a stable form and an 'active' form. The former, P_r, changes into the latter, P_{fr}, when irradiated with light of wavelength 6600 Å (6600×10^{-10} metres) or after a brief period of daylight. Similarly, P_{fr} reverts to P_r on exposure to light in the far-red part of the spectrum of wavelength 7350 Å, but it also reverts slowly and spontaneously in darkness. As we have already indicated, phytochrome is chemically active in the P_{fr} from, presumably catalyzing some biochemical reaction on which certain crucial stages in the plant's life history depend.

For a while it was thought that the rate at which the active form of phytochrome spontaneously reverts to the stable form provides the plant with a 'clock' for measuring the dark period. However, it was eventually realized that, although the role of phytochrome in photoperiodism is of great importance, that idea that it alone provides an effective measure of time for the plant is inadequate. In many cases light reactions connected with other pigments, most of which are still unknown, are thought to be involved. Moreover, it has been found that the light offered to the plant may reveal different action spectra depending on when within the dark

† The many-sided effect of phytochrome on plants indicates that it is connected with some general metabolic link or links and is not confined to regulating photoperiodic reactions (Borthwick and Hendricks 1961). A useful textbook on phytochrome and plant growth is that by Kendrick and Frankland (1976).

period the plant is subjected to it (Bünning 1973, p. 217). Although it is clear that a 'biological clock' exists in plants influencing the timing of the flowering process, it has not yet been discovered how it controls the plant's responsiveness to light, except that it seems to be coupled with the environment mainly through the phytochrome system.

In the case of animals, red light of wavelength between 4400 and 8000 Å has been found to have a marked effect on different bird species, but the photoreceptor pigments have not been identified. As in the case of plants, the action spectra depend on when within the dark period the light is offered. Adaptive evolution of photoperiodically controlled seasonal processes has led to certain thresholds, depending on the species concerned, becoming the decisive factors in any photoperiodic reaction. In the strongly photoperiodic species of northern latitudes the threshold is sufficiently critical to make differences in photoperiods of only a few minutes a day sufficient to initiate the relevant process, for example gonadal recrudescence in birds or induction of diapause in insects. As in the case of plants, it is only when the animal experiences light at a specific time each day (the 'photoinducible phase') that the particular process is initiated. Light experienced at other times is ineffective. The actual timing of the photoinducible phase varies from species to species, and in some cases even within a given species if it has a wide geographical distribution.

The use of daylight by living organisms as an initiating stimulus is not surprising, at least in middle and high latitudes, since it is the most regular of the seasonal changes that occur in those regions of the Earth's surface. Photoperiodism may well have become important during the evolutionary conquest of land, where temperature is a less reliable indicator of change than in the sea. Since natural selection favours those organisms that produce offspring at the most propitious time of year, species have tended to evolve responses in those environmental changes that provide the most stable source of predictive information.

3.4. Biological rhythms in animals and plants (i)

So far we have considered a variety of evidence suggesting that many living organisms are able to make surprisingly accurate time 'measurements'. Nevertheless, it has not been possible to show how they are able to do this. We have considered animal navigation and plant and animal photoperiodism. Now we turn our attention to the third field of research that has provided evidence for the existence of biological clocks, the study of periodic rhythms† in the behaviour and physiological activity of living organisms.

The discovery that an organism can display precise cyclic or rhythmic behaviour even where no perceptible change occurs in its environment was first made by a French astronomer, Jean de Mairan, in 1729. He was interested in the leaf movements of plants. Many plants extend their leaves in the hours of daylight and fold them at night. Mairan observed that these movements will continue even when the plants are kept in constant darkness. It was as if the plant sensed the sun without being exposed to it. Some thirty years later Henri-Louis Duhamel du Monceau (1758) repeated Mairan's experiment and confirmed that it was as if the plant, in the absence of any light clue, was able to tell the time. He also found that its ability to do this was independent of temperature. Later other scientists studied similar phenomena, and Darwin discussed them in his last important book, *The power of movement in plants*, published in 1880. Darwin concluded that the folded, or 'sleeping', attitude of plants was such as to expose the smallest amount possible of leaf surface to the night sky. He realized that, owing to natural selection, the plants concerned possessed an ideal means of protecting themselves against the cold night air, but he had no idea of the nature of the underlying mechanism. Although the great German botanist Pfeffer‡ and others thought that the continuation of the daily leaf movements in conditions of total darkness might be an after effect of previous experience of alternating day and night, Darwin believed in the intrinsic nature of the periodicity, which he thought was 'to a certain extent inherited' (Darwin and Darwin 1880). He said that he was unable to follow Pfeffer's reasoning. Years later Pfeffer altered his views on the cause of leaf movements and came to believe in the existence of hereditary diurnal rhythms. In a notable series of experiments he prevented leaf movements by attaching weights and by other means. When the leaves were later released so as to resume movement, he found that they did so without any phase shift compared with the controls (Pfeffer 1911). This led him to reject his original simple feedback hypothesis and to postulate diurnal internal controlling processes that continue even when the movements themselves are prevented.

† In biology, the term *rhythm* is used to denote any periodic fluctuation, or oscillation, in the physiology or behaviour of an organism.

‡ W. F. Pfeffer (1845–1920) is regarded as one of the greatest of experimental botanists and as a precursor of modern molecular biologists. He introduced exact physicochemical methods into plant physiology and was famous in his day for his pioneer studies of osmosis. Erwin Bünning, a successor to Pfeffer's chair in Tübingen, has recently written a biography of Pfeffer (Bünning 1975).

Any controlling process of this type can only be regarded as a biological clock if it is actually used by the organism concerned to measure time. The mechanism involved may be either *exogenous*, i.e. external or environmental, or else *endogenous*, i.e. internal or self-sustaining. Two types of endogenous clock have been considered as possible mechanisms of biological chronometry: the *rhythmic type* that automatically oscillates with a constant frequency, and the *hour-glass type* that measures a definite lapse of time and then has to be reset to measure a further lapse. For many years biologists refused to believe that living organisms could contain any endogenous clocks. Instead various attempts were made to find an exogenous 'factor X' which was assumed to cause the diurnal periodicity in physiological processes. The crucial discovery that eventually led to the abandonment of this view was made in 1930 by Erwin Bünning, who found that in conditions of constant darkness and uniform temperature the cycle of leaf movement in the bean *Phaseolus* has a period that was only approximately 24 h, being in fact between 25 and 26 h.† Since this period did not coincide with any known change in the environment, Bünning concluded that the origin of the rhythmic movement of its leaves must be intrinsic to the plant itself and depend on an internal clock (Bünning and Stern 1930).

In recent years the useful term 'circadian', from the Latin *circa diem* meaning 'about a day', introduced by F. Halberg in 1959 has come into general use to denote all biological rhythms that deviate somewhat from precise 24-hour periodicity. According to Bünning (1973), the term should be restricted to rhythms that can continue in constant temperature and in the absence of diurnal light–dark cycles. Under these uniform ambient conditions, circadian cycles are found to have periods between 22 and 28 hours, and usually between 23 and 25 hours. Only a few cases are known in which, under such conditions, it is difficult to detect significant deviations from the 24-hour period. In natural conditions, however, the period of an endogenous circadian rhythm in an organism is usually forced to match that of the 24-hour cycle of day and night. The endogenous rhythm is then said to be *entrained* by the external cycle.

In general, an external cycle can entrain an endogenous oscillation if their periods do not differ too much. The signal responsible for the entrainment is called the *Zeitgeber* (time giver) or *synchronizer*. When the rhythm considered is not entrained it is said to be *free running* (Pittendrigh 1958) and oscillates with its natural period, e.g. 22 hours approximately for Candolle's *Mimosa pudica*. Besides differences between

† Bünning (1960) has pointed out that deviations from a 24-hour period had been discovered previously, in particular as long ago as 1832 by de Candolle who discovered that the leaf movements of *Mimosa pudica* showed a period of 22–23 hours in darkness (Candolle 1832).

species, individuals within a species differ (genetically) in their free-running periods.

Bünning's discovery of circadian rhythmicity was the result of his experiments on the effect of red light on leaf movement in plants. He undertook these experiments after studying important but neglected researches by Rose Stöppel. Stöppel (1916) had carefully rechecked Pfeffer's work on plant rhythms and had discovered the remarkably accurate timing of the leaf movements of the plant *Calendula*. She could not believe that such precise timing from day to day was possible without some cue from the environment, but she attributed it to some subtle factor X. She had worked with the aid of red illumination, which in those days was thought to be 'safe', i.e. ineffective as a *Zeitgeber* or synchronizer. Bünning and Stern (1930) discovered that, on the contrary, red light is the most important synchronizer for higher plants, even if given for only a minute or two. They found that in genuinely constant ambient conditions plant rhythms tend to drift out of phase with the earth's rhythm of diurnal rotation.

It was soon found that circadian rhythms are not limited to periodic leaf movements and also that such rhythms survive the subjection of the organism to abnormal light–dark cycles in the laboratory. The Dutch scientist Anthonia Kleinhoonte (1929, 1932) put seeds of the large jack bean (*Canavalia ensiformis*) to sprout in abnormal light–dark cycles (8 hours of each) for 17 days. They then showed a rhythm corresponding to a 16-hour day. Next she submitted them to continual light, so that they had no clue at all concerning the length of the day. Nevertheless, although these plants had never seen true day and true night from the time when they had sprouted and had been forced to accommodate themselves to a completely unnatural light–dark cycle, they gradually shifted from the abnormal cycle to the normal cycle in phase with the earth's daily rotation, despite the absence of any external cue. This showed that the diurnal rhythm was not imprinted by diurnal fluctuations of the environment during an early stage of the plant's development. Similar results were obtained for the normal time sense of bees even when they were brought up under constant conditions of light or darkness (Wahl 1932). One of the most thorough investigations of this kind was that made by Bünning (1935) on the fruit fly *Drosophila*. The particular rhythm studied was that of eclosion, or emergence of the adult fly from the pupal case. This occurs regularly at a precise time close to dawn.† It was found that the exposure of *Drosophila* to constant conditions of light

† It is advantageous to emerge then since the insects are not exposed immediately to dehydration by sunlight, which could be damaging. Just after they emerge their cuticula may not be waterproof.

and temperature for fifteen consecutive generations made no difference, thereby indicating the endogenous nature of the diurnal rhythm involved.

In 1936 Bünning proposed the hypothesis that the mechanism responsible for photoperiodic time measurement in plants was the endogenous diurnal rhythm controlling its leaf movements. He suggested that each oscillation comprises two alternating half-cycles differing in their sensitivity to light. The first 12 hours Bünning called the *photophile*, or light-requiring, half-cycle, and the second the *scotophile*, or dark-requiring, half-cycle. Bünning postulated that each half-cycle, or phase, has different biochemical properties and different sensitivities to light. For 'short-day' plants flowering is induced when light is restricted to the photophile phase, but for 'long-day' plants flowering depends upon the extension of light into the scotophile phase.†

Of the two types of endogenous clock that might be involved in plant photoperiodism, only the rhythmic, or oscillator, type was believed by Bünning to be concerned. This was contrary to prevailing opinion at the time, and it was not until 1960 that Bünning's ideas made much impact outside Germany or on others besides plant physiologists. In that year Bünning and Joerrens (1960) introduced the idea of the endogenous rhythmic clock into insect physiology in place of the hour-glass model previously assumed by most workers in the field. Since then, despite some exceptions to it, Bünning's hypothesis has tended to dominate the whole field of photoperiodism.

According to Bünning's hypothesis, photoperiodic regulation means that some process is linked with a means of time measurement, the organism concerned having an inherited internal time scale at its disposal. Day by day the actual duration of daylight is compared with this scale. As soon as the actual length of day (or night) exceeds,or becomes less than, a critical duration some physiological change (photoperiodic reaction) occurs. Consequently, photoperiodic reactions result from the fact that the circadian clock controls the quantity and quality of responsiveness to light. Maximum responsiveness to light breaks occurs at what Bünning calls the *subjective midnight point*, generally about 16 to 18 hours after sunrise.

Bünning has stressed the importance of circadian rhythms in natural

† More recently, Pittendrigh and his colleagues (Pittendrigh 1960, Pittendrigh, Rosenweigh, and Rubin 1959, Pittendrigh and Bruce 1959, Pittendrigh and Minis 1964, 1971) have found it necessary to introduce the concept of the *inducible phase*, defined as that part of the scotophile phase when light (in some way not yet understood) interacts with some component of the biological clock involved to cause (or prevent) photoperiodic induction, depending on whether the organism has a long-day (or short-day) response.

selection,† because of the necessity for many organisms to perform certain functions at certain times of day, for example the hatching time of *Drosophila* and the time memory of bees. He pointed out that in the case of bees it is obviously important that the clock operates by oscillations and not by non-repetitive processes, so that information about the best visiting hours for regions with specific plant species can be stored on days with unfavourable weather. The bees can then stay home for a day or two and still be able to remember the often closely defined time of day at which a particular species secretes its nectar. 'These are only a few examples, and they can be supplemented at will. Thus we may say that diurnally periodic oscillations in plants and animals are not luxuries but are really used as clocks'. (Bünning 1960, p. 3.)

For selection to have occurred there must have been a variety of possibilities from which a choice could be made. In fact, there is abundant evidence that *oscillation is a fundamental property of all living things.*‡ Originally, however, these oscillations need not have been diurnal or even circadian and their periods may have differed significantly from 24 hours.§ Oscillations were necessary to keep cells alternating between extreme physiological states, anabolic processes predominating in the one

† Until recently it was generally believed that circadian leaf movements belonged to that small group of complicated physiological processes for which no adaptive value could be found. Darwin's foresight in this matter has, however, now been justified, for Bünning (1971) has shown by experiment that these mysterious leaf movements are important for the precision of photoperiodic time measurement. Threshold values for the plant's perception of light involved in photoperiodism are often as low as 0.1 lux. This low value means that moonlight could disturb photoperiodic time measurement, since it often reaches 0.5 lux and sometimes more, the highest values occurring under full moon at midnight, the time when plants are most responsive to light (subjective midnight point). However, the night position of the leaves reduces the intensity of moonlight coming from the zenith and reaching the surface of the leaves to about 0.05–0.1 lux. As Bünning (1971, p. 208) remarks, 'Darwin was correct in his view that leaf movements have an adaptive value'. It is possible, that circadian rhythms when originally evolved ensured maximum utilization of solar energy. Once evolved they became a kind of inbuilt mechanism, just like the chromosomes carrying the genetic 'blue print' and the energy-releasing enzymes.

‡ Persistent rhythmicity often continues in organisms even when they are removed from the particular environment with which these periodic changes enable them to cope. Besides rhythms with periods that match characteristic rhythms of the environment there is a whole array of biological phenomena with a wide variety of periods, e.g. from the beats of cilia in protozoa that tend to be too fast for the eye to follow (typically 80 milliseconds) to the flowerings of large tropical bamboos that often occur only every 30–40 years, so that in some parts of Asia a man is said to be so old that he has seen the bamboo flower twice. To most of these periodic phenomena there are no matching environmental variables. Such rhythms represent the raw material from which evolution has had to select those that approximate periodicities in the environment (Sweeney 1969, Chapter 6).

§ A point which Bünning does not mention is that in the remote geological past the terrestrial day was, in all probability, significantly shorter than 24 hours.

and catabolic in the other. If these extreme states are not reached certain functions cannot occur. This is clearly seen in the unexpected injuries suffered by many organisms when kept in constant light. The reason why these symptoms are relieved by an occasional exposure to darkness, say once a week, is because this allows an oscillation to start whereby a physiological extreme can be reached. Although it is not strictly necessary that the oscillation should keep in step with the rotation of the earth, such an adaptation to the earth's rotation has caused our circadian rhythms to evolve from oscillations with widely diverging periods such as are still to be found in many fungi. (Bünning 1960)

Circadian rhythms are exhibited by most plants and animals from unicellular organisms† to man. Indeed, we now recognize a fairly sharp division between those organisms that display circadian rhythms and those that do not. Organisms which lack a discrete nucleus (bacteria and most algae) have no need of circadian rhythms because their lifespan is less than 24 hours. During that time a bacterial culture divides several times. There is consequently no selective advantage in a bacterium 'knowing' what time of day it is! For organisms living longer than 24 hours it is advantageous to be prepared for the cyclic changes that occur in that time, and the more complex the life of the organism the greater the advantage.

Not all biological rhythms that have external correlates are circadian. Some marine organisms show rhythms in their behaviour that are clearly associated with high and low tide. For example, green flatworms (*Convoluta*) come to the surface of the sand at high tide and then bury themselves in the sand as it dries. It has been found that this rhythm continues when they are placed in an aquarium where there are no tides (Bohn 1903). Lunar cycles also occur in some marine organisms. The palolo-worm of the Pacific and Atlantic oceans is the best known. It reproduces only during the neap tides of the last quarter moon in October and November. Also the brown alga *Dictyota* discharges most eggs 9 days and 15–16 days after exposure to moonlight (Bünning 1973, p. 183). Since a number of species continue their periodic behaviour under laboratory conditions when they are exposed neither to tidal conditions nor to moonlight, it is clear that the rhythms concerned must be endogenous. Lunar and semilunar cycles may be due to the interaction of diurnal rhythms with tidal rhythms, certain phases of the two coinciding so as to produce regular reinforcements at intervals of about 15 or 29 days.

† The most thoroughly studied unicellular organism in which rhythms can be detected is *Euglena gracilis*. These motile green flagellates collect together more rapidly in a light beam by day than by night. Circadian rhythms have also been found in the individual cells of multicellular organisms.

This possibility has been studied in detail by F. A. Brown and his colleagues working at Woods Hole, Massachusetts, in the case of the common fiddler crab *Uca*. This animal inhibits the inter-tidal zone of the seashore and spends much of its time in a burrow that it digs for itself. When the tide ebbs and exposes the burrow it comes out to feed. It exhibits a striking 24-hour cycle of variation in colour. During daylight the black pigment in its skin cells disperses through them to make it dark and so protect it from the bright sun and predators. Towards nightfall the crab becomes paler as the pigment concentrates at the centres of the cells, and at dawn the cycle begins all over again. It has also been found that the epoch of maximum darkening tends to occur some 50 minutes later each day. This epoch is correlated with the time of low tide which advances at this rate from day to day. It is therefore clear that there must be another cycle of about 12 hours 25 minutes as well as the 24-hour one.

The 24-hour rhythm of pigment changes in *Uca* is so precise that, in Brown's view, it cannot be described as circadian. He has argued, in opposition to most other investigators of biological rhythms, that organisms under natural conditions possess no intrinsic daily rhythmicity but are rhythmic in response to subtle pervasive geophysical factors, e.g. fluctuations in the earth's magnetic field. In his opinion, biological rhythms are essentially exogenous rather than endogenous, even if some internal timing mechanisms exist. He believes that the many deviations from precise 24-hour periodicity are due to daily phase shifts, or perturbations, a concept that he has termed *autophasing* (Brown 1965).†

That the nature of biological rhythmicity in animals is more diverse than in plants is evident from the detailed investigations by A. D. Lees of the bean aphid *Megoura viciae*, an insect to which he believes Bünning's

† According to Brown, the influence of temperature and the intensity of light, even when they are kept constant, vary with fluctuations in the sensitivity of responsiveness of the organism to external influences, and shifts of phase and period can result.

To decide whether rhythms are truly endogenous or whether Brown is right, Hȧmner and his associates tested organisms on a turntable at the South Pole. He argued that, if a turntable placed there rotated once in every 24 hours in the sense opposite to the earth's rotation, an organism placed on it would remain at rest with respect to a fixed direction in space. The diurnal rhythms of the organisms so tested were not found to be influenced either by location at the Pole or by any rotation of the turntable. He concluded that although it is possible that some external periodic stimulus may regulate the biological clock, it does not arise from any factor associated with the earth's rotation. (Hamner *et al.* 1962). When Hamner mentioned at the Cold Harbor Symposium on Biological Clocks in 1960 this experiment that he was then planning, Brown (1960, p. 70) commented, 'The experiment of Professor Hamner sounds very interesting. Positive results would be most exciting, but I would caution him against interpretation of negative results in view, for example, of the known complex diurnal movements of the magnetic, relative to the geographical, poles'.

For a clear and not highly technical account of Brown's ideas, by a former pupil of his, see Bennett (1974) *passim*. An excellent debate on Brown's views can be found in Brown, Hastings, and Palmer (1970).

hypothesis cannot be applied. His investigations have led him to conclude that the photoperiodic clock in *Megoura*, although endogenous, is not an oscillatory rhythm but an interval timer of the hour-glass type. It is started by the onset of darkness, is stopped by dawn, and effectively measures the duration of the dark period. Lees has been careful to point out, however, that a simple hour-glass model will not account adequately for all the facts (Lees 1965, 1973). Pittendrigh (1966) maintained that the two types of timer, oscillator and hour-glass, are not necessarily mutually exclusive, and in a recent detailed survey Saunders (1976b, pp. 158 *et seq.*) concluded that the biological clocks in many insects are a combination of both hour-glass and oscillator, clocks with at least two components being necessary to account for all aspects of insect photoperiodism.†

The existence of circadian rhythms throughout the animal and plant world has led biologists to inquire whether there may be any physiological rhythms with periods of about a year. This is more difficult to determine since it clearly requires a patient year-by-year investigation.‡ The first positive evidence bearing on the question was discovered by K. C. Fisher and E. T. Pengelley, who were studying animal hibernation (Pengelley and Fisher 1963). A species of ground squirrel that inhabits the Rocky Mountains was kept in a small windowless room at freezing point and supplied with ample food and water. From August until October it ate and drank normally and maintained a body temperature of 37 °C, despite its cold surroundings. In October, as would have happened if it had been in its natural surroundings, it stopped eating and drinking and began to hibernate, its body temperature falling to just above freezing point. In April it became active again and resumed its normal behaviour and body temperature. Similar experiments with other ambient temperatures gave the same result, the period of each complete cycle being a little less than a year. The usual criteria for the existence of an endogenous rhythm were satisfied: its period was not precisely a year, it did not synchronize with any periodic external signal, and it was independent of the ambient temperature.

The *circannual rhythm*, as it is now called, was even manifested when these animals were kept at a uniform temperature so close to their normal body temperature that it was impossible for them to hibernate. Although

† All models, however, have one feature in common: long days have a positive effect in that they give rise to development or prevent an insect going into diapause, whereas short days are considered to be neutral. It is probable that long days lead to a production of a chemical substance that accumulates until it can stimulate the release of a hormone in the brain which initiates the sequence of events associated with egg development, moulting, or metamorphosis (Saunders 1976a, p. 121).

‡ According to Sweeney (1969, p. 87), a rigorous demonstration of an annual rhythm in plants has probably never been achieved. One difficulty is that environmental conditions of light and temperature should be kept constant for three cycles, i.e. three years.

food and water were available, they reduced their consumption and lost weight during the winter and then reverted to normal in the spring. As E. T. Pengelley and S. J. Asmundson (1971), who made this experiment, have commented, 'There could hardly be a more convincing statement of the existence of an internal clock operating independently of the environmental conditions'.

The discovery of a circannual clock in hiberators has led to the search for similar rhythms in other animals, particularly birds. We have seen (p. 137) that, by altering the daily rhythm of light and dark, Rowan influenced the timing of the characteristic restlessness that initiated the birds' migration. He suspected that the length of the day was probably not the only factor that influenced the urge to migrate. His belief has since been confirmed by the discovery of a circannual clock in some migratory birds. Experiments on warblers, some being kept under natural conditions and others under a constant ambient temperature and a daily cycle of 12 hours light and 12 hours darkness, have shown that differences in environmental conditions have little effect on the birds' migratory urge. It did not matter whether they were kept in Europe, where they normally spend the summer, or in Africa, where they usually winter. E. Gwinner, who conducted these experiments, concluded that 'an internal annual periodicity participates in the regulation of seasonal rhythms in warblers' (Gwinner 1971, p. 419).

Another cycle that has been similarly studied is the annual growth and shedding of antlers by deer (R. J. Goss *et al.* 1974). Tropical deer, when confined to zoos in temperate latitudes, maintain the same annual cycle despite the difference in day-length pattern. Notable evidence for believing that antler growth is controlled by a circannual rhythm is provided by a blind elk that was kept under observation for six years at Colorado State University. Despite the absence of any light cues, it shed and regenerated its antlers on time throughout this period (Pengelley and Asmundson 1971, p. 77).

The adaptive value of a circannual clock in a hibernator is that it enables the animal to prepare well in advance for a future environmental condition that would be difficult to cope with if the animal waited until its onset, e.g. cold weather (Pengelley and Asmundson 1974, pp. 138 *et seq.*). Such a clock can give an animal a warning that it may not always obtain from its environment. Birds wintering in tropical conditions near the equator can seldom receive much of a signal from their surroundings to inform them that it is time to migrate to their breeding places in more temperate latitudes. Gwinner (1971, p. 420) believes that circannual rhythms are fairly widespread among long-lived animals, but Farner (1971) has made the point that, although it is possible to trace back the development of circadian periodicity to a very early stage in evolution, 'two thirds of the evolution of life was over before a circannual periodicity could have any adaptive significance because animals didn't live for as long as a year'.

Stress has been laid on the deviation of circadian and circannual rhythms from the exact day and exact year respectively. If, however, the external environment played no part in regulating these rhythms, they would become increasingly out of phase with the day–night cycle and with the seasons. Consequently, the organism concerned must depend on some cues or signals from its environment to correct its clock and keep it running more or less in phase, just as we use time signals from our national observatories to regulate our clocks and watches. In the case of circadian rhythms, the necessary cues (leading to entrainment) appear to be provided by the daily variations in light and temperature. These variations, and possibly other influences too, may be involved in the regulation of circannual rhythms.

3.5. Biological rhythms in animals and plants (ii)

Circadian rhythms are remarkably resistant to chemical manipulation. Of the many potentially disruptive chemical agents that have been applied experimentally only deuterium oxide, ethyl alcohol, valinomycin and lithium ions have been found to have any influence on the periods of rhythms. Particular interest attaches to the first of these, for the substitution of deuterium oxide (D_2O) for part of the intracellular water (H_2O) lengthens the period of various rhythms, e.g. leaf movement in *Phaseolus* (Bünning and Baltes 1963), phototaxis in *Euglena* (Bruce and Pittendrigh 1960), and circadian rhythms in several species of birds (Palmer and Dowse 1969). As has been remarked by Sweeney (1969, p. 121), the substitution of a heavier atom for hydrogen might well be expected to slow a rhythm down if it is generated by an endogenous pacemaker, be it chemical or physical, but it could hardly alter the period of an exogenous 'clock'. This can be regarded as a further argument in favour of Bünning's rather than F. A. Brown's point of view.

One of the most important characteristics of circadian rhythms in both animals and plants is the temperature independence of their periods.†

† Although the period (and hence the frequency) of the rhythm is independent of temperature, the amplitude is not and often varies considerably. In the words of Sweeney and Hastings (1960), it was in 'the elegant paper' by Pittendrigh (1954) on *Drosophila* that the problem of temperature independence 'was first fully and forcefully placed in what we now consider its proper perspective', for it was the discovery by Pittendrigh that the periods of circadian rhythms are temperature compensated that first convinced biologists generally that they could be regarded as clocks synchronizing internal biological processes with external physical time.

This may not seem particularly surprising in the case of warm-blooded animals, because they maintain their metabolic processes at a controlled temperature. In cold-blooded animals, however, the body temperature fluctuates with the environmental temperature, and the rate of metabolism increases and decreases as the latter rises and falls. Nevertheless, the rates of the biological 'clocks' in these animals also tend to remain remarkably constant, at least within certain ranges of temperature.

A good example of independence of ambient temperature, irrespective of whether the rhythm concerned is circadian or not, is provided by the fiddler crab *Uca*, studied by F. A. Brown and his colleagues (see p. 147). A number of these crabs were placed in a dark room that was maintained at a uniform temperature. It was found that different temperatures between 26 and 6 °C had no appreciable effect on the colour rhythm. Although at the lower temperature the actual expansion of the pigment cells was far less than at the higher, the period of the rhythm of colour changes neither gained nor lost more than a few minutes in two months. When, however, the temperature was lowered to about 0 °C the rhythm disappeared. When it was raised again, the rhythm was restored but was out of phase by the appropriate amount. For example, when the low temperature was maintained for six hours the restored rhythm was out of phase by a quarter of a cycle, whereas if the low temperature persisted for 24 hours the restored rhythm was in phase.

In all types of rhythms within a certain ambient temperature range, temperature independence (i.e. comparative independence) seems to be the rule. In the single-celled marine plant *Gonyaulax polyedra*, responsible for sea phosphorescence or induced flashing, there is a small but consistent increase of period with increase of temperature (Hastings and Sweeney 1957). There is a maximum period, which occurs at about 28 °C. According to Sweeney (1972, p. 145), this feature suggests that 'temperature compensation accounts for the temperature independence of the period and that it is not quite perfect'. However, generally speaking, the variation of period with temperature is no greater in unicellular than in multicellular organisms.

It is thought that temperature compensation results from the interaction of distinct chemical processes that respond differently to changes of temperature. The effect of temperature on period is usually expressed in terms of the ratio

$$Q_{10} = \frac{P(T-10)}{P(T)} = \frac{\nu(T)}{\nu(T-10)}$$

where P denotes the period and ν the corresponding frequency, as a function of the temperature T measured in degrees Celsius. Table 1,

TABLE 1

Plant	Rhythm	Period	Q_{10}
Gonyaulax	Induced Luminescence in dim light	23 h at 16 °C 27 h at 26 °C	0.85
	Spontaneous luminescence in darkness	23.3 h at 20 °C 24.7 h at 25 °C	0.9
Phaseolus	Leaf movement in darkness	28.3 h at 15 °C 28.0 h at 25°C	1.01
Kalanchoe	Petal movement in darkness	21.9 h at 15 °C 21.3 h at 25 °C	1.03
Neurospora	Zonation of growth in dim red light	22 h at 24 °C 21.6 h at 31 °C	1.03
Bryophyllum	CO_2 output in darkness	23.9 h at 16 °C 22.4 h at 26 °C	1.06

which relates some plant circadian rhythms, is based on one given by Jones and Mansfield (1975, p. 108).†

Low sensitivity to changes of temperature is what we would expect of a biological clock if it is to be at all effective, for no clock is of much use if it is very inaccurate; it is, therefore, understandable that, if they are to function as biological clocks, circadian rhythms should be largely independent of ambient temperature and also of chemical changes in their environment. Indeed, if they were temperature sensitive, they would be thermometers rather than clocks! From their temperature insensitivity, however, it is reasonable to conclude that the rhythms concerned are not metabolic.

In contrast to their comparative independence of ambient temperature, the free-running periods of circadian rhythms, under otherwise constant conditions, depend on the intensity of illumination. In general, the frequency increases (or decreases) linearly with the logarithm of the intensity. For animals, Aschoff has formulated a general criterion that has come to be called *Aschoff's Rule:* as light intensity is increased the free-running period of activity in a day-active animal decreases, whereas

† $Q_{10} = 1$ signifies complete temperature independence, i.e. complete temperature compensation. $Q_{10} \sim 1$ for many rhythms, periods often being slightly shorter (and hence $\nu(T)$ slightly greater) at higher temperatures, so that $Q_{10} > 1$.

that of a night-active animal increases (Aschoff 1960, p. 14). As Aschoff himself took care to point out, this is a rule to which some exceptions may be expected, and in fact comparatively few insects obey it.

There is considerable experimental evidence that circadian oscillations are not of the simple harmonic type associated with the interaction of properties analogous to inertia and elasticity. If a plant which displays a circadian rhythm that is effectively temperature independent between, say, 10 and 30 °C is chilled to temperatures near 0 °C for a sufficiently long time, e.g. for more than 10 hours, the same interval of time at subsequent normal temperature always tends to elapse before a new maximum of the cycle is reached. This indicates that prolonged chilling does not fix the oscillator in the phase that prevailed at the outset of chilling but causes it to 'relax' to its 'zero position', indicating that the cycle must be regarded as a 'relaxation oscillation'.† This is confirmed by imposing the abnormally low temperature in different phases of the cycle; there is one phase of several hours which cannot be delayed very much by chilling (the relaxation phase), whereas chilling in the other phase (the tension, or energy-supply, phase) causes the oscillator to relax to its 'zero value'.

† Relaxation oscillations depend on a discharge taking place when some limiting potential or intensity is reached that leads to instability. A good example is provided by the flashes of a neon lamp with a parallel condenser in series with a resistance and a source of electromotive force. Discharge occurs at regular intervals, whenever the potential difference across the condenser reaches a critical value (the flashing potential). Discharge is usually relatively rapid and relaxation oscillations (e.g. as manifested by voltage–time curves or current–time curves in the case of the above example) consequently tend to be markedly asymmetric and non-sinusoidal. In short, relaxation oscillations have two distinct and characteristic phases: one during which energy is stored up slowly (e.g. in a spring or condenser) and another in which the energy is discharged rapidly when a certain critical state has been reached. The term *relaxation oscillation* is due to Balthasar van der Pol (1926). He analyzed the concept mathematically and found that many relaxation oscillations can be based (with simplifying choice of units x and t) on the famous one-parameter non-linear differential equation named after him:

$$\ddot{x} + \mu(x^2 - 1)\dot{x} + x = 0$$

with $\mu \gg 1$. (Simple harmonic sinusoidal oscillations, e.g. pendulum oscillations, are approximated to when $\mu \ll 1$.) He applied this non-linear differential equation to many natural phenomena, including the heartbeat (van der Pol and van der Mark 1929) and biological rhythms (van der Pol 1940).

The frequency of a pendulum oscillation is insensitive to external disturbance, but its amplitude is not. By contrast, the amplitude of a relaxation oscillation is very stable, whereas its frequency can be changed easily. A relaxation oscillation can, therefore, be synchronized much more readily than can a pendulum oscillation.

The mathematical analysis of van der Pol's equation is given in books on non-linear oscillations and differential equations, e.g. Minorsky (1962) and Davies and James (1966). An important property of the equation is that it has a unique periodic solution (*limit cycle* in the x, y plane, where $\dot{x} = y$, $\dot{y} = -x - \mu(x^2 - 1)y$) that is approached asymptotically (as $t \to \infty$) by every other (non-trivial) solution.

As Bünning (1960) has pointed out, this interpretation enables us to understand how the 'clock' can be regulated by a change in the light or the ambient temperature. If light (e.g. a light break in the dark phase) or high temperature is offered in the tension phase, the tension is increased and the phase is lengthened, e.g. by an hour or two. However, if this treatment is repeated for several days, the adjustment process is also repeated until the light or high temperature no longer occurs in the tension phase. Consequently, the phases ultimately occur several hours later in the day than before the treatment. On the other hand, if the light or high temperature is offered in the relaxation phase, this phase is not completed since these treatments tend to promote tension and the relaxation process is terminated an hour or two earlier than usual in favour of the tension phase. Consequently the phase is advanced and the 'clock' accelerated. The clock may be stopped by long exposure to constant conditions, particularly constant illumination. However, a single stimulus, such as interruption by a dark period, will set the clock running again.

Although the non-metabolic biological 'clocks' are more or less temperature independent within certain limits of temperature, they can nevertheless, as we have seen, make use of temperature as a synchronizing clue to adjust their setting and bring it into phase with some significant external condition. It follows that one of the most important problems now facing biologists is to discover a physiological mechanism that can respond to temperature in the one respect and yet be independent of it in the other. The problem has been studied biochemically, but there is no sign that any enzyme fluctuation is involved in such a 'clock' mechanism (Bünning 1973, p. 135). The biochemical oscillations investigated seem to be for the most part manifestations of the *working* of the clock (the 'hands of the clock', to quote Bünning's apt metaphor) rather than its inner mechanism. Indeed, the problem may be a *biophysical* rather than a *biochemical* one.†

A good example of the need to distinguish between the 'hands' of a biological clock and its basic mechanism is provided by the fiddler crab *Uca.* If its temperature is lowered to 10 °C, all its locomotor activity ceases. When, however, the temperature is restored again to its normal value, not only is the locomotor rhythm resumed but it is found to be in phase with that of control crabs that have not been chilled. This is a clear indication that the chilled crab's clock continues to run accurately even when no rhythm is in evidence. Consequently, the clock and the actual

† A 24-hour rhythm is not easily derived from the relatively rapid changes encountered in biochemical processes, those studied experimentally tending to have periods of less than 15 minutes (Pye 1969).

processes that produce the locomotor rhythm must be distinct. They can be uncoupled and later recoupled.

At one time there was a tendency to look for 'master clocks' in living organisms. The first claim made for the identification of such a mechanism was made by G. P. Wells (1955), who located one in the oesophagus of the lug worm *Arenicola marina* that lives on beaches of muddy sand. He found that it controlled the feeding-movement rhythm of this organism which consists of 3-minute bursts and 1-minute rests, irrespective of whether or not food is present, followed by locomotory movements (and ejection of sand) every 40 minutes. The rhythm is similar to that of the heart, except that it is transferred through the entire nervous system of the animal, affecting every aspect of its physiology and behaviour.

Some years later the location of another master clock was claimed by Dr. Janet Harker of Cambridge, in the cockroach *Periplaneta Americana.* If kept in a standard light–darkness cycle, this insect shows a definite circadian rhythm in its foraging activities, being most active at the onset of darkness. However, if it has been kept in continual light for a long time, it ceases to display any measurable rhythm in its activities. A cockroach with a good rhythm was immobilized by the removal of its legs and was then grafted by Harker on to the back of one with no such rhythm but able to move about. The blood systems of the two insects were joined by means of a capillary tube to form a single circulation. Harker (1964) found that the lower insect, although still in continual light, soon developed the same circadian rhythm as had previously been shown by the upper one. Moreover, and this she regarded as the crucial discovery, the rhythmical cockroach imparted the *phase* of its activity to the other. This was interpreted as a strong indication that the rhythm is due to the periodic release of some hormone into the blood stream.

From transplantation experiments Harker concluded that the source of this secretion is the sub-oesophageal ganglion forming the ventral part of the brain. This organ contains specialized nerve cells which secrete hormones under the influence of light that enters the eyes of the insect. When these cells were removed and implanted in the body cavity of another cockroach that had previously had its head removed, they went on functioning as before. The headless insect continued for several days to run around at the time of day to which the phase of secretion of the implanted cells had been set by the donor cockroach's previous experience of light and darkness.

Further experiments, however, indicated that a second circadian process is involved in the foraging activities of the cockroach. This second clock influences the setting of the 'master clock' (preventing it from being reset by random light flashes such as transitions from shade into sunshine), but it is situated outside the oesophageal ganglion. This discovery and the failure of other investigators to corroborate Harker's findings† makes it seem unlikely that there is any simple solution to the problem of the precise nature and location of the complete mechanism controlling circadian and other rhythms in animals.

There is, however, some evidence other than that studied by Wells and Harker that there are control centres in the brains of certain organisms. For example, Lees (1960, 1964) has shown that the photosensitive area, and presumably the 'clock', of *Megoura* lies within the brain. Recently it was claimed that a possible biological clock appeared to have been identified in the golden hamster (Stetson and Watson–Whitmyre 1976). Stetson and Whitmyre found that, if a part of its brain called the suprachiasmatic nucleus of the hypothalamus is removed, the hamster's activity cycle, which normally clocks on when darkness begins and clocks off at dawn, is completely upset and the animal is active in the light and shows no rhythmic variation in the dark. Also the normal timing of sex gland activity, which ensures that breeding occurs only at the most favourable time of year, is disrupted. The suprachiasmatic nucleus is close to the main visual pathway. It is therefore well placed for monitoring light signals and hence for the entrainment of motor and hormonal activities on the light–dark cycle. Whatever the ultimate verdict on Stetson and Whitmyre's discovery may be, it does not solve the problem for animals such as insects that have no nucleus suprachiasmaticus.

Pittendrigh (1976) has drawn attention to some recent experiments on silk moths and sparrows. Removal of the midbrain in the former and the pineal gland in the latter destroys rhythmicity. When a midbrain is implanted into a headless moth and a pineal gland into the anterior chamber of the eye of a pinealectomized sparrow the rhythms are restored. Clearly the brain of a silk moth and the pineal gland of a sparrow *contain* pacemaking systems, but whether the pacemaker consists of cells, a single cell, or part of a cell remains unknown.‡ Pittendrigh

† For a recent general review of the results of Harker and her critics, see Saunders (1976b, pp. 205–10).

‡ At the XVIIth International Ornithological Congress held in West Berlin in June 1978, M. Menaker of the University of Texas, Austin, described some experiments with house sparrows which seem to indicate that they have circadian master clocks in the pineal gland. (This tiny light-sensitive neurosecretory organ in the brain may have functioned as an eye in some ancestral vertebrates.) When this gland is removed and the birds are kept in constant darkness the normal endogenous (free-running) rhythm of activity keeps down. The pineal gland's influence on circadian clocks is transmitted *chemically* (by way of the blood stream) and not neurally. One of the hormones produced in the pineal gland is melatonin and there are various indications that it is involved in circadian rhythmicity. It is possible there are some subordinate self-sustained oscillators which are synchronized by the pineal gland.

makes the important point that we must be cautious in attributing all of a system's temporal organization to a discrete pacemaker. Moreover, we must not confuse pacemakers and 'slave' processes controlled by them.

At the present time many rhythm biologists believe that in most organisms there are, instead of single master clocks, a large number of independent clocks which tend to become desynchronized when the organism is isolated and deprived of external time cues. Unlike animals, plants show no signs at all of any central regulator of periodicity. This has led Bünning and others to conclude that they have a clock in every cell. This view has received some remarkable experimental support from an investigation by Beatrice Sweeney (1960). Diurnal rhythms are usually measured in many cells simultaneously either in multicellular organisms or in populations of unicellular organisms. Sweeney devised an elegant technique for successfully measuring the diurnal rhythm of oxygen produced by photosynthesis in a single *Gonyaulax* cell.

From time to time it has been suggested that the cell cycle might be the generator of circadian rhythms, but there are good reasons for doubting this. Not only does this cycle tend to be too temperature sensitive, but chloramphenicol halts cell division and yet has no effect on circadian rhythms (Sweeney 1972, p. 149). It is, therefore, unlikely that the cell cycle is the basic oscillator responsible for circadian rhythms in general. Indeed, it may well be that cell division is controlled by the circadian oscillator. It is possible that each living cell contains a multitude of 'clocks', some of which are replicating themselves while others are causing cell processes to be rhythmic. In the search for a possible central controlling clock within cells their nuclei might have been expected to yield the secret, but experiments with plants have shown that rhythms persist not only after nuclear metabolism has been inhibited but even after the nucleus itself has been removed† (Sweeney and Haxo 1961).

The suggestion, inspired by the advent of molecular biology, that the circadian 'clock' mechanism is associated with the control system for nucleic acid metabolism has been investigated by Ehret (Ehret and Trucco 1967). He advanced the hypothesis that there are sub-units of chromosomes, which he called 'chronons',‡ that play the role of a cellular

† Some specific rhythms, however, seem to depend upon metabolic processes in the nucleus. As has been remarked by Bellamy (1970, p. 114), there may be many different processes that give rise to clock-like phenomena. 'The fact that most rhythms at a higher level have been found to be independent of temperature suggest that they are generated either by physicochemical reactions on membranous structures within the cell, or by diffusion processes associated with ion movement, but at the moment there is no evidence as to the exact molecular mechanism'.

‡ Not to be confused with the very different 'chronon' concept discussed on pp. 201–4 in Chapter 4!

chronometer, each chronon consisting of a number of genes in a definite linear order. Ehret postulated that they are transcribed *serially* in this order, the complete transcription of one chronon taking about 24 hours. During that time the concentrations of specific messenger RNA's would vary and so would those of the proteins whose syntheses are directed by the specific molecules of RNA. At one end of a long segment of DNA there would be an initiator gene and at the other end a terminator. The messenger RNA of the latter would somehow stimulate the production of a protein that would reactivate the initiator to repeat the process all over again.

Sweeney (1969, p. 131) has argued cogently that it is premature to construct theories in which the generation of circadian rhythms is attributed to the transcription of genetic messages, as has been assumed by Goodwin (1963) as well as by Ehret and Trucco. More recently, Lewin (1975) has also criticized Ehret's model: 'A few years ago, when gene expression was surrounded by an almost naïve enthusiasm in molecular biology, the "magic" formula drifted into circadian rhythm, in the form of a chronon model. This proposed a circadian transcription of genes as the driving mechanism of the clock . . . this system would probably be too inflexible, and it does not now stand up to experimental test'.

The currently most favoured hypothesis concerning the circadian clock is that it is a *membrane clock*. As previously mentioned, a 24-hour rhythm is not easily derived from the rapid processes of biochemistry; experimentally studied systems tend to have periods of less than 15 minutes (Pye 1969). Membrane models by drawing on slow processes, such as lateral diffusion of proteins within membranes, could lead to periods of approximately 24 hours.† Njus, Sulzman, and Hastings (1974) have investigated the hypothesis that the biological clock is basically a feedback mechanism of this type. They find that temperature compensation can be readily accounted for if the liquid composition of membranes is controlled to maintain a definite fluidity independent of temperature. As Sweeney (1969, p. 134) has commented, 'Perhaps as membranes become better understood the solution of the problem of the mechanism of cellular clocks will become apparent'.‡

† Pavlidis (1973) has suggested that much longer periods than those of a single oscillator can be obtained by systems involving couplings of more than one such oscillator. 'From a biological point of view', he writes, 'one should expect that any rhythm regulator at the cellular level should consist of a group of coupled oscillators'. It would produce a more reliable system because of redundancy. Moreover, coupled oscillators *as a group* can have much longer periods than any of the oscillators individually. Synchronization among the various biological oscillators can occur when the coupling is not direct but through sensory organs, e.g. organisms which flash periodically. See Hanson *et al.* (1971) and Buck and Buck (1976).

‡ For recent ideas on membranes as components in circadian oscillators and some promising experimental approaches, see Sweeney (1976) and Njus (1976).

We have seen that, despite overwhelming evidence for the existence of biological clocks in both animals and plants, no generally accepted identification of a single such clock has yet been made. This situation is, however, not quite so paradoxical as it appears for, although to function properly an organism must be able to control the timing of its physiological processes and therefore biological clocks must exist for this purpose, we cannot expect them to be readily identifiable, particularly as dissecting techniques are more likely to destroy than to reveal the processes concerned. As Hans Kalmus (1964) has remarked, 'When dissecting an animal a biological clock, like the mind, is just not among the organs, though—again like the mind—time-keeping must have an affinity to special organs'. Moreover, there is the difficulty to which we have already referred, that we may easily mistake the 'hands' of the clock for the clock itself. Nevertheless, despite the elusiveness of the temporal mechanisms controlling them, it is clear that biological rhythms are a universal and fundamental characteristic of life. Instead of rigidly adhering to Claude Bernard's concept of homeostasis, based on the conviction that an organism strives to maintain many of its physiological parameters at a constant level, we must recognize that life depends on processes that vary rhythmically in time.

3.6. Biological rhythms in man

Biological rhythms tend to play a less prominent role in the life of man than in that of animals and plants, presumably because in the course of evolution he has become less dependent upon them. Nevertheless, there are two biological 'clocks' that still have a major influence on human life, the menstrual 'clock' and the diurnal 'clock' that normally causes us to spend part of the 24-hour day asleep. It is the generally held view of sleep physiologists that the peremptory need of man and most other multicellular animals to sleep for a while during each 24 hours is probably not due to fatigue but rather to the activity of a circadian clock whose function is to stimulate activity in daylight and inactivity in darkness. This view reflects our dependence on vision as the primary sense controlling our activities and hence on the importance for us, and for most other animals, of light as a requirement for efficient functioning.† Before the invention of artificial light we needed some means of ensuring that we were alert and active only when we could see what was going on around us and inactive at night when darkness put us at a disadvantage. The fact that we still sleep, although we now have artificial illumination whenever we require it, does not contradict the hypothesis that sleep has been

† For nocturnal animals, such as bats, that rely on hearing, and rats, that rely mainly on smell and touch, the advantages of light and dark are reversed.

nature's solution to this problem for, even if the biological utility of sleep is no longer evident, we cannot easily undo the effects of hundreds of millions of years of evolution and automatically dispense with it.

Evidence for the influence on our sleeping habits played by the alternation of light and darkness is provided by communities living in high latitudes. Their sleeping habits differ markedly in the perpetual daylight of summer and in winter darkness (Lobban 1960, p. 326). One of the arguments for the view that sleep is not primarily due to fatigue is that even in lower latitudes there are people who require very little sleep and seem to be none the worse for it. There are also some animals, for example a species of porpoise, that have never been observed to sleep, presumably because they live in conditions that obviate the need for it (Oatley 1975). Moreover, even if each day there is an accumulation of products of fatigue that our bodies eventually have to deal with, this does not conflict with the hypothesis that sleep is primarily controlled by an internal clock since recuperation can take place when we are least active.

Sleep is not the only one of our physiological states that varies with the 24-hour cycle for, even though we normally do not notice such variations, many of our physiological functions exhibit definite circadian rhythms. For example, it is well known that our internal temperature is normally at its highest in the evening and lowest early in the morning. Since this phenomenon was first described by A. Gierse in 1842, it has been found that there is apparently no organ and no function in the human body that does not display a similar daily rhythmicity (Aschoff 1965). The occurrence of free-running rhythms departing slightly from a period of exactly 24 hours has been observed on many occasions in man (Conroy and Mills 1970, p. 122). As we have seen, such a departure from a precise 24-hour period is generally regarded as good evidence for the existence of an endogenous circadian clock.

Next to the diurnal patterns of sleep and temperature variations, our daily pattern of excretion from the kidneys is the most readily testable of our circadian rhythms. Since the kidney has no known information channel from the outside world, it must presumably be subject only to rhythmic influences from within the body (Mills 1973, p. 57). Some urinary rhythms, for example excretion of calcium and magnesium, depend on the pattern of meals. On the other hand, excretion of potassium is not greatly affected by the pattern of living. This has been strikingly illustrated by experiments on humans subjected to abnormal time schedules. For example, Mary Lobban (1960) and her associates took a group of volunteers to Spitsbergen to live in the continual summer light in isolated camps in uninhabited surroundings with no external cues for telling the time. They were given wrist watches that had been manipulated so as to encourage them to live either a 21-hour or a 27-hour day.

At the end of six weeks, although some functions and urinary excretions had adjusted themselves to the artificial day, the well-marked rhythm of potassium excretion showed no such change but still adhered to the 24-hour pattern. This suggests that the excretion of potassium is an indicator of the existence of an endogenous clock.

Adaptation of an endogenous rhythm to a phase shift is much easier than to a change of period. An abrupt change of phase of external rhythms occurs whenever we fly across several time zones. In recent years, with the general increase in rapid transport by air, we have become aware of a new kind of illness, called 'dysrhythmia' or 'jet-lag fatigue', suffered by many people when they make a long journey by air east or west. The first systematic observations of the disruptions caused by flying east or west were reported by Strughold (1952). He discovered that many people, especially older people, experienced physiological discomfort on such flights, becoming hungry and sleepy and feeling awake at the wrong times. He later reported (Strughold 1965) that pilots who had to cross and recross several time zones frequently in the month were prone to nervous stress. Measurements of sodium, potassium, and other levels in urinary samples taken following rapid time-zone displacement indicate that biochemical excretions are initially in phase with the original environment. Gradually, however, they become synchronized with the new environment, although this readjustment often takes several days.

Blatt and Quinlan (1972) suggest that the malaises reported after rapid transmeridian flight may be due to three different factors. First, there is a phase shift between the activity cycle of the individual and the activity schedule of the environment. Where the former is at a resting level the latter may be flooding him with stimuli, and this discrepancy can cause psychological distress and lead to less efficient performance. The second factor is sleep loss, either because the individual arrives at a time when he feels like retiring but is unable to do so owing to the activity going on all around him, or because he awakens at his biological awakening time and finds it is the middle of the night. The third factor, in their view, is the effects of the desynchronization of the individual's circadian rhythms which occurs as these rhythms begin to shift to the new time zone. The shifts take place at different rates and this variation leads to a desynchronization of the phase relationships among the different rhythms. The psychological effects of this transitional loss of synchronization have not yet been extensively studied, although it seems likely that they enhance the feelings of irritability and dysphoria caused by loss of sleep. The considerable differences between individuals in the ease and speed with which they adjust themselves to shifts of time zone depend on personality differences, particularly as regards the flexibility of their normal sleeping patterns and their responsiveness to internal physical cues as distinct from environmental factors. For example, those whose consumption of food is controlled more by 'clock time' than by feelings of hunger are less affected by dysrhythmia.

Another source of information about human circadian rhythms is provided by people who work at night or on rotating shifts. Many studies on night-workers and shift-workers have concentrated on working efficiency and psychological factors such as job satisfaction and dislike. The first investigators of ill health, accidents, and errors among these workers attributed such effects to the work itself, whereas it has since been realized that they often stem from the phase shifts that occur at weekends and when workers have time off. Luce (1971, 1972) has drawn attention to the findings of Soviet scientists that the abnormal physiological rhythms displayed by night-workers on the Moscow Underground railway were not due to working at night but rather to a poorly organized routine of daytime rest. Also, in Sweden, it has been found that meter readers in a gasworks made most errors on night shift, fewer during afternoon shift, and fewest in the morning. Andlauer and Metz (1953), in a survey of a thousand industrial workers in the Rhône valley, found that nearly half could not adjust to a 7-day rotation and a third could not tolerate a 2-day rotation. Body-temperature rhythms did not adapt to either schedule, and Dr. Andlauer recommended a rearrangement that offered stable work hours.

Abnormal time schedules are not the only source of support for the idea that many of our physiological rhythms are endogenous. Certain illnesses occur periodically, for although some diseases, such as malaria and gout, are recurrent without being periodic, there are others in which the same pathological symptoms reappear at regular intervals over many years, often with no obvious effects on health between attacks. The most significant of these are periodic peritonitis, periodic fever, periodic oedema, and periodic joint disorder. Periodic peritonitis, a recurrent inflammation of the lining of the stomach or intestines, is an inherited disease due to a recessive gene. It is largely, but not entirely, confined to certain ethnic groups in the Mediterranean area. The most effective way to diagnose the disease is by recording the symptoms in a diary. Reimann (1963) has pointed out that, like other periodic diseases, this disorder often appears at intervals of seven days or integral multiples of seven days. This would indicate that, in addition to a circadian clock, a weekly clock is also involved.† Periodic fever, without any obvious cause, can

† According to Reimann (1963), the seventeenth century antiquarian and writer John Aubrey appears to have suffered from this condition for several years. He had attacks of fever, vomiting, and abdominal pain lasting 12 hours, at first every fortnight, then monthly, then quarterly, and eventually at intervals of 6 months until they disappeared.

occur in an otherwise healthy person, and again the period is often 7 days or an integral multiple thereof. No pathological symptoms have been discovered in the individuals subject to this complaint who, after death from other causes, have come for autopsy. Periodic oedoma, by far the most dangerous of the periodic diseases, seems to be linked with a dominant gene. It may affect any part of the body, the great danger being death from asphyxia due to laryngeal oedoma. In one series, almost half the descendants of a person known to have been a victim of this disorder also suffered from it. There is again a tendency for attacks to occur at intervals of 7 days or an integral multiple thereof. Similarly, attacks of periodic arthrosis, or swelling of one or more joints, particularly the knees and elbows, often occur at similar intervals. Greatly increased amounts of fluid in the synovial spaces produce the swelling. Patients often experience a sudden rush of fluid into the joints at the start of an attack. Other periodic disorders, besides the principal four enumerated, affect the glands, the skin, the bone marrow, etc.

One of the most famous and curious forms of periodic illness is purpura, or periodic bleeding. There are many accounts of mystics who, like St. Francis, showed the stigmata, or legendary wounds of Christ, regularly on Fridays or on specific days of the Christian calendar.† Generally, purpura gives rise to various kinds of internal bleeding, often into the skin, but it can be fatal when it occurs in vital organs. In the case of the mystics, an emotional factor is clearly involved. Recurrent emotional and mental illnesses occur with a variety of periods. Charles Lamb's sister, Mary, suffered a cyclic psychosis for half a century from the age of thirty. During one of her attacks she killed her mother, of whom she was very fond. She was placed in the care of her brother; whenever she had an attack, and this occurred regularly, he rushed her to hospital and she was put in a straitjacket. On recovery, she resumed her normal life of writing stories and enteraining literary friends until the next attack. Despite the long series of attacks, she showed no signs of mental or physical deterioration other than those of old age. It is well known that manic depression is a recurrent illness, although only some victims have a cycle of clock-like periodicity. Manic depressives with periods ranging up to two and possibly ten years have been cited by Richter (1965, pp. 60–71). He has also drawn attention to a case of remarkably precise timing in a Parkinsonian patient. Following an attack of *encephalitis lethargica* when a child, she eventually became completely incapacitated.

† Perhaps the most famous modern bearer of stigmata was Theresa Neumann, born in 1898 near Munich. She displayed her first stigmata at 28, bleeding from the left breast on Fridays. She gradually developed more stigmata, these occurring regularly on Thursday nights and Fridays, accompanied by periods of ecstasty. She died in 1962.

For most of the day she was bedridden with marked rigidity and tremors of arms and legs. Although mentally alert, she could not speak clearly and her handwriting was indecipherable, but at 9 p.m. each evening she underwent a sudden change of personality and for two or three hours she could walk about, look after herself, and write and talk clearly. There have also been cases of patients who display two distinct personalities on alternate days, tasks left unfinished at the end of one 24-hour period being resumed 24 hours later.

Richter thinks that in these and other periodic disorders that can afflict man, a number of biological rhythms may be involved which affect different parts of the body. Pathological rhythms should, however, be dissociated from regular physiological rhythms. He has distinguished what he calls 'peripheral clocks' from central clocks and also from homeostatic clocks involving feedback mechanisms. Peripheral clocks are outside the nervous system, are quite independent of any feedback process, and have their own inherent rhythms. They can be found, for example, in the joints. In some patients with intermittent hydrarthrosis several joints may exhibit the same period, e.g. a week, but be out of phase. In considering how such clocks might work, how shock and other agents can elicit cyclic responses, and similar problems, it occurred to Richter that a clue was provided by the observation (see p. 143) that the pupae of *Drosophila* have an inherent 24-hour emergence rhythm, so that if kept in a 12-hour light, 12-hour dark regime they will all emerge simultaneously, but when kept in constant darkness, and hence deprived of any external stimuli, they will emerge at all times of the day and night. If, however, a shock—such as a brief flash of light or thermal stimulus—is administered to them (when they are kept in otherwise constant darkness) they will all be brought into phase and will emerge only at 24-hour intervals (Pittendrigh 1954). Richter argued that 'what happened to the individual members of a colony of flies in response to a strong shock might also happen to the individual units of an organ that are clearly bound together into an integrated whole' (Richter 1960, p. 1527). He was thus led to formulate what he called his *shock-phase hypothesis.* According to this, the functioning units of any organ in the body have an inherent rhythm characteristic of the organ, but under normal conditions these units are out of phase. Under sudden shock of trauma, however, they may be synchronized and give rise to a periodic disturbance.

Richter's shock-phase hypothesis assumes that individual units of an organ may function quite independently of one another, and he has cited some evidence in favour of this assumption (Richter 1965, p. 86). The action of transmitter mechanisms, such as the hormones, is to synchronize the rhythms of the different components of the body rather than to impose a rhythm on a non-rhythmic tissue. Rhythmicity may exist at many levels, including the cellular, but whether there is a 24-hour master clock in the human body is still far from being decided. It is unlikely to be in the cortex, since rhythms persist even in the absence of cortical function (Conroy and Mills 1970, p. 126). The hypothalamus has been tentatively proposed as a more likely site. Reimann (1963) has suggested that many periodic illnesses may be due to sudden excitations in this or other mid-brain regions. Partial confirmation of the role of this part of the brain in controlling the circadian clock has been obtained by Richter in experiments on rats. He found that he could not eradicate their inherent 24-hour rhythms by depriving them of oxygen, suspending their heart beats, action of drugs, etc., but only by partial destruction of the hypothalamus.

More recently, Richter (1968) experimented with squirrel monkeys. They have the advantage of being primates and therefore much closer than rats to man. Moreover, they have larger and more highly developed brains. He found that these animals appear to have a clock that measures both 24-hour and 12-hour periods, and that it is more accurate, reliable, and constant than that of any other animal previously tested. This was the first experimental evidence for the possession of such a clock in a primate, but it is likely that all primates possess it. Richter also discovered evidence of a monthly cycle that occurred in males as well as in females. It was revealed by activity-distribution patterns that did not deviate greatly from a period of 24 hours 50 minutes. This result coincides with the fact that, on the average, the moon and the tides appear 50 minutes later each day. Since the animals had been blinded,† Richter concluded that either the times of onsets of activity were determined by effects of the moon other than its light, or else by an inherent lunar (approximately 30-day) clock to which the 24-hour clock had become entrained. Deviations from the 30-day average led Richter to regard this clock as independent of the moon cycle. Instead, he concluded that these monkeys have an inherent 30-day clock that was built into them far back in their evolutionary history by the tides or the monthly recurrences of light from the full moon or both. Lunar cycles are associated with reproductive activity, and so the ability to adjust to them and also to the recurrence of the tides presumably had survival value. The lunar clock has probably long since become independent of the tides and light of the full moon, just as the inherent 24-hour clock no longer depends on the daily change of light and darkness.

Normal man gives little overt indication of possessing any inherent

† They were blinded to demonstrate the presence of a 24-hour clock that was not entrained by light but was free running.

biological clocks. In this respect he differs even from the primates, those that are active in the light being almost totally inactive in the dark, so that we can only conclude that the 24-hour clock plays an important part in their survival. Since, under certain pathological conditions, the 24-hour clock may be revealed as clearly in man as in animals, it must normally exist within him in what Richter (1975) has described as a 'submerged' condition. With the discovery of fire some hundreds of thousands of years ago, man ceased to have the same need for his activities to be dominated by the circadian clock since he was no longer dependent for his safety on natural illumination. Richter argues that, because in the squirrel monkey there is clear evidence of a 12-hour rhythm associated with the 24-hour cycle, the circadian clock must have had its evolutionary origin in a survival process in the tropics, where day and night do not differ greatly in their duration. Originating as an inherent timing mechanism with survival value for the organism, this clock has become independent of all external and internal influences and functions independently of light. Although the 29–30-day clock still manifests itself in normal human females, it too has become otherwise submerged, to appear only under pathological conditions associated with various psychiatric disturbances. It presumably had its origin when survival depended on the ability to adjust to the light of the full moon or else to the tides. Although little is yet known about the existence of an inherent annual clock in man, pathological conditions in a few psychiatric patients have given some indications of yearly cycles. If such a clock has been built into the brain, it may have been by annual variations of temperature, and this would indicate an origin in temperate parts of the world rather than in the tropics.

Although we do not yet know for sure where the master clocks, if any, that control our submerged biological rhythms are located, it is possible that in the course of evolution inhibitory controls on these rhythms may have developed in the higher centres of our brains. Some support for this hypothesis has come from experiments by Richter, Jones, and Biswanger (1959), who found that thyroid-deficient rats which failed to show cyclic changes displayed them after removal of the frontal lobes of the brain. Whether the hypothesis is valid or not, it is evident that in the course of the last few hundred thousand years man's survival has come to depend less on particular rhythms and more and more on his ability to perform equally well throughout the day, the month, and the year. Nevertheless, our cognitive time sense, however much it may be controlled by social and psychological factors, is superimposed on the rhythms of the biological clocks that beat within us far below the level of consciousness. These in turn have been selected because of their close chronometric relation to external influences of an astronomical nature associated with 'universal time'.

References

ADRIAN, E. D. (1947). *The physical background of perception*, Chap. V. Clarendon Press, Oxford.

ANDLAUER, P. and METZ, B. (1953). Variations nychtémerales de la fréquence des accidents du travail continu. *Arch. Mal. prof. Méd. Trav. Secur. Soc.* **14** (6), 613.

ASCHOFF, J. (1960). Exogenous and endogenous components in circadian rhythms. In *Cold Spring Harbor Symp. 25, Biological clocks*, pp. 11–27. The Biological Laboratory, Cold Spring Harbor, New York.

—— (1965). Circadian rhythms in man, *Science* **148,** 1427–32.

BELL, C. R. (1965). Time estimation and increases in body temperature. *J. exp. Psychol.* **70,** 232–4.

—— (1966). Control of time estimation by a chemical clock. *Nature* (*Lond.*), **210,** 1189–90.

—— (1975). *J. exp. Psychol.* **27,** 531.

BELLAMY, F. (1970). Animal rhythms. *Sci. Prog.* **58,** 99–115.

BENNETT, M. F. (1974). *Living clocks in the animal world.* Charles C. Thomas, Springfield, Ill.

BLATT, S. J. and QUINLAN, D. M. (1972). The psychological effects of rapid shifts in temporal referents. In *The Study of Time* (ed. J. T. Fraser, F. C. Haber, and G. H. Müller), pp. 506–22. Springer-Verlag, Berlin, Heidelberg, New York.

BLISS, V. L. and HEPPNER, F. H. (1976). Circadian activity rhythm influenced by near zero magnetic field. *Nature* (*Lond.*) **261,** 411–2.

BOHN, G. (1903). *C. R. Acad. sci. Paris* **37,** 576–8.

BORTHWICK, H. A. and HENDRICKS, S. B. (1961) *Encyclopedia of plant physiology* (ed. W. Ruhland), Vol. 16, pp. 299–330. Springer, Berlin.

BORTHWICK, H. A. and PARKER M. W. (1952) The reaction controlling flower initiation. *Proc. nat. Acad. Sci.* U.S.A. **38,** 929–34.

BRADFIELD, J. R. G. (1955). Fibre patterns in animal flagella and cilia. *Symp. Soc. exp. Biol.* **9,** 306–31.

BROWN, F. A. (1960). Response to pervasive geophysical factors and the biological clock problem. In *Cold Spring Harbor Symp. 25, Biological clocks,* pp. 57–72. The Biological Laboratory, Cold Spring Harbor, New York.

—— (1965) A unified theory for biological rhythms. In *Circadian clocks* (ed. J. Aschoff), pp. 231–61. North-Holland, Amsterdam.

BROWN, F. A., HASTINGS, J. W., and PALMER, J. D. (1970). *The biological clock: two views.* Academic Press, New York, London.

BRUCE, V. G. and PITTENDRIGH, C. S. (1960). An effect of heavy water on the phase and period of a circadian rhythm in *Euglena*. *J. cell. comp. Physiol.* **56,** 25–31.

BUCK, J. and BUCK, E. (1976). Synchronous fireflies. *Sci. Am.* **234**(5), 74–85.

BÜNNING, E. (1935). Zur Kenntnis der endogenen Tagesrhythmik bei Insekten und Pflanzen. *Ber. dtsch. bot. Ges.* **53,** 504–623.

—— (1936). Die endogene Tagesrhythmik als Grundlage der photoperiodischen Reaktion. *Ber. dtsch. bot. Ges.* **54,** 590–607.

—— (1960). Opening address: biological clocks. In *Cold Spring Harbor Symp. 25, Biological clocks*, pp. 1–9. The Biological Laboratory, Cold Spring Harbor, New York.

—— (1971). The adaptive value of circadian leaf movements. In *Biochronometry* (ed. M. Menaker), pp. 203–11. National Academy of Sciences, Washington, D.C.

—— (1973). *The physiological clock* (revised 3rd edn.). The English Universities Press, London.

—— (1975). *Wilhelm Pfeffer, Apotheker, Chemiker, Botaniker, Physiologe: 1845–1920.* Wissenschaftliche Verlagsgesellschaft, Stuttgart.

Bünning, E., and Baltes, J. (1963). Zur Wirkung von schwerem Wasser auf die endogene Tagesrhythmik. *Naturwissenschaften* **50,** 622–3.

Bünning, E., and Joerrens, G. (1960). Tagesperiodische antagonistische Schwankungen der Blauviollet and Gelbrot-Empfindlichkeit als Grundlage der photoperiodischen Diapause-Induktion bei *Pieris brassicae. Z. naturf.* **15,** 205–13.

Bünning, E., and Stern, K. (1930). *Ber. dtsch. bot. Ges.* **48,** 227–52.

Butler, W. L., Norris, K. H., Siegelman, H. W. and Hendricks, S. B. (1959). Detection, assay and preliminary purification of the pigment controlling photoresponsive development of plants. *Proc. Natn. Acad. Sci. U.S.A.* **45,** 1703–8.

von Buttel-Reepen, H. (1915). *Leben und Wesen der Bienen.* Vieweg, Braunschweig.

de Candolle, A. P. (1832). *Physiologie végétale.* Béchet Jeune, Paris.

Comfort, A. (1968). Feasibility in age research. *Nature* (*Lond.*) **217,** 320–2.

Conroy, R. and Mills, J. N. (1970). *Human circadian rhythms.* Churchill, London.

Darwin, C. and Darwin, F. (1880). *The power of movement in plants*, pp. 407–8. John Murray, London.

Davies, T. V. and James E. M. (1966). *Nonlinear differential equations.* Addison-Wesley, Reading, Mass.

Dempsey, E. W. and Morrison, R. S. (1943). The electrical activity of a thalamo-cortical relay system. *Am. J. Physiol.* **138,** 283.

Duhamel du Monceau, H.-L. (1758). *La physique des arbres*, Vol. II, p. 158. Guerin et Delatour, Paris.

Ehret, C. F. and Trucco, E. (1967). Molecular models for the circadian clock. I. The chronon concept. *J. theor. Biol.* **15,** 240–62.

Emlen, S. T. (1975). The stellar-orientation system of a migratory bird. *Sci. Am.* **233,** 102–111.

Farner, D. S. (1971). Contribution to discussion on paper by Eberhard Gwinner. In *Biochronometry* (ed. M. Menaker), pp. 426–7. National Academy of Sciences, Washington, D.C.

Forel, A. H. (1906). Mémoire du temps et association des souvenirs chez les abeilles. *Bull. Inst. gén. Psychol.* **6,** 257–9.

François, M. (1927). Contribution à l'étude du sens du temps interne comme facteur de variation de l'appréciation subjective des durées. *Année Psychol.*

28, 188–204.

VON FRISCH, K. (1967). *The dance language and orientation of bees* (transl. L. E. Chadwick). Harvard University Press, Cambridge, Mass.

GARNER, W. W. and ALLARD, H. A. (1920). Effect of the relative length of day and night and other factors of the environment on growth and reproduction in plants. *J. agric. Res.* **18,** 553–606.

GOODDY, W. (1958). Time and the nervous system: the brain as a clock. *Lancet,* No. 7031, 1139–41.

GOODWIN, B. C. (1963), *Temporal organization in cells.* Academic Press, London, New York.

GOSS, R. J., DINSMORE, C. E., NICHOLS-GRIMES, E., and ROSEN, J. K. (1974). Expression and suppression of the circannual antler growth cycle in deer. In *Circannual clocks* (ed. E. T. Pengelley), pp. 393–422. Academic Press, New York, London.

GWINNER, E. (1971). A comparative study of circannual rhythms in warblers. In *Biochronometry* (ed. M. Menaker), pp. 405–427. National Academy of Sciences, Washington, D.C.

HAMNER, K. C. (1960). Photoperiodism and circadian rhythms. In *Cold Spring Harbor Symposium 25, Biological clocks,* p. 270. The Biological Laboratory, Cold Spring Harbor, New York.

—— (1966). Experimental evidence for the biological clock. In *The voices of time* (ed. J. T. Fraser), p. 287. Braziller, New York.

HAMNER, K. C. and BONNER, J. (1938). Photoperiodism in relation to hormones as factors in floral initiation and development. *Botan. Gaz.* (Chicago) **100,** 388–431.

HAMNER, K. C., FINN, J. C., SIROHI, G. S., HOSHIZAKI, T. and CARPENTER, B. H. (1962). The biological clock at the South Pole. *Nature* (*Lond.*) **195,** 476–80.

HANSON, F. E., CASE, J. F., BUCK, E., and BUCK, J. (1971). Synchrony and flash entrainment in a New Guinea fire-fly. *Science* **174,** 161–4.

HARKER, J. E. (1964). *The physiology of diurnal rhythms.* Cambridge University Press, Cambridge.

HART, D. S. (1951). *J. exp. Biol.* **28,** 1–12.

HASTINGS, J. W. and SWEENEY, B. M. (1957). On the mechanism of temperature independence in a biological clock. *Proc. nat. Acad. Sci. U.S.A.* **43,** 804–11.

HOAGLAND, H. (1933). The physiological control of judgements of duration: evidence for a chemical clock. *J. gen. Psychol.* **9,** 267–87.

—— (1966). Some biochemical considerations of time. In *The voices of time* (ed. J. T. Fraser), p. 329. Braziller, New York.

HOLUBÁŘ, J. (1969). *The sense of time: an electrophysiological study of its mechanisms in man* (transl. J. S. Barlow). M. I. T. Press, Cambridge, Mass.

HOWES, E. L. and HARVEY, S. C. (1932). Age factor in velocity of growth of fibroblasts in the healing wound. *J. exp. Med.* **55,** 577.

JONES, M. B. and MANSFIELD, T. A. (1975). Circadian rhythms in plants. *Sci. Prog.* **62,** 103–25.

KALMUS, H. (1964). Circadian organization (a review of the first edition of *The physiological clock*, by E. Bünning). *Nature* (*Lond.*) **204,** 411.

KEETON, W. T. (1974). The mystery of pigeon homing. *Sci. Am.* **231,** 103.

KEILIN, D. (1959). The problem of anabiosis or latent life: history and current concept (The Leeuwenhoek Lecture, 1958). *Proc. R. Soc. B* **150,** 149–91.

KENDRICK, R. E. and FRANKLAND, B. (1976). *Phytochrome and plant growth* (Institute of Biology's Studies in Biology No. 68). Edward Arnold, London.

KLEBS, G. (1913). *Sitzungsber. Heidelb. Akad. Wiss. Abt. B* **3,** 47.

KLEINHOONTE, A. (1929). Über die durch das Licht regulierten autonomen Bewegungen der *Canavalia*-Blätter. *Arch. néerl. Sci. exactes nat., Ser 3B* **5,** 1–110.

—— (1932). Untersuchungen über die autonomen Bewegungen der Primärblätter von *Canavalia ensiformis.* Jahrb. Wiss. bot., **75,** 679–725.

KRAMER, G. (1950). *Naturwissenschaften,* **37,** 188.

—— (1952). Experiments on bird orientation. *Ibis* **94,** 265–85.

LEES, A. D. (1960). Some aspects of animal photoperiodism. In *Cold Spring Harbor Symposium 25, Biological Clocks,* pp. 261–8. The Biological Laboratory, Cold Spring Harbor, New York.

—— (1964). The location of the photoperiodic receptors in the aphid *Megoura viciae* Buckton, *J. exp. Biol.* **41,** 119–23.

—— (1965). Is there a circadian component in the *Megoura* photoperiodic clock? In *Circadian clocks* (ed. J. Aschoff), pp. 351–6. North-Holland, Amsterdam.

—— (1973). Photoperiodic time measurement in the aphid *Megoura viciae. J. Insect Physiol.,* **19,** 2279–316.

LEWIN, R. (1975). In search of the biological clock. *New Sci.*, 13 November, 386–7.

LINDAUER, M. (1960). Time-compensated sun orientation in bees. In *Cold Spring Harbor Symp. 25, Biological clocks,* pp. 371–7. The Biological Laboratory, Cold Spring Harbor, New York.

—— (1971). *Communication among social bees.* Harvard University Press, Cambridge, Mass.

LIPPOLD, O. (1970). Origin of the alpha rhythm. *Nature* (*Lond.*) **226,** 616–8.

—— (1973). *The origin of the alpha rhythm.* Churchill Livingstone, Edinburgh, London.

LOBBAN, M. C. (1960). The entrainment of circadian rhythms in man. In *Cold Spring Harbor Symp. 25, Biological clocks,* pp. 325–32. The Biological Laboratory, Cold Spring Harbor, New York.

LOCKE, J. (1690). *An essay concerning human understanding* (ed. A. I. Pringle-Pattison), p. 107. Clarendon Press, Oxford.

LUCE, G. G. (1971). *Biological rhythms in human and animal physiology.* Dover, New York.

—— (1972). *Body time.* Temple Smith, London.

MACLEOD, R. B. and ROFF, M. T. (1936). An experiment in temporal disorientation. *Acta psychol.* **1,** 381–423.

MATTHEWS, G. V. T. (1968). *Bird navigation,* p. 159. Cambridge University Press, Cambridge.

MILLS, J. N. (1973). Transmission processes between clock and manifestations. In *Biological aspects of circadian rhythms*, pp. 27–84. Plenum, London, New York.

MINORSKY, N. (1962). *Nonlinear oscillations.* Krieger, New York.

NJUS, D. (1976). Experimental approaches to membrane models. In *The molecular basis of circadian rhythms* (ed. J. W. Hastings and H. G. Schweiger), pp. 283–94. Dahlem Konferenzen, Abakon Verlagsgesellschaft, Berlin.

NJUS, D., SULZMANN, F. M., and HASTINGS, J. W. (1974). Membrane model for the circadian clock. *Nature* (*Lond.*) **248,** 116–20.

DU NOÜY, LECOMTE (1936). *Biological time*, p. 160. Methuen, London.

OATLEY, K. (1975). Clock mechanisms of sleep. *New Sci.* 15 May, 371–4.

PALMER, J. D. and DOWSE, H. B. (1969). Preliminary findings on effect of D_2O on the period of circadian activity rhythms (Abstract). *Biol. Bull.* **137,** 388.

PAVLIDIS, T. (1973). *Biological oscillators: their mathematical analysis*, Chapter 8. Academic Press, New York, London.

PENGELLEY, E. T. and ASMUNDSON, S. J. (1971). Annual biological clocks. *Sci. Am.* **224**(4), 72–9.

——, and —— (1974). Circannual rhythmicity in hibernating mammals. In *Circannual clocks* (ed. E. T. Pengelley), pp. 95–160. Academic Press, New York, London.

PENGELLEY, E. T. and FISHER, K. C. (1963). The effect of temperature and photoperiod on the yearly hibernating behaviour of captive golden-mantled ground squirrels. *Can. J. Zool.*, **41,** 1103–20.

PFEFFER, W. F. P. (1911). *Abh. math.-phys. Kl. K. Sächs. Gesl. Wiss.* **32,** 163–295.

—— (1915). *Abh. math.-phys. Kl. K. Sächs. Ges. Wiss.* **34,** 1–154.

PIÉRON, H. (1923). Les problèmes psychophysiologiques de la perception du temps. *Année Psychol.* **24,** 1–25.

PITTENDRIGH, C. S. (1954). On temperature independence in the clock controlling emergence time in *Drosophila. Proc. Natn. Acad. Sci. U.S.A.* **40,** 1018–29.

—— (1958). Perspectives in the study of biological clocks. In *Symp. on Perspectives in Marine Biology*, pp. 239–68. University of California Press, Berkeley, Calif.

—— (1960). Circadian rhythms and the circadian organization of living systems. In *Cold Spring Harbor Symp. 25, Biological clocks*, pp. 159–82. The Biological Laboratory, Cold Spring Harbor, New York.

—— (1965). On the mechanism of the entrainment of a circadian rhythm by light cycles. In *Circadian clocks* (ed. J. Aschoff), pp. 277–97. North-Holland, Amsterdam.

—— (1966). The circadian oscillations in *Drosophila pseudoobscura pupae:* a model for the photoperiodic clock. *Z. Pflanzenphysiol.*, **54,** 275–307.

—— (1976). Circadian clocks: What are they? In *The molecular basis of circadian rhythms* (ed. J. W. Hastings and H. G. Schweiger), pp. 283–94. Dahlem Konferenzen, Abakon Verlagsgesellschaft, Berlin.

PITTENDRIGH, C. S. and BRUCE, V. G. (1957). An oscillator model for biological clocks. In *Rhythms ans synthetic processes in growth* (ed. A. Rudnick), p. 75. Princeton University Press, Princeton, N.J.

—— (1959). Daily rhythms and coupled oscillator systems and their relation to thermoperiodism and photoperiodism. In *Photoperiodism and related phenomena in plants and animals* (ed. R. B. Withrow), pp. 475–505. American Association for the Advancement of Science, Washington, D.C.

PITTENDRIGH, C. S. and MINIS, D. H. (1964). The entrainment of circadian oscillations by light and their role as photoperiodic clocks. *Am. Nat.* **98,** 261–94.

—— (1971). The photoperiodic time measurement in *Pectinophora gossypiella* and its relation to the circadian system in that species. In *Biochronometry* (ed. M. Menaker), pp. 212–47. National Academy of Sciences, Washington, D.C.

PITTENDRIGH, C. S., ROSENWEIGH, N. S. and RUBIN, M. L. (1959). A biological clock in neurospora, *Nature (Lond.)*, **184,** 169–70.

PYE, E. K. (1969). Biochemical mechanisms underlying the metabolic oscillations in yeast. *Can. J. Bot.* **47,** 271–85.

REIMANN, H. A. (1963). *Periodic diseases.* Blackwell, Oxford.

RENNER, M. (1955). *Naturwissenschaften* **42,** 540–1.

RICHTER, C. P. (1960). Biological clocks in medicine and psychiatry: shock-wave hypothesis. *Proc. Natn. Acad. Sci. U.S.A.* **46,** 1506–30.

—— (1965). *Biological clocks in medicine and psychiatry.* Charles C. Thomas, Springfield, Ill.

—— (1968). Inherent twenty-four hour and lunar clocks of a primate—the squirrel monkey. *Commun. behav. Biol. A*, **1,** 305–32.

—— (1975). Astronomical references in biological rhythms. In *The study of time II* (ed. J. T. Fraser and N. Lawrence), pp. 39–53. Springer-Verlag, Berlin-Heidelberg- New York.

RICHTER, C. P., JONES, G. S. and BISWANGER, L. (1959). Periodic phenomena and the thyroid. I. Abnormal but regular cycles in behaviour and metabolism produced in rats by partial radiothyroidectomy. *Arch. Neurol. Psychiat.* **81,** 233.

ROWAN, W. (1926). On photoperiodism, reproductive periodicity and the annual migrations of birds and certain fishes. *Proc. Boston Soc. nat. Hist.* **38,** 147–89.

SAUER, E. G. F. (1957). *Z. Tierpsychol.* **14,** 29–70.

SAUER, E. G. F. and SAUER, E. M. (1960). Star navigation of nocturnal migrating birds. The 1958 planetarium experiments. In *Cold Spring Harbor Symp. 25 Biological clocks*, pp. 463–73. The Biological Laboratory, Cold Spring Harbor, New York.

SAUNDERS, D. S. (1976a). The biological clock of insects. *Sci. Am.* **234**(2), 114–21.

—— (1976b). *Insect clocks.* Pergamon Press, Oxford.

SCHÄFER, E. A. (1907). *Nature (Lond.)* **77,** 159–63.

SPIEGEL, E. A., WYLIS, H. T., ORCHINIK, C. W., and FREED, H. (1955). The thalamus and temporal orientation. *Science* **121,** 771.

STETSON, M. H. and WATSON-WHITMYRE, M. (1976). Nucleus suprachiasmaticus: the biological clock in the hamster? *Science* **191,** 197–9.

Stöppel, R. (1916). *Z. Bot.* **8,** 609–84.

Strehler, B. L. (1977). *Time, cells and aging*, p. 287–8, Academic Press, New York.

Strughold, H. (1952). Physiological day–night cycle after global flight. *J. Aviat. Med.* **63,** 464–73.

—— (1965). The physiological clock in aeronautics and astronautics. *Ann. N. Y. Acad. Sci.*, **134,** 413–22.

Sweeney, B. M. (1960. The photosynthetic rhythm in single cells of *Gonyaulax polyedra.* In *Cold Spring Harbor Symp. Biological clocks*, pp. 145–7. The Biological Laboratory, Cold Spring Harbor, New York.

—— (1969). *Rhythmic phenomena in plants.* Academic Press, London, New York.

—— (1972). Circadian rhythms in unicellular organs. In *Circadian rhythmicity*, pp. 137–55. Centre for Agricultural Publishing and Documentation, Wageningen, Netherlands.

—— (1976). Evidence that membranes are components of circadian oscillators. In *The molecular basis of circadian rhythms* (ed. J. W. Hastings and H. G. Schweiger), pp. 267–81. Dahlem Konferenzen, Abakon Verlagsgesellschaft, Berlin.

Sweeney, B. M. and Hastings, J. W. (1960). Effects of temperature upon diurnal rhythms. In *Cold Spring Harbor Symp. 25, Biological clocks*, pp. 87–103. The Biological Laboratory, Cold Spring Harbor, New York.

Sweeney, B. M. and Haxo, F. T. (1961). *Science* **134,** 1361.

Tournois, M. J. (1912). *Ct. R. Acad. Sci. Paris* **155,** 297–300.

—— (1914). *Ann. Sci. Nat. Bot. Ser. IX* **19,** 49–191.

van der Pol, B. (1926), On 'relaxation oscillations' *Phil. Mag.* **2,** 978–92.

—— (1940). Biological rhythms considered as relaxation oscillations. *Acta Med. Scand. Suppl.* **108,** 76–88.

van der Pol, B. and van der Mark, J. (1929). The heartbeat considered as relaxation oscillation, and an electrical model of the heart. *Arch. néerl. Physiol.* **14,** 418–43.

Wahl, O. (1932). Neue Untersuchungen über das Zeitgedächtnis der Bienen. *Z. vergl. Physiol.* **16,** 520–89.

Wells, G. P. (1955). *The sources of animal behaviour* (Inaugural lecture delivered at University College, London, 5 May 195u), pp. 4–7. H. K. Lewis, London.

Wiener, N. (1958). *Scientia* **93,** 199–205.

4

MATHEMATICAL TIME

4.1. Time and number

The abstract mathematical idea of time as a geometrical locus—the so-called 'spatialization of time'—is one of the most fundamental concepts of modern science. Its psychological origin lies in our intuitive conception of time as one-dimensional. Our instinctive recognition of this linear property may be due to the previously mentioned fact that, strictly speaking, we can consciously attend to only one thing at a time, and that we cannot do this for long without our attention wandering. Our idea of time is thus directly linked with our 'train of thought', i.e. with the fact that the process of thinking has the form of a linear sequence. This linear sequence, however, consists of discrete acts of attention. Consequently, in the first instance, time is more naturally associated with counting, and hence with number, than with the linear continuum of geometry. We have already stressed the importance of rhythm in the development of the idea of time. Since counting is the simplest of all rhythms—a series of units, each considered as exactly similar to the rest, being freely combined into groups—we may attribute the power to form numbers to an elementary rhythm of the attention. It is surely no accident that the words 'number' and 'rhythm' come from two Greek terms (*ἀριθμός* and *ἀυθμὸς*) which were derived from a common root (*ῥεῖν*, to flow).

The peculiarly close relation between time and counting† has been emphasized both by philosophers of time and by philosophers of mathematics. For example, Aristotle, in endeavouring to distinguish between time and motion, came very close to reducing time to number. On the other hand, L. E. J. Brouwer in developing his celebrated

† Experiments on the pre-linguistic ability of birds to 'count' have revealed that this relation has very deep roots indeed, for there is evidence that the 'internal counting, which O. Koehler (1951) has shown to be a dormant capability of birds, is based upon the memory of a series of previous actions performed successively in time. A jackdaw, trained to raise one after another the lids of boxes in a row and secure five baits, found one in the first box, two in the second, and one in the third. He then went back to his cage, but later returned to the boxes, bowed his head once before the first, twice before the second, and once before the third. He then opened the fourth box and, finding nothing in it, went on to the fifth and took out the single bait which it contained. He left the remaining boxes untouched. The 'intention bowing' before the first three boxes presumably indicated that the bird 'counted' by remembering its previous actions.

As for human beings, the celebrated mathematician and lightning calculator A. C. Aitken (1959) stated that when engaged in mental arithmetic he did not visualize much but that 'the auditory–rhythmic impulse' was strong.

'intuitionist' theory of mathematics in the first two decades of the present century based his construction of the natural numbers on the conceptual multiplicity of the intervals of time, which he regarded as the primary intuition of the human intellect. Brouwer's doctrine derives from Kant who argued that 'arithmetic produces its concepts of number through successive addition of units in time' (Kant 1783). Although Kant did not regard arithmetic as the science of time in the way in which he regarded geometry as the science of space, because arithmetical relations are independent of time, he regarded both space and time as universal forms of our intuition, or conception, of phenomena and therefore as *a priori* or inherent in human reason. Consequently, any modification of our ideas of space and time was not only unnecessary but 'unthinkable'. It is well known that from the Kantian standpoint space is essentially unique and Euclidean, even though it is not something inherent in nature but only in our conception of nature. Similarly, time must be unique too, although Kant—a notoriously obscure writer—does not appear to be so explicit on this point.

Kant's highly original theory of space and time made a profound impression upon one of the greatest mathematicians of the first half of the nineteenth century, William Rowan Hamilton, who some thirty years after Kant's death read a paper before the Royal Irish Academy in which he argued that, just as geometry is the pure mathematical science of space, so there must also be a pure mathematical science of time and that this must be algebra (Hamilton 1837). Dissatisfied with the formalist approach of Peacock, who regarded algebra as a 'system of signs and their combinations', he demanded a more 'real' foundation. He sought this foundation in our intuition of time, but his object was to derive algebra from this intuition rather than to use algebra to elucidate it. He began with three fundamental principles: (i) that the notion of time is connected with existing algebra; (ii) that the notion or intuition of time may be developed into an independent pure science; (iii) that the science of pure time thus developed is co-extensive and identical with algebra, so far as algebra is a science. However, if algebra is to be based on time, which Hamilton regarded as a one-dimensional continuum of point-like instants, a difficulty is encountered when we come to consider the roots of a quadratic equation if they are imaginary. His memoir is largely devoted to his attempt to overcome this difficulty by his theory of moment couples (A_1, A_2), where A_1 is a primary moment and A_2 a secondary moment, irrespective of whether A_2 follows, precedes, or coincides with the primary moment. From this conception he developed an algebraic theory of number couples which led to an algebraic (as distinct from a geometrical) conception of complex numbers involving the square root of minus one. At the end of his second paper on this topic he refers to a paper by Graves on the logarithms of complex numbers and concludes with the following eloquent justification of his standpoint: 'But because Mr. Graves employed in his reasoning the usual principles respecting Imaginary Quantities, and was content to prove the symbolical necessity without showing the interpretation, or inner meaning of his formulae, the present Theory of Couples is published to make manifest that hidden meaning: and to show, by this remarkable instance, that expressions, which seem according to common views to be merely symbolical and incapable of being interpreted, may pass into the world of thoughts, and acquire reality and significance, if Algebra be viewed as not a mere Art or Language, but as the Science of Pure Time'.

Although the square root of minus one was not a number in the traditional sense, it obeyed all the formal algebraic rules of classical numbers and was therefore a new arithmetic entity rather than an element of a new algebra. Hamilton's algebraic investigations culminated, however, some eight or nine years later in his famous discovery of quaternions, the first example of a non-commutative algebra. Thus, the ultimate conclusion of his train of thought was the revelation that algebra is not unique, a view very difficult to reconcile with his Kantian conception of its nature and a powerful argument in favour of the formalist philosophy of mathematics to which he was so resolutely opposed. As for his peculiar association of algebra with the concept of time, the ultimate verdict was passed half a century later by the great algebraist Cayley in his Presidential Address to the British Association in 1883. After pointing out that Hamilton used the term algebra in a very wide sense, so as to include the differential calculus, he confessed that he could not himself recognize the connection of algebra with the notion of time. 'I would go further:' he said, 'the notion of continuous variation is a very fundamental one, made a foundation in the Calculus of Fluxions (if not always so in the Differential Calculus) and presenting itself or implied throughout in pure mathematics: and it may be said that a change of any kind takes place only in time; it seems to me, however, that the changes which we consider in mathematics are for the most part considered quite irrespectively of time. It appears to me that we do not have in Mathematics the notion of time until we bring it there'.

In the same year, in a fundamental paper on the meaning of the mathematical continuum, Georg Cantor (1883) argued that we cannot begin to determine this concept precisely by appealing either to the idea of time or that of space, for these ideas themselves can only be clearly explained by means of a continuity concept which must be more primitive and independent of them. The neo-Kantian philosopher Ernst Cassirer therefore re-interpreted Kant's theory of arithmetic as the study of 'series' which find concrete expression in the time sequence. He maintained that Kant himself was primarily concerned with the 'transcendental' determination of time as the prototype of an ordered sequence, and he argued that, instead of our intuition of time being the foundation of the logical concepts of sequence and order from which the laws of arithmetic can be derived, our idea of time implicitly depends on these concepts.

This point of view has been rejected by Brouwer who followed Kronecker in his criticism of Cantor and returned to Kant's original standpoint as regards time, although rejecting his theory of space. During the nineteenth century, philosophers and mathematicians came to be sharply divided on the latter, following the discovery about twenty years after Kant's death of non-Euclidean geometry by Lobatchewski and independently by Bolyai, for, although men of science generally continued to regard Euclidean geometry as the unique form of physical space, many mathematicians believed that other geometries were 'thinkable', i.e. logically permissible, whereas most philosophers did not. The acceptance of these other geometries greatly strengthened the standpoint of the formalist in their dispute with the intuitionists concerning the nature of pure mathematics. Nevertheless, in his famous inaugural lecture at Amsterdam, Brouwer (1913) maintained that 'However weak the position of intuitionism seemed to be after this period of mathematical development, it has recovered by abandoning Kant's apriority of space but adhering the more resolutely to the apriority of time'. He argued that the 'falling apart of moments of life into qualitatively different parts to be reunited only while remaining separated by time'—in other words the psychological fact that our minds operate by successive acts of attention—was the fundamental phenomenon of the human intellect which by a process of abstraction became the basis of all mathematical thinking—'the intuition of the bare two-oneness'. By repetition this process gives rise to all finite ordinals, and by its indefinite repetition the smallest infinite ordinal ω may also be generated. Also, the same basic intuition gives rise to 'the intuition of the linear continuum, i.e. of the "between" which is not exhaustible by the interposition of new units and which therefore can never be thought of as a mere collection of units'. Brouwer concluded that 'In this way the apriority of time does not only qualify the properties of arithmetic as synthetic *a priori* judgments, but it does the same for those of geometry, and not only for elementary two- and three-dimensional geometry, but for non-Euclidean and n-dimensional geometries as well. For since Descartes we have learned to reduce all these geometries to arithmetic by means of the calculus of coordinates'.

The past half-century has seen the intuitionists holding their ground against both the formalists and those who regard mathematics as an extension of logic. The attention of Brouwer and his followers has been focused on problems concerning the nature and foundations of pure mathematics, but their achievements make it all the more imperative for us to examine their fundamental presupposition concerning the apriority of time. We can be guided in this by Helmholtz's criticism of Kant's doctrine of space. Helmholtz pointed out that the latter can be divided into two parts: (i) space is a pure form of intuition; (ii) Euclidean geometry is the unique science of space and holds *a priori*. He argued that the second proposition is not a necessary consequence of the first and, in fact, he rejected it. However, he accepted the first proposition, although in his opinion nothing could be inferred from it beyond the fact that all things in the world have spatial extension. Can we adopt a similar attitude to Kant's doctrine of time? For Brouwer's purpose, it is only necessary to assume that time is a 'pure form of intuition' in the generally accepted sense that our experience is characterized by temporal succession based on the two-term relation of before and after. It is not necessary to accept Kant's view that the attribution of temporal characteristics to the universe itself necessarily leads to logical antinomies and that time is therefore nothing more than the form of our inner sense.

4.2. Time, geometry, and functional variation

The intuitionist outlook is associated both with the idea of time and with mathematical 'construction', as has been emphasized by Kant and by Brouwer. Indeed the association of the ideas of mathematical construction and time has even led Brouwer to reject the logical principle of 'the excluded middle', at least as applied to the concept of mathematical existence. For Brouwer, the existence of mathematical entities is synonymous with the possibility of their construction, and a particular theorem is neither true nor false until we have a construction for deciding the question. On the other hand, the formalist and logistic philosophies of mathematics depend on belief in the timeless character of mathematical 'existence'. This idea of mathematics may be regarded as the ultimate product of a line of thought which began with Plato.

Plato's philosophy of form was based on a critical analysis of the Pythagorean philosophy of number. According to Pythagoras and his school, the essence of things must be sought in number. Numbers were, however, represented geometrically, by patterns of 'points' or units having position. Moreover, many of the more powerful mathematical techniques employed to obtain numerical results were geometrical in character. The purely arithmetical technique of counting was not only less powerful, i.e. less general. than the geometrical method, because of the complete absence of anything corresponding to modern algebraic symbolism, but it led to notorious difficulties, notably the incommensurability of the diagonal of a unit square. Although these difficulties could be obviated by the kinematic, or fluxion, technique of moving points and lines, according to which the point flows into the line, and so on, Plato was too strongly influenced by the arguments of Parmenides and Zeno (see §4 of this chapter) to adopt this method. On the contrary, he purged the Pythagorean mathematics of its more strictly 'arithmetical' content. This was associated with time, process, and generation, for the Pythagoreans regarded numbers as generated by successive additions of the 'one' or arithmetical unit. (Their theory of the unit and the dyad is very similar to that of Brouwer.) Consequently, although Plato regarded time as an essential feature of the sensible world, he rigorously excluded it from the science of pure geometry which he associated with, and only with, the eternal world of ideal forms. As a result, he was resolutely opposed to mathematical 'construction'. In a significant passage in *The Republic* he complained of the mathematicians who 'constantly talk of "operations" like "squaring", "applying", "adding" and so on, as if the object were to *do* something, whereas the true purpose of the whole subject is knowledge—knowledge, moreover, of what eternally exists, not of anything that comes to be this or that at some time and ceases to be' (Cornford 1941).

Once one realizes that Plato's objection to mathematical 'construction' is due to his dislike of introducing temporal considerations into pure geometry, a flood of light is thrown on his curious insistence that so-called 'mechanical solutions' of the famous problems of the squaring of the circle, the trisection of any angle, and the duplication of the cube are not permissible. It is often stated that his objection was to the practical use of actual mechanical instruments, but this argument as it is usually presented fails to make sense, for it is said, for example by Cajori (1919), that he rejected the ingenious solutions of Archytas, Eudoxus, and Menaechmus because 'they required the use of other instruments than the ruler and compasses'. Surely the objection to the use of mechanical instruments should extend to all without exception? The crucial Platonic distinction, as I see it, is that whereas the bisection of an angle, for example, is associated with a particular configuration of straight lines and circular arcs which can be contemplated *statically*, i.e. without reference to time, the trisection of an angle, as performed by Hippias, was a construction involving a *moving* configuration of lines and therefore dependent upon temporal considerations.

In the solution of Hippias a curve known as the *quadratrix*, which was also applied to the squaring of the circle, was generated in the following manner. One side AB of a square ABCD rotates uniformly about the vertex A through a right angle to the position AD. Meanwhile the adjacent side BC slides uniformly between AB and CD so as to arrive at the position AD in the same time. The curve BEF, the quadratrix generated by the point of intersection, has the property that the perpendicular from any point on it to AD is proportional to the angle between AD and the line joining A to that point. (This can be seen most easily by considering the motions concerned in reverse.) To trisect the angle EAD, all that is necessary is to trisect the line EG and then draw lines from the points of trisection parallel to AD so as to intersect the quadratrix in H and K. The lines AH and AK divide the angle into three equal parts (see Fig. 4.1).

The essential features of this construction are that the motions are uniform and that they begin and end at the same time. As presented by its author, it is definitely kinematic. Although this geometry of movement, or generative geometry, was regarded by Plato as inapplicable to the world of ideal forms, Greek mathematicians, notably Archimedes in his book *On Spirals*, studied the pure geometry of curves defined kinematically. Moreover, kinematic geometry was applied by Plato's pupil Eudoxus to the analysis of the planetary motions, probably following elementary pioneer attempts by the Pythagoreans. This association of geometry of movement with astronomy was one of the most original and far-reaching intellectual achievements of the ancient Greeks. The two other ancient civilizations that made the most elaborate investigations in mathematical astronomy, the Seleucid Babylonians and the Mayas of Central America, developed only arithmetical techniques.

Despite his profound interest in problems of motion and change, Aristotle was insistent on the rigid separation of mathematics from physics. Nevertheless, in the seventh book of his *Physics* there is a prolix discussion of uniform motion in which time is treated as if it were a

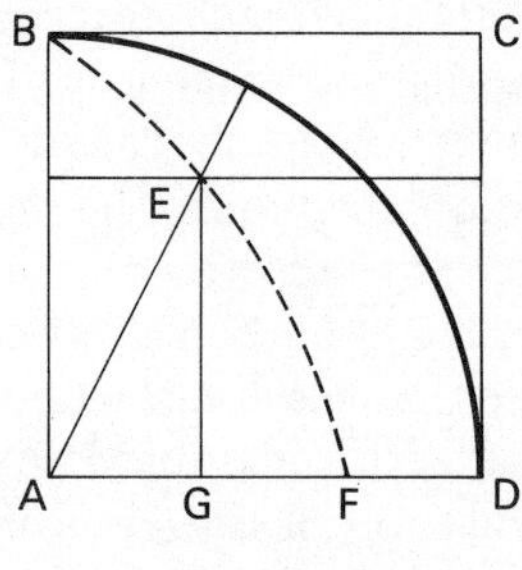

FIG. 4.1

geometrical quantity similar to space and like the latter capable of infinite divisibility. Indeed, there is considerable evidence of a geometrical point of view in this chapter (in particular, Aristotle denotes an interval of time by the same notation, for example ZH, as the Greek geometers used for a line segment), although his discussion is much less precise than any mathematical argument in the writings of Archimedes.

Marshall Clagett (1956, p. 77) has drawn attention to the fact that the Greek geometers of motion tended to give comparative rather than metric definitions, either comparing the *distances* traversed in two uniform motions when the times are assumed to be the same, or the *times* when the distances are the same. These comparisons were true proportions in the Euclidean sense, since they were between quantities of the same nature. Consequently, scarcely any of the Greek authors arrived at the idea of velocity as a number or magnitude representing the ratio of the unlike quantities, distance and time.

The oldest kinematic treatise of the Latin West known to us is the *Liber de motu* of Gerard of Brussels, an obscure geometer of the first half of the thirteenth century. In this curious work, although the author does not define velocity as a ratio of unlike quantities, he assumes that the speed of a motion can be assigned some number or quantity which is neither distance nor time (Clagett 1956, p. 152). Composed when knowledge of Greek geometry was first becoming widespread in the Latin West, this treatise is full of elementary mathematical errors. Nevertheless, it exerted a major influence on the school of philosophers at Merton College, Oxford, in the following century, stimulating them to study the kinematics of non-uniform, or accelerated, motion. The modern concept of acceleration, which we now recognize as indispensable for the formulation of a satisfactory dynamics, was never even imagined by the Greeks, let alone discussed and analysed.† In groping their way towards this concept, which was extremely difficult to formulate before the invention of the differential calculus, the Schoolmen of the fourteenth century were greatly handicapped by their lack of algebraic symbolism. Their discussions were purely verbal and tediously prolix. Nevertheless, they led to one of the greatest advances in knowledge ever made.

Before the kinematic concept of acceleration could be correctly formulated, two other ideas were necessary: (i) the idea of time as the *independent*, and space as the dependent, variable; (ii) the idea of instantaneous velocity.

† The first explicit treatment of acceleration, in the sense of motion that becomes quicker, appears to have been given by Strato of Lampsacus who became head of the Lyceum in 287 B.C., but his treatment was unsatisfactory because he had no clear-cut concept of instantaneous velocity.

The general mathematical notion of the variable was gradually formulated by the late Scholastic philosophers following the great condemnation of Aristotle's philosophy in 1277 by Bishop Tempier of Paris and Archbishop Kilwardby of Canterbury. We have seen that Aristotle made a rigid distinction between mathematics and physics, the former being concerned with 'things' which do not involve motion' and the latter with things that do. Moreover, since in terrestrial motions, unlike celestial, there appeared to be no general uniformity, physical motion was regarded not as a 'quantity' but rather as a 'quality' that does not increase or decrease through the joining together of parts (Duhem 1909). Duns Scotus, who died in 1308, was among the first who broke with this tradition and considered the general problem of the variability of qualities, or 'latitude of forms' as it was called. The problem arose out of the need to account for the observed fact of the variation in the intensity of qualities, despite the axiomatic Aristotelian principle of the immutability of substantial forms. For example, if we increase or diminish its intensity, a ray of light will become more or less luminous without anything being added to its nature or subtracted from it, for this is simply light itself; intensity is, therefore, a mode, or intrinsic property, of light. The term *latitudo* to denote the range in which the intensity of a quality may vary seems to have been introduced by a slightly earlier philosopher, Henry of Ghent, who died in 1293. He was one of the advisers of Bishop Tempier in 1277. According to Henry, the '*intensio*' of a quality consists in approach to a certain terminus in which the quality achieves its full perfection (Maier 1939, pp. 10, 27–9). Duns Scotus and his followers held that increase in intensity occurs by addition, the proper analogy being not the addition of stone to stone, but of water to water; the new individual quality formed by such addition contains the previous one (Maier 1939, pp. 32–8, 45–9). William of Ockham (died 1349) and the nominalists generally followed Scotus in regarding *intensio* as an additive increase, and attention came to be focused on the logical problem of how to denominate a subject in which the intensity of a quality varies from one point to another. This problem merged into the mathematical one of describing the various possible modes of spatial or temporal variation of intensity. The term *latitudo* came to refer to a configuration or particular mode of variation of intensity in space or time.

The pioneer figure in the mathematical, as distinct from the dialectical, development of the idea of the variable was Thomas of Bradwardine, whose *Tractatus de Proportionibus* was written in 1328. A recent writer has drawn attention to the fact that Bradwardine's work set the stage for modern physics based on Galileo's wedding of mathematics and experimental observation. 'Bradwardine used mathematics for the systematic and general expression of theory; Galileo used it for the systematic generalization of experimental observation' (Crosby 1955, p. 17). Bradwardine's work is noteworthy for introducing into mathematics functions more complex than simple linear proportionality. Among other leading figures of the Merton mathematical school of the first half of the fourteenth century were William Heytesbury, who defined acceleration as the velocity of a velocity, John of Dumbleton, and Richard Swineshead, who was called Calculator after his main work which was not, however, concerned with calculation, as we understand the term, but with the verbal and arithmetical theory of uniform and non-uniform rates of change. Incidentally, it may be mentioned that the terms *fluxus* and *fluens* which he used in this context were destined to be employed some three hundred years later by Newton when he spoke of a variable as a fluent and its rate of change as a fluxion.

Despite the pioneer achievements of the Mertonians, the principal mathematical advance in the study of the variable in the fourteenth century was made in France by Nicole of Oresme, who was born about 1323 and died in 1382 as Bishop of Lisieux. One of the greatest mathematicians in the later Middle Ages, he was no less outstanding as a writer on natural philosophy and economics and was unquestionably one of the most versatile men of his day. He appears to have been the first to use fractional indices, or powers, systematically. His treatise *De Configuratione Qualitatum*, written probably before 1361, was remarkable for the fact that, following the Greek tradition which regarded number as discrete and geometrical magnitude as continuous, he abandoned the dialectical Mertonian discussions of variation in terms of number and instead associated continuous change with a geometrical diagram. A horizontal line (*longitudo*) was drawn to represent the extension in space or in time of a given form, the properties or 'qualities' of which, for example, colour, density, etc., were to be determined. This line was divided into equal divisions called degrees. An intension, or rate of change, by which a form acquired a quality, was represented by a vertical line (*latitudo*) drawn on a uniform scale through the corresponding *longitudo*. When all the latitudes had been plotted, a line drawn through their summits yielded a geometrical figure† which Oresme called the *linear* configuration of the quality in question.

Oresme's work also marked a significant advance in the development of the idea of instantaneous velocity. He was apparently the first to represent an instantaneous rate of change by a straight line: '*sed punctualis velocitas instantanea est ymaginanda per lineam rectam*' (Wieleitner 1914,

† Although Oresme appears to have been the first to apply the graphical technique systematically to represent the idea of functional variation, and thereby played a decisive role in the geometrization of time, he certainly did not invent the idea of graphical representation. The oldest known graph dates probably from the tenth century.

p. 226). The idea of an instantaneous velocity had been expressly rejected by Aristotle, and until the modern theory of limits was developed a satisfactory definition was unattainable. The distinction between velocity as a simple fraction of distance and time ($v = s/t$) and as an 'instantaneous' quality of motion ($v = \mathrm{d}s/\mathrm{d}t$, in modern notation) was drawn by Bradwardine in his discussion of dynamical motion (Crosby 1955, p. 44). The importance of the intuitive idea of instantaneous velocity was further enhanced as a result of the revival in France of the physical idea of 'impetus': that a body once set in motion will continue to move because of an internal tendency which it then possesses. This anti-Aristotelian theory, which may be regarded as a vague anticipation of the Newtonian principle of inertia, goes back to Joannes Philoponos in the sixth century A.D. It was revived by Peter John Olivi, who died in 1298, and more especially by Jean Buridan, Rector of the University of Paris, who died about 1358.

In spite of his explicit assertion that an instantaneous velocity is to be represented by a straight line, Oresme followed Aristotle in maintaining that every velocity persists throughout a time: '*omnis velocitas tempore durat*' (Wieleitner 1914, p. 225). In attempting to clarify the concept of instantaneous velocity, Oresme stated that the greater this velocity, the greater would be the distance covered, if motion were to continue uniformly at this rate. The Merton mathematician, William Heytesbury, made the same point (Wilson 1956). Although both the Merton school and Oresme had the correct kinematic conception of acceleration, *with time as the independent variable*, much confusion over this issue persisted until the time of Galileo. Historically, this confusion can be traced back to Aristotle's two definitions of 'the quicker': (i) as that which traverses equal space in less time; (ii) as that which traverses greater space in equal time. The former led historically to the erroneous conclusion that in the naturally accelerated motion of falling bodies velocity increases uniformly with respect to *distance*. This conclusion was supported, in one way or another, by Strato, Alexander of Aphrodisias, Simplicius, Albert of Saxony, and even by Galileo before he arrived at the correct formulation.

Modern researches have shown that Galileo was closer to his fourteenth-century predecessors in mathematical kinematics than is generally realized, Mach was quite wrong when he stated that Galileo had to create for us the entirely new concept of acceleration (Mach 1883). He was anticipated in his concept of kinematic acceleration by the Mertonians and by Oresme, and, in his correct application of the concept to the law of falling bodies, to some extent by Dominico Soto, a Spanish Dominican who died in 1560. After defining uniform acceleration correctly, Soto said that this was the kind of motion proper to freely falling bodies and to projectiles. Thus, although Galileo went much further than his predecessors in formulating a complete kinematics and in applying it to naturally occurring motions, he was by no means so original as had been generally assumed. In particular, he was not the first to use the geometrical concept of time.

4.3. Time and the calculus

In the course of the seventeenth century, the geometrization of time led to remarkable developments in mathematics, owing to the success of kinematic methods. Thus, the invention of logarithms by Napier, published in 1614, was based on the comparison of two moving points, as shown in Fig. 4.2. The point P moves along a straight line AB while another point Q moves along a line of unlimited length starting at C. Both points have the same speed at the start when P leaves A and Q leaves C, but whereas Q maintains the same speed all the time the speed of P at any instant is proportional to the distance PB. Napier defined the logarithm of the number measured by the distance PB as the number measured by the distance CQ.

The outstanding mathematical achievement associated with the geometrization of time was, of course, the invention of the calculus of fluxions by Newton. Newton's concept of fluxion was based on a tacit appeal to our intuition of motion. He was greatly influenced, and partially anticipated, by his predecessor in the Lucasian chair, Isaac Barrow. Both Barrow and his famous, but mathematically incompetent, contemporary Thomas Hobbes rejected the arithmetization of mathematics advocated by John Wallis, Savilian Professor at Oxford. Instead, they stressed the fundamental significance of continuous geometrical quantity. Nevertheless, there was an important difference between their points of view.

Hobbes, whose criticism was embittered by Wallis's rejection of his naïve attempt to 'square the circle', castigated the *Arithmetica Infinitorum* of Wallis as a 'scurvy book' (Hobbes 1656, p. 283). and referred to the arithmetization of geometry in Wallis's *Tractatus de Sectionibus Conicis* as a 'scab of symbols' (Hobbes 1656, p. 361). Hobbes was greatly impressed by Galileo's success in basing dynamics on the idea of changing velocity and endeavoured to make motion the foundation of his whole philosophy.

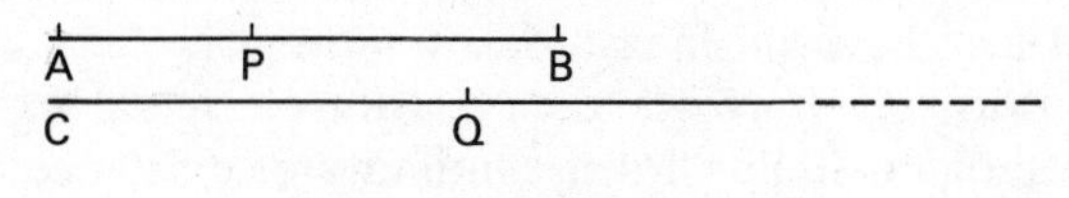

Fig. 4.2

He introduced the concept of *conatus* as the beginning of geometrical extension. Motion at a point he regarded as motion in a minimum indivisible interval. Time he defined as merely the 'phantasm', or decaying image in the mind, of before-and-after in motion. He did not regard it as the measure of motion, for 'we measure time by motion and not motion by time'. Moreover, in his view, 'The present only has a being in nature; things past have a being in the memory only, but things to come have no being at all, the future being but a fiction of the mind applying the sequels of actions past to the actions that are present' (Hobbes 1651).

Barrow, although opposed to the arithmetical and algebraic tendencies of Wallis (and some continental mathematicians) and in agreement with the view held by Hobbes that mathematics should be identified with geometry, nevertheless appreciated the significance of time. In this respect he must be regarded as something of a pioneer, for to most thinkers of his age space was a far more important concept. Thus, even Descartes, despite his appreciation of the usefulness of algebra and his profound interest in problems of motion, was so obsessed by the concept of geometrical extension that for him time was relatively unimportant. Whereas he regarded extension as the principal attribute of physical things, time was only a mode of our thinking about them (Descartes 1644).

Barrow's view of the nature of time is not only of great interest in itself, but also because of his influence on Newton. Indeed, just as Newton's philosophy of space derives from the Cambridge Platonist Henry More, so his philosophy of time similarly derives from Barrow, whose lectures he attended whilst a student. In his *Lectiones Geometricae*, Barrow maintained that 'because mathematicians frequently make use of time, they ought to have a distinct idea of the meaning of the word, otherwise they are quacks' (Barrow 1735, p. 4). Although he believed that 'there is a great affinity and analogy between space and time', he was careful to distinguish between them. He criticized Hobbes, because 'he feared not to compare *lines* and *times* together, as homogeneous quantities obtaining a mutual proportion, though things of the most distant nature'. He was profoundly impressed by the kinematic method in geometry which had been studied with great effect by Galileo's pupil Torricelli, and he realized that to understand this method it was necessary to study time. Although time is measurable by motion, he was careful to point out, as was Plotinus in his criticism of Aristotle, that it cannot both be a measure of motion and also be itself measured by motion.

'Time', according to Barrow, 'denotes not an actual existence, but a certain capacity or possibility for a continuity of existence; just as space denotes a capacity for intervening length. Time does not imply motion, as far as its absolute and intrinsic nature is concerned; not any more than it implies rest; whether things move or are still, whether we sleep or wake, Time pursues the even tenour of its way' (Barrow 1735, p. 35). We see here the origin of Newton's famous definition: 'Absolute true, mathematical time, of itself, and from its own nature, flows equably without relation to anything external'. Barrow continues, 'Time implies motion to be measurable; without motion we do not perceive the passage of Time. We evidently must regard Time as passing with a steady flow; therefore, it must be compared with some handy steady motion, such as the motion of the stars, and especially of the Sun and Moon . . .'. Barrow does not, however, leave the matter there. To the question of how one knows that the Sun is carried by an equal motion and that one day, or year, is exactly equal to another, 'I reply that, if the sundial is found to agree with motions of any kind of time-measuring instrument, designed to be moved uniformly by successive repetitions of its own peculiar motion, under suitable conditions, for whole periods or for proportional parts of them; then it is right to say that it registers an equable motion. It seems to follow that strictly speaking the celestial bodies are not the first and original measures of Time; but rather those motions, which are observed round about us by the senses and which underlie our experiments, since we judge the regularity of the celestial motions by the help of these. Not even is Sol himself a worthy judge of time, or to be accepted as a veracious witness, except so far as time-measuring instruments attest his veracity by their votes'.

Barrow answers the question of the ultimate relation of time and motion in the following way: 'Time may be used as a measure of motion; just as we measure space from some magnitude, and then use this space to estimate other magnitudes commensurable with the first; i.e. we compare motions with one another by the use of time as an intermediary'. He regarded time as essentially a mathematical concept which has many analogies with a line 'for time has length alone, is similar in all its parts and can be looked upon as constituted from a simple addition of successive instants or as from a continuous flow of one instant; either a straight or a circular line'. This clear statement is perhaps the earliest explicit formulation of the concept of geometrical time, for just as Euclid only speaks of straight line segments and never of the complete straight line in our sense, so Galileo used only such segments to denote definite time intervals. Nevertheless, as already mentioned, Barrow did not *identify* time with a line. Time, in his view, was 'the continuance of anything in its own being', and in a passage to which we shall return in Chapter 5, he remarks, 'nor do I believe there is anyone but allows that those things existed equal times which rose and perished together' (Barrow 1735, p. 5).

In his discussion of the analogy between time and a line, Barrow pointed out that the latter can be regarded either as composed of points or as the trace of a moving point. Similarly, he argued, time can be thought of as either an aggregation of instants or as the continuous flow of one instant. Mathematically, his kinematic method was extremely fruitful. If he had not adhered so resolutely to the synthetic style of the ancient geometers and at the same time deliberately neglected algebraic methods, he might have anticipated Newton in his development of the calculus of fluxions as a powerful analytical tool. Both Barrow and Newton were brought face to face with the subtle problems of the continuum and of the nature of instantaneous velocity.

Barrow's views on these questions were far from precise. 'To every instant of time, or indefinitely small particle of time, (I say instant or indefinite particle, for it makes no difference whether we suppose a line to be composed of points or of indefinitely small linelets; and so in the same manner, whether we suppose time to be made up of instants or indefinitely minute timelets); to every instant of time, I say, there corresponds some degree of velocity, which the moving body is considered to possess at that instant' (Barrow 1735, p. 38). Again, in his argument that the area under a velocity time curve represents distance, he maintained that a surface may be represented by an aggregate of straight lines; although he realized that, strictly speaking, very narrow rectangles ought to be substituted for the lines, he claimed that 'it comes to the same result whichever way you take it' (Barrow 1735, p. 39).

Newton's approach was more subtle. Unlike Barrow, he tended to follow Wallis in breaking away from the idea of number as merely a collection of units. He came close to anticipating the modern concept of limit in his idea of the 'ultimate ratio' of 'evanescent increments'. Indeed, it has been suggested that, if he had devoted more of his time to clarifying this idea, he might have anticipated the 'rigorous' method developed in the nineteenth century by Cauchy (Boyer 1949, p. 196). Nevertheless, in Newton's writings (and also for that matter in those of Leibniz) we do not find that a limit is regarded explicitly as a number in its own right, but as a quotient of two quantities. In this respect Newton was a traditionalist, for in the Euclidean view ratios of geometrical magnitudes took the place which we now assign to the so-called 'real numbers'.

Newton seems to have regarded mathematics as being primarily a method for the solution of physical problems; for example, in the preface to the *Principia*, he says that geometry is but a 'part of universal mechanics'. It is therefore not surprising that his idea of a limit was closely associated with his geometrical and temporal intuitions, particularly the latter since he tended to regard time as the standard independent variable. In his famous memoir *The Analyst*, published in 1734, the philosopher Berkeley criticized Newton's definition of fluxion as the ultimate ratio of evanescent increments, since it appeared that the latter were neither finite quantities nor zero but 'the ghosts of departed quantities'. Newton himself had not been unaware of the difficulty and endeavoured to circumvent it by the following argument, to be found in the *Scholium*, following the lemmas of Book I of the *Principia*, in which the idea of time plays a central role: 'Perhaps it may be objected that there is no ultimate proportion of evanescent quantities; because the proportion, before the quantities have vanished, is not the ultimate, and when they are vanished, is none. But by the same argument it may be alleged that a body arriving at a certain place, and there stopping, has no velocity; because the velocity, before the body comes to the place, is not its ultimate velocity, when it has none. But the answer is easy: for by the ultimate velocity is meant that with which the body is moved, neither before it arrives at its last place and the motion ceases, nor after, but at the very instant it arrives, that is, that velocity with which the body arrives at its last place, and with which the motion ceases. And in like manner, by the ultimate ratio of evanescent quantities is to be understood the ratio of the quantities not before they vanish, nor afterwards, but with which they vanish . . .' (Cajori 1934).

In direct opposition to this point of view, it is now generally agreed by mathematicians that the difficulties concerning the foundations of the calculus first exposed by Berkeley were not adequately resolved until last century when Cauchy, Dedekind, Cantor, Weierstrass, and others imparted a far greater precision to the basic concepts than had obtained hitherto. These mathematicians all adhered to the formalist view of the nature of the subject. In particular, they rejected the Newtonian conception of the calculus as a scientific description of the generation of magnitudes. Consequently, *the gain in precision which they achieved was accompanied by the elimination of temporal concepts.* For example, the modern definition which identifies the limit of an infinite sequence with the sequence itself has deprived the *mathematical* problem of whether a variable reaches its limit of all meaning. As a result, the limit concept has been finally divorced from its intuitive dependence on the concept of motion. Thus, although the ideas of time and motion played so central a role in the rise of the new mathematical analysis in the seventeenth century, two hundred years of discussion concerning its foundations ultimately culminated in the paradoxical result that 'the very aspect which led to its rise was in a sense again excluded from mathematics with the so-called "static" theory of the variable which Weierstrass had developed' (Boyer 1949, p. 288). According to this view, 'The variable does not represent a progressive passage through all the values of an interval, but the disjunctive assumption of any one of the values in the interval. Our vague intuition of motion, although remarkably fruitful in having suggested the investigations which produced the calculus, was found, in the light of further elaboration in thought, to be quite inadequate and misleading'.

Thus, the wheel has turned full circle and mathematical analysis is now characterized by its neo-Platonic 'elimination of time'.

4.4. Zeno's paradoxes

The emancipation of the theory of the mathematical variable and of the continuum from all temporal considerations cannot be overlooked when claims are made that this theory automatically resolves the notorious paradoxes of time and motion associated with the name of Zeno of Elea. For example, the paradox of the flying arrow—according to which the arrow cannot move because at each instant of its flight it occupies a space equal to itself and therefore has no room to move—*cannot* be resolved, as Boyer claims, merely by pointing out that it 'involves directly the conception of the derivative and is answered immediately in terms of this' (Boyer 1949, p. 25), for this mathematical concept has been freed of all reference to the ideas of time and motion, and these are the very ideas with which the paradox is concerned.

Although we cannot be sure of Zeno's object in formulating his paradoxes nor even of their original wording, philosophical interest in them has been sustained for twenty-four centuries and shows no sign of abating. Zeno, a native of Elea in southern Italy, flourished about the middle of the fifth century B.C. He was a pupil of Parmenides, the pioneer of logical argument in philosophy. Parmenides regarded the senses as deceptive and believed that reality is indivisible and timeless. He may have been originally a Pythagorean, for like them he was a believer in the spherical shape of the world. His pupil Zeno used his great logical gifts to advance his master's doctrine by showing that the ideas of plurality and change lead to logical antinomies. In particular, he criticized the concept of time in four paradoxes concerning motion. These paradoxes fall into two groups according as it is assumed that time (and correspondingly space) is either discrete or continuous, i.e. either composed of indivisible units of short but finite duration or else infinitely divisible.

The paradox of the arrow is directed against the contention that time is composed of indivisible instants. It is supplemented by an ingenious argument known as the 'stadium' which is obscurely expressed by Aristotle (Lee 1936, p. 55), whose *Physics* is our primary surviving source of

A 1 2 3   A 1 2 3
S 1 2 3   S   1 2 3

FIG, 4.3

reference on Zeno's teachings. The gist of Zeno's argument, which also involves the Pythagorean concept that space is composed of discrete points, seems to be given by the following. A row of points A passes a fixed stadium S, also composed of points, at the rate of one point per instant. The two diagrams in Fig 4.3. represent A in relation to S at successive instants. Suppose also that another row of points B moves at the same minimum rate as A but in the opposite direction. Then at successive instants we have the situations represented in Fig. 4.4, from which we see that at *successive instants* B1 is in line with A1 and then with A3. Zeno argues that this is absurd, for there must be an instant as A moves one way and B the other at which B1 is in line with A2, the point intermediate between A1 and A3. This contradicts the idea that the two instants previously considered are successive. Hence, there are no successive instants and so time (and correlatively space) in infinitely divisible.

Despite its ingenuity, this is one of the easiest of Zeno's arguments to answer. For, *if* space and time are composed of discrete units, then relative motions must be such that the situations typified by the diagrams of Fig. 4.4 *can* occur at successive instants. Zeno's rejection of this possibility is not based on any logical rule but simply on a fallacious appeal to 'common sense'. Indeed, Zeno is in fact guilty of a logical error himself when he makes this appeal, for he is tacitly invoking a postulate of continuity which is incompatible with the hypotheses adopted at the beginning of the argument.† Strange as it may seem, if we adopt such hypotheses then motion will be a discontinuous succession of distinct configurations, as in a roll of cinema film, and at no time will intermediate configurations occur. The passage of an electron from one orbit to

† The same mistake is also implicit in the criticism by A. Grünbaum (1968, pp. 117–20) of this analysis of Zeno's stadium paradox, for he objects to the conclusion that whether a given vertical alignment of the rows of numbered points of A and B qualifies as an event or not depends on the relative velocity of the rows concerned. He argues that 'none of our present day kinematic knowledge even gives a hint of the possibility of the aforementioned dependence of event-status on relative motion'. This remark shows that he has missed the point, for, *if* time is composed of indivisible instants, then what Grünbaum calls 'event-status', i.e. whether the alignment of, say, A2 or B1 is a possible event or not, must necessarily depend on the relative motion of A and B. This conclusion cannot be rejected by an appeal to 'present-day kinematic knowledge', which is based on the assumption of the mathematical continuity of time. Of course, from our practical experience it is clear that, if time consists of successive instants, these must be extremely small, because the assumption of continuity works so well.

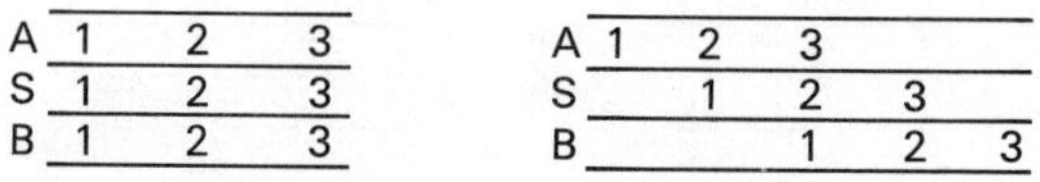

A	1	2	3
S	1	2	3
B	1	2	3

A	1	2	3		
S		1	2	3	
B			1	2	3

FIG. 4.4

another as envisaged in the elementary Bohr theory of the atom is of this type.

Having rejected Zeno's paradox of the stadium, we are left with that of the arrow as an argument against the existence of temporal instants. An amusing variant of this paradox is given in the article on Zeno in Bayle's famous *Dictionnaire*, published in 1696. He recalls a story told by Sextus Empiricus of the sophist Diodorus who lectured against the existence of motion. Having dislocated his shoulder, he went to have it set. 'How?' said the doctor. 'Your shoulder dislocated! That cannot be; for, if it moved, it did so either in the place where it was, or in the place where it was not. But it did not move, either in the place where it was, or in the place where it was not, for it could neither act nor suffer in the place where it was not!'

Zeno's paradox of the arrow raises profound issues concerning the nature of motion. The American philosopher Charles Peirce, whose writings have attracted much more attention in recent years than in his lifetime (he was born in 1839 and died in 1914), has reformulated this paradox in the form of the following syllogism (Peirce 1934):

Major premise: No body in a place no larger than itself is moving;
Minor premise: Every body is a body in a place no larger than itself;
Conclusion: Therefore, no body is moving.

The error, in his opinion, lies in the *minor premise*, which is only true in the sense that during a time sufficiently short the space occupied by a body is as little larger than itself as you please. He concludes that all can be inferred from this is that during no time a body moves no distance. Although this particular form of the argument is interesting, it is incomplete because it fails to take account of the conception of motion inherent in the *major premise.*

Bertrand Russell, on the other hand, has confined his discussion of the paradox to this premise. In his opinion, Zeno assumes that when a thing is changing its position there must be some internal state of change in the thing; in other words, a moving body is in 'a state of motion' which is qualitatively different from a state of rest. 'He then points out', says Russell, 'that at each instant the arrow simply is where it is, just as it would be if it were at rest. Hence, he concludes that there can be no such thing as a *state* of motion, and therefore, adhering to the view that a state of motion is essential to motion, he infers that there can be no motion and that the arrow is always at rest' (Russell 1946). This argument raises important issues, but the boot is on the other foot. I submit that in this paradox Zeno is adopting Russell's own view that a moving body is *not* qualitatively different from a static one and that motion can be recognized only by change of position. In a temporal instant no change of position can occur and hence, Zeno argues, there can be no motion. If there were some *intrinsic* change in a body due to motion, then the major premise formulated above would collapse. Instead, the whole force of Zeno's argument, as I see it, derives from the intuitive conviction, expressed in this premise, that motion is analysable only into states of motion and not into states of rest. Motion, in other words, can be compounded only of motions and not of immobilities.

It follows that there are two alternative ways of escape from Zeno's conclusion: either we can distinguish at any instant a moving body from a stationary one by some descriptive feature other than change of position, for that, as Zeno correctly points out, cannot be instantaneous, except in the ideal case of infinite velocity—which is not in question, or we can boldly accept the explanation which Zeno rejects as paradoxical, that motion can be compounded of immobilities. Russell agrees with Zeno in rejecting the first possibility and comes very near to agreeing with him over the second. 'Weierstrass, by strictly banishing all infinitesimals', he writes—referring to his rigorous arithmetization of mathematical analysis and the calculus—'has at last shown that we live in an unchanging world, and that the arrow, at every moment of its flight, is truly at rest. The only point where Zeno probably erred was in inferring (if he did infer) that, because there is no change, therefore the world must be in the same state at one time as another' (Russell 1937).

Russell's point can be made in a less provocative fashion. If we adopt the view that motion merely signifies change of position in the sense that a body is at different positions at different times, then, however strange it may seem, there is nothing *illogical* in asserting that, since at each instant the body is in a unique position, at that instant it is indistinguishable from a stationary body in the same place. A sequence of photographs of an arrow in flight when viewed separately show it in a succession of quasi-stationary states. Owing to the phenomenon of retention of vision, when these pictures are run sufficiently rapidly through a cinema projector the arrow appears to be moving. The difference between the two interpretations depends entirely on how rapidly the photographs succeed each other in our vision, i.e. merely on the temporal relations of one photograph to another. If we look upon this as an exact analogy and regard motion as a phenomenon necessarily referring to different instants, Zeno's paradox collapses, because in the syllogistic form above the phrase 'is moving' strictly means 'is moving at a given instant', and this is meaningless.

Although this argument resolves the paradox of the arrow from a purely logical and semantic point of view, it does not entirely dispose of the question from the point of view of physics and natural philosophy. The definition of motion which we have adopted, however natural it may seem to us, is by no means an obvious one. Indeed, it is highly sophisticated. This is clear historically. For example, in the scholastic discussions of motion in the fourteenth century, Duns Scotus maintained that motion was a *forma fluens*, a continual flow which cannot be divided into successive states,† whereas Gregory of Rimini argued that motion is a *fluxus formae*, or 'flux of form', a continuous series of distinguishable states. Gregory said that during motion the moving body acquired from instant to instant a series of distinct attributes of place (Crombie 1961). In this view he was partially influenced by the nominalist philosopher William of Ockham who denied that movement was due to the real existence of some form or flux of form in the moving body. Instead, it was sufficient to regard the moving body as having at different instants different spatial relationships to some other body. This idea, that *motion is a relation and not a quality*, was also adopted by Nicholas of Autrecourt. His definition of motion has been expressed by Weinberg in the form '"x is moved" means "x is at a at time t, x is separated from point b at time t, x is at b at time t_1 and separated from a"' (Weinberg 1948). This is the conception of motion which we adopted above.

The idea that motion is a relation rather than a quality is a necessary presupposition of the law of inertia, although of course one must be careful not to read a conscious anticipation of the latter into early formulations of the former, particularly as the law equates only states of uniform rectilinear motion with those of rest. According to the principle of relativity of uniform motion in classical mechanics, a body in uniform motion is in all respects identical with itself when at rest; its state of motion in no way modifies it, except as regards position. However, with the rise of the Special Theory of Relativity (which will be discussed in Chapter 5), there came a subtle modification of this conception. Although a body in uniform motion is still regarded as *intrinsically* identical with itself when at rest, it is not so according to the observer with respect to whom it is moving; its relative spatial extension in the direction of motion

† In recent times the best-known advocate of this point of view was the philosopher Henri Bergson. He argued that in discussing motion we must distinguish between the space traversed and the act by which it is traversed. He maintained that the former can be subdivided but the latter cannot, for 'it is quite possible to subdivide an *object*, but not an *act*' (Bergson 1910).

is shortened by a factor depending on its speed. Although this Fitzgerald–Lorentz contraction, as it is called, does not conflict with our argument for rejecting Zeno's paradox of the arrow, it has an unexpected repercussion on the detailed formulation of this paradox, for, instead of a moving body occupying either a place no larger than itself or a place as little larger than itself as we please, according as we consider it at an instant or over an interval of time sufficiently short, we now have to reckon with the possibility of the moving body seeming to occupy a *smaller* place than itself, i.e. a smaller place when moving than when at rest!

Having considered Zeno's two arguments for proving that if time is composed of indivisible instants motion cannot occur, we now turn to his two further arguments by which he claims that motion is equally impossible if time (and, correlatively, space) is divisible *ad infinitum*. Whereas the stadium and the arrow are independent paradoxes, the two which we shall now examine—the Dichotomy and the Achilles—are intimately related: each involves an infinite sequence in time, in the one case a regression into the past and in the other a progression into the future. According to the former, motion can never begin, for before an object can traverse a distance (however small) it must first traverse half that distance, and similarly before it can move through half it must cover a quarter, and so on *ad infinitum*. Therefore, if it is to traverse any distance whatsoever in a finite time it must have completed an infinite number of operations in that time. Zeno rejects this as impossible. By the Achilles paradox, on the other hand, Zeno claims to show that, granted the possibility of motion, 'the slowest runner will never be overtaken by the swiftest, since the pursuer must first reach the point from which the pursued started, and so the slower must always be ahead' (Lee 1936, p. 51). Incidentally, in this concise statement of Aristotle there is no mention of the 'tortoise', but the commentator Simplicius (who lived in the sixth century A.D.) in a fuller account of the paradox wrote, 'The argument is called the Achilles because of the introduction into it of Achilles, who, so the argument says, cannot possibly overtake the tortoise he is pursuing. For the overtaker must, before he overtakes the pursued, first come to the point from which the pursued started. But, during the time taken by the pursuer to reach this point, the pursued advances a certain distance; even if this distance is less than that covered by the pursuer, because the pursued is the slower of the two, yet nonetheless it does advance, for it is not at rest... And so, during every period of time in which the pursuer is covering the distance which the pursued moving at its lower relative speed has already advanced, the pursued advances a yet further distance; for even though this distance decreases at each step, yet, since the pursued is also definitely in motion, it does advance some positive distance. And so by taking distances decreasing in a given proportion *ad infinitum* because of the infinite divisibility of magnitudes, we arrive at the conclusion that not only will Hector never be overtaken by Achilles, but not even the tortoise' (Lee 1936, p. 51).

It is remarkable how views on the significance of this paradox have differed. Peirce, for example, said that 'this silly little catch presents no difficulties at all to a mind adequately trained in mathematics and in logic' (Peirce 1934). On the other hand, Russell has described Zeno's four paradoxes concerning motion as 'all immeasurably subtle and profound', despite the fact that 'the grossness of subsequent philosophers pronounced him to be a mere ingenious juggler, and his arguments to be one and all sophisms' (Russell 1937). In view of the perennial power of the problem to attract minds of the highest quality†, unlike the equally ancient problem of 'squaring the circle' which, in its original form, now only attracts cranks, one cannot help feeling that those who despise it may have missed the essential point. This concerns the limit of a function as its argument 'tends' to a certain fixed value, the argument in this case being the distance described by Achilles and the function being the time. By simple arithmetic we calculate where and when Achilles should catch the tortoise, and we then ask whether the limit of the function, in the sense envisaged by Zeno, is 'attained', i.e. is the corresponding value of the function calculated. As we have already remarked, Newton in Lemma 1 of Book I of the *Principia* seems to have claimed that limits of functions are always 'attained'. The subtleties of the question were not fully apparent to him. As we saw above, the clarification of the problem in the nineteenth century deprived the *mathematical* question of whether a variable 'attains' its limit of all meaning. The temporal concepts which are inevitably associated with terms such as 'tend to' and 'attain' are now explicitly excluded from pure mathematics. The question therefore becomes one of the relevance of the mathematical formulation of the problem to the actual problem of time and motion considered by Zeno.

This was partially understood by Georg Cantor, but not completely; in any case, his primary concern was with the pure mathematical concept of the continuum and not with problems involving time and motion. For centuries thinkers had tried to elucidate the idea of the linear continuum, but no-one before Cantor had succeeded in characterizing it as a linear set with a specific structure. Indeed, it seemed that it must either be thought of as a primitive notion incapable of further logical and mathematical analysis or be based on the extra-logical and 'non-mathematical' concept of time (Fraenkel 1953). In discussing the meaning

† An interesting and unexpected discussion of the Achilles paradox is to be found at the beginning of Chapter XXII, Book Twelve, of Tolstoy's *War and Peace*.

of the concept of the continuum, Cantor came to the conclusion that it should be regarded as a more primitive concept than time or space or any independent variable. He argued that we cannot begin with time or space, for these concepts themselves can only be explained by means of a continuity concept which must be independent of them (Cantor 1883). Acceptance of this point of view does not, however, compel us to agree with him when he goes on to argue that specific consideration of time is not needed at the critical point where the ability of Achilles to overtake the tortoise is under consideration. Taking, for purposes of illustration, the speed of Achilles to be ten times that of the tortoise, the sum of the successive distances that Achilles covers in coming up at the end of each time-interval considered to the place where the tortoise was at the beginning of that interval is given by an infinite series of the form

$$\text{(A)} \qquad 10+1+\tfrac{1}{10}+\tfrac{1}{100}+\tfrac{1}{1000}+\ldots.$$

The sum of the corresponding time-intervals is given by another infinite series of the form

$$\text{(T)} \qquad 1+\tfrac{1}{10}+\tfrac{1}{100}+\tfrac{1}{1000}+\ldots.$$

Cantor argued that, if the A-series converges to a finite limit, then so does the T-series. The convergence of the A-series is independent of temporal considerations. It does converge, and *therefore* Achilles catches the tortoise.

This argument has been accepted by Russell, Whitehead, and Broad, to name only three of the most distinguished. However, whereas Russell paid due tribute to Zeno, Whitehead dismissed the paradox with the derisory comment that Zeno was guilty of a mathematical fallacy owing to his ignorance of infinite numerical series (Whitehead 1929). Broad, while accepting Russell's solution as set out in his *Principles of Mathematics* (1903), felt that it did not meet the specific difficulty 'which many intelligent persons feel' that at no point given by the construction has Achilles reached the tortoise. In a short note, published in 1913, he pointed out that, although the number of points given by the construction is infinite, they do not constitute all finite points of the line and there is nothing in Zeno's argument to show that Achilles and the tortoise do not meet at some point not given by the construction. The sum of the A-series corresponds to such a point (Broad 1913).

Although Broad hoped that this argument would settle the controversy finally, since the matter was urgent 'because it and Zeno's other paradoxes have become the happy hunting-ground of Bergsonians and like contemners of the human intellect', his hope has not yet been fulfilled. Nor is this surprising, because, on the assumption of continuity of space and time which is at issue, Achilles must go through all the points of the construction *before* he can meet the tortoise, and this is the root of the difficulty, for, if Achilles goes through all these points in the way prescribed, he performs an infinite sequence of acts. *The fact that the total time interval allotted to him for this feat has a finite measure does not automatically ensure that he can actually complete the sequence.* As William James (1909) rightly pointed out, the argument (still accepted by Whitehead twenty years later) that because the infinite series of time intervals involved has a finite sum *therefore* Achilles must catch the tortoise 'misses Zeno's point entirely. Zeno would have been perfectly willing to grant that, if the tortoise can be overtaken at all, he can be overtaken in (say) twenty seconds, but he would still have insisted that he can't be overtaken at all'.

To bring out more clearly the fundamental point at issue, let us consider an alternative to the Achilles paradox which is less artifical and more typical of the problems of mathematical physics.† It is the 'paradox of the bouncing ball'. Imagine a ball projected vertically upwards from a horizontal floor with initial velocity v against uniform gravity, the downward acceleration being denoted by g and the coefficient of restitution between the ball and the floor being e. The time t until the first bounce is $2v/g$, and the total time that elapses from initial upward projection until the ball finally comes to rest on the floor, assuming that each bounce is instantaneous, is given by the infinite series

$$t = \frac{2v}{g}(1+e+e^2+\cdots) = \frac{2v}{g}\left(\frac{1}{1-e}\right).$$

If, for example, we take $32\,\text{ft}\,\text{s}^{-2}$ for g, $16\,\text{ft}\,\text{s}^{-1}$ for v, and $e=\frac{3}{4}$, we find that $t=4$ seconds. In these four seconds, the ball is assumed to have bounced an infinite number of times. Is this possible?

To answer this question let us label the succession of events when the ball bounces by the sequence

$$A_0, A_2, A_4, \ldots, A_{2n}, \ldots$$

where A_0 refers to the initial projection of the ball upwards, and the succession of events when the ball has zero velocity, i.e. is at the top of an upward path, by the sequence

$$A_1, A_3, A_5, \ldots, A_{2n+1}, \ldots.$$

† Moreover, it is not open to Bergson's objection that motion, unlike space, cannot be regarded as divisible into a sequence of distinct stages that can be arbitrarily chosen by us, as assumed by Zeno (Bergson 1910, 1911).

The ball thus describes an up-and-down sequence of paths in the successive intervals of time

$$A_0A_1, A_1A_2, A_2A_3, \ldots, A_{2n}A_{2n+1}, A_{2n+1}A_{2n+2}, \ldots.$$

After the elapse of four seconds from when the ball was first projected upwards, the number of paths described by it is infinite and is given by the cardinal number aleph-zero. However, each time the ball hits the ground the number of paths described is even. The distinction between even and odd does not apply to aleph-zero. Hence the only possibility is that the top points, attained at the events A_{2n+1}, eventually coincide with the bottom points, attained at the events A_{2n}, and this is in fact what must happen when the ball is at rest and no longer bouncing. Nevertheless, if we consider the time intervals $A_{2n}A_{2n+1}$ when the ball is going up and $A_{2n+1}A_{2n+2}$ when the ball is going down, however short they become they never actually reduce to zero. Consequently, although eventually (when four seconds have elapsed from the start) we should reach the stage when both the upward and the downward intervals are all of zero duration, nevertheless *this stage cannot be attained by traversing the sequence considered.*

In the physical world bounces are not strictly instantaneous; moreover, the paths considered in the above mathematical analysis would eventually become shorter than the diameters of molecules, atoms, and elementary particles. Instead, in any actual experiment of the type discussed only a finite number of bounces would occur before the ball ceases to move up and down. The mathematical analysis leading to an infinity of bounces is clearly an idealization, but even in the ideal case of instantaneous bounces the resulting paradox reveals a serious difficulty *in principle* of the hypothesis that time is infinitely divisible.

Both in the bouncing ball paradox and in the Achilles paradox we are concerned with an open set of instants preceding, respectively, the instant when the ball comes to rest and the instant when Achilles draws level with the tortoise. In the dichotomy paradox we are concerned with another open set of instants of the form

$$\ldots \frac{1}{2^n}, \ldots, \tfrac{1}{8}, \tfrac{1}{4}, \tfrac{1}{2}, 1,$$

and the relation of this infinite set to the immediately preceding instant, that would be labelled zero. In this case, the difficulty revealed by the paradox is that, if time is infinitely divisible, motion would appear to be impossible because it cannot start! Putting it another way, the dichotomy paradox can be applied to time itself, i.e. to any clock, for it is equally difficult to understand how time can advance from a given 'initial' instant, here labelled zero, into the open set of instants given above. On the other hand, in the Achilles and bouncing ball paradoxes we encounter the corresponding difficulty of understanding how time can advance through an open set of events, or instants, to a given subsequent event, or goal. All the sequences concerned are convergent, and the isolated event in each case is the associated limit, occurring either immediately before or immediately after the open end of the sequence. The difficulty is one of temporal *transition*, which has no analogue in pure mathematics (analysis and geometry), from which all temporal concepts like transition have long been rigorously excluded.

There are two ways in which these paradoxes can be resolved. One is to abandon the concept of 'becoming' or temporal transition† and to regard events as similar to points in space except for the ordering relations 'earlier than', 'later than', and 'simultaneous with', a point of view which will be discussed in Chapter 7. If, however, the concept of temporal transition is retained, the paradoxes can be resolved only by abandoning the assumption of the infinite divisibility of time, and hence of the concept of the point-like instant. Rejection of the latter is in accord with the point of view expressed by A. N. Whitehead (1938): 'There is no nature apart from transition, and there is no transition apart from temporal duration. This is why an instant of time, conceived as a primary simple fact, is nonsense'.

To sum up this discussion of Zeno's paradoxes concerning time and motion: it appears that the two based on the concept of indivisible temporal instants are on a different footing from the other two that depend on the assumption of the infinite divisibility of time. The former do not involve any logical antinomies, although they may seem to conflict with common sense. The latter, however, reveal that definite logical antinomies result if we try to combine the hypothesis of continuity, and hence of infinite divisibility, with that of the transitional nature of time. Consequently, the concept of temporal transition implies that there is a limit to the divisibility of time and that, although the hypothesis that time is truly continuous has definite *mathematical* advantages, it is an *idealization* and not an actual characteristic of physical time.

4.5. Temporal atomicity

For most of us nowadays the concept of temporal atomicity is difficult to imagine because of a natural tendency to believe in the continuity of our

† This is the solution adopted by Grünbaum, who remarks that 'the disavowal of becoming as an attribute of elementary physical events is a necessary condition for the meaningful affirmation of the denseness of physical time' (Grünbaum 1968, p. 51).

own existence. On the other hand, the concept of material atomicity, which signifies that there are absolute minimal parts of matter, is now universally accepted. Similarly, since the rise of the quantum theory, it has become commonplace to regard energy as being ultimately atomic. Whether physical length should also be pictured in the same way is still a moot point, although it would seem to be in general accord with the trend of modern ideas to postulate a lower limit to spatial extension in nature. Closely linked with this concept is the hypothesis of minimal natural processes and changes, according to which no process can occur in less than some atomic unit of time, the *chronon.* Acceptance of the ideas of spatial and temporal atomicity in physics does not, of course, preclude us from applying mathematical concepts of space and time involving numerical continuity in our calculations, but the infinite divisibility associated with these concepts will then be purely mathematical and will not correspond to anything physical.

A consistent treatment of such questions is not easy.† As we have seen, even Zeno implicitly appeals to continuity in his discussion of the stadium paradox, although it is based on the hypothesis of spatial and temporal atomicity. A logically consistent treatment of the problem must be based on the idea of successive discrete states between which no other states can be interpolated. Conversely, in a temporal continuum there would be no successive states, for between any pair we could interpolate an infinity of intermediate states. The two concepts of physical discontinuity and mathematical continuity must therefore be kept sharply distinct when discussing the subtleties implicit in the problems raised by Zeno. It must not be assumed that in some sense the second concept underlies the first, as, for example, so acute a thinker as Professor Grünbaum does implicitly when he maintains that, since space and time are 'extensive magnitudes whose values are given by real numbers' and the set of real numbers in pure mathematics is a continuum, therefore '"atoms" of space and time presuppose logically all the constituent parts of which they can be regarded to be the sum' (Grünbaum 1955). When he goes on to argue that in each of the finite indivisible elements of space the moving body traverses a continuum, and hence a dense infinity of points and instants, he appears to have fallen into the same trap as Zeno. Instead, *on the hypothesis of temporal atomicity*, Achilles will not traverse anything in an indivisible element of time; at each instant he will be in a definite place, at one instant he will be here, at the next instant there, and that is all there is to it.

After three hundred years during which mathematical physics has been

† For useful discussions of this topic see the papers by Hamblin (1972) and Čapek (1972).

dominated by the continuous geometrical time of Galileo, Barrow, and Newton, the idea that time is atomic, or not infinitely divisible, has only recently come to the fore as a daring and sophisticated hypothetical concomitant of recent investigations in the physics of atoms and elementary particles. Yet, in the Middle Ages† the atomicity of time was maintained by various thinkers, notably by the Jewish philosopher Maimonides who lived in the twelfth century and wrote in Arabic. In the most celebrated of his works. *The Guide for the perplexed*, he wrote, 'Time is composed of time-atoms, i.e. of many parts, which on account of their short duration cannot be divided.... An hour is, e.g., divided into sixty minutes, the minute into sixty seconds, the second into sixty parts, and so on; at last after ten or more successive divisions by sixty, time-elements are obtained which are not subjected to division, and in fact are indivisible ...' (Maimonides 1190).

It has been suggested that the Arabic writers of the Middle Ages drew not only on ancient Hellenic and Hellenistic science and philosophy but also on the theories of the Indian philosophers. The Sautrânkitas, a Buddhist sect which originated in the second or first century B.C., formulated a metaphysical theory of the momentariness of all things, according to which everything exists but for a moment and in the next moment is replaced by a facsimile of itself, cinematographically as it were. This theory, which presupposes the resolution of time into 'atoms', was presumably devised in order to account for the perpetual changes that take place in the physical world (Jacobi 1909). Whether or not the origin of Maimonides' concept of time is to be sought so far afield, he was almost certainly quoting from previous writers, for the same idea appears in the *Etymologiae* of Isidore of Seville, who died in A.D. 636, and also about a century later in the *De Divisionibus Tempororum* of the Venerable Bede, who died at Jarrow in the year 735. According to Tannery (1922), the idea that time is composed of individual instants was transmitted to the Middle Ages by Martianus Capella, the Roman author of an encyclopaedic work which he wrote in Carthage about A.D. 470. The same idea occurs again in the ninth book of the popular encyclopaedia *De Proprietatibus Rerum* of Bartholomew the Englishman (a Franciscan monk of the French province who may have been a pupil of Grosseteste at Oxford), written *ca.* 1230–1240. In this work, which although methodical and comprehensive was rather antiquated in outlook, astronomy was presented from the point of view of Macrobius and Martianus Capella. We find that Bartholomew divided the day into 24 hours, each hour into 4 points or 40 moments, the moment into 12 ounces, and the ounce into

† In antiquity, the idea of indivisible atoms of time may have been advocated by Zenocrates, a pupil of Plato (see Sambursky 1959).

47 atoms,† thus giving only 22 560 atoms in an hour, as compared with Maimonides' figure of 60^{10} or more! Tannery has indicated a modern philological relic of this ancient concept of temporal atomicity in the Italian word *attimo*, signifying *instant*.

The theory of Maimonides that time is composed of time atoms and that the universe would exist only for one of them were it not for the continual intervention of God was also held by Descartes. In his view, since a self-conserving being requires nothing but itself in order to exist, self-conservation must be the unique prerogative of God. Consequently, a material body has only the property of spatial extension and no inherent capacity for endurance, God by his continual action recreating the body at each successive instant. For 'the conservation of a substance, in each moment of its duration, requires the same power and act that would be necessary to create it, supposing it were not yet in existence; so that it is manifestly a dictate of the natural light that conservation and creation differ merely in respect of our mode of thinking and not in reality' (Descartes 1641). Hence, Descartes was compelled to postulate that the instants at which creaturely beings exist must be discontinuous, or atomic. Temporal existence was therefore, in his view, like a line composed of separate dots, a repeated alternation of the state of being and the state of non-being. Descartes' contemporary Torricelli also regarded time as 'granular', a succession of discrete segments which he called 'instants' (Westfall 1971).

The idea of temporal atomicity does not necessarily imply that there must be gaps between successive instants. The essential criterion for atomicity is that there is a limit to the division of any duration into constituent parts. In other words, time would be like a line which can be divided into a finite number of adjacent segments with no intervals between them. As previously mentioned, it would mean that, from the temporal aspect, there are minimal processes in nature, no process occurring in less than some shortest unit of time, or *chronon*.

An objection often cited by philosophers to the chronon concept is that no interval of time, however short, is thinkable without its limits (or 'end-points'), and that these must be durationless instants. This objection applies, however, only to our usual way of thinking about time nowadays, i.e. to our abstract spatialization of time, and in particular to pursuing too far the above analogy of chronons with adjacent linear segments. For, if in

† The introduction of the curious factor 47 was presumably connected with the Metonic cycle, named after the astronomer Meton who flourished in Athens in the time of Pericles (about 432 B.C.), which was based on the discovery that 19 solar years are almost equivalent to 235 (i.e. 5×47) lunar months. In order to express the month as an integral number of time atoms, the factor 47 was introduced.

nature time is a succession of minimal processes that cannot, *even in principle*, be subdivided, then these processes *cannot* have beginnings and endings, point-like or otherwise.

Modern speculations concerning the chronon have often been related to the idea of a smallest natural length. One suggestion was that this is given by the effective diameter of the proton and electron, that is to say about 10^{-15} metres. If this were a shortest natural length and we divided it by the fastest possible speed, that of light *in vacuo* (3×10^8 metres per second), the resulting interval of time would be about 10^{-23}–10^{-24} seconds. A time of this order characterizes the normal decay processes of nucleons, i.e. processes involving the so-called 'strong interactions' between protons and neutrons, and also the lifetime of the most transient elementary particles. It is therefore possible that if the chronon exists it is of this order of magnitude.† A purely theoretical unit of length much shorter than 10^{-15} metres can, however, be constructed from the three fundamental constants, G, h, and c (constant of gravitation, Planck's constant, and velocity of light). It is $\sqrt{(Gh/c^3)}$, and is of the order of 10^{-34} metres. If this is divided by the velocity of light, it gives a time of the order of 10^{-43} seconds. It is possible that this time, sometimes called the 'Planck time', might be the chronon.

So far, however, most physicists have felt no need to adopt the concept of temporal atomicity, the quantum concept of *stationary states* being reconciled with a continuous time variable through quantum mechanics. If temporal atomicity is not adopted, the only other alternative to the hypothesis of the infinite divisibility of time is that it is wholly indivisible. This view was enthusiastically adopted by Bergson as a means of escaping the difficulties raised by Zeno concerning both temporal continuity and atomicity without abandoning belief in the reality of time. Bergson's philosophical merit lay not in the particular ideas which he formulated but rather in the originality of his emphasis on those characteristics of time which are truly temporal and not quasi-spatial. Unfortunately, in attacking the geometrization (or spatialization) of time he went too far and argued that, because time is essentially different from space, therefore it is fundamentally irreducible to mathematical terms. This conclusion is quite unfounded. Because time is not spatial it does not follow that it is wholly indivisible and not measurable, any more than temperature or hardness. Until there is general agreement concerning the chronon, the

† As a by-product of their experiment at CERN to measure the lifetime of high-velocity muons (see p. 259), Bailey *et al.* (1977) have calculated that, if there is a fundamental quantum of length (smallest length measurable), it cannot be more than about 10^{-17} metres. However, this does not necessarily mean that the associated estimate for the magnitude of the chronon must be correspondingly reduced, for an independent method of calculating it on the basis of Heisenberg's uncertainty principle leads to a similar value (see p. 283).

concept of mathematical time underlying physical science, including microphysics, will continue to be based on the hypothesis of continuity or infinite divisibility.

4.6. Mathematical time as a type of serial order

Since the mathematical instant of zero duration is the precise analogue of the geometrical point, it cannot be regarded as the theoretical correlate of the 'now' of our sensory awareness which we have seen is definitely not durationless. Moreover, our examination of Zeno's paradoxes has led to the conclusion that for motion to be possible the point-like instant must be regarded as a logical fiction. It follows that we can accept this concept only as a mathematical device which is employed simply as an aid to calculation.

There is nothing unusual in this procedure. Indeed, mathematical physics abounds with instances where the logically fictitious character of the devices employed is far more blatant. Quantities which by their very nature must be discrete are repeatedly denoted by differentials, despite the fact that differentiability presupposes continuity. For example, in statistical mechanics the symbol $\mathrm{d}N$ denotes a number of particles and therefore, strictly speaking, should be integral. Or again, in electrical problems $\mathrm{d}q$ denotes an element of charge, despite our knowledge of the quantum nature of electricity. However objectionable these apparently self-contradictory procedures may seem to the logically minded, they are never queried by physicists who, if challenged, would argue that they are justified because of the smallness of $\mathrm{d}N$ compared with the total number of particles under consideration and of $\mathrm{d}q$ relative to the total charge in question. These are, of course, necessary conditions for the practical efficacy of such devices, but their justification is mathematical facility. Similarly, in the case of time (and also of space) physicists adhere to the hypothesis of continuity because differential equations are usually far more tractable than difference equations (analytically at least; the increasing use of computational techniques and machines which depend on discrete numerical processes may lead to some modification of this consideration). Thus the primary reason why physicists have clung to this hypothesis has been mathematical convenience.

In the last fifty or so years, however, several distinguished philosophers and pure mathematicians have not been content to leave the matter at that. Just as in the previous half-century Dedekind and others had felt dissatisfied with the lack of any logical definition of irrational numbers, such as $\sqrt{2}$, although mathematicians had long operated successfully without any definition, so Whitehead, Russell, and others similarly came to the conclusion that dimensionless points and instants ought to be 'constructed' and not merely postulated. Dedekind defined irrational numbers in terms of rational numbers which in turn are constructed from the positive integers. Whitehead and his colleagues and followers sought to define the dimensionless instants of mathematical time in terms of perceptible events of finite duration and their perceptible temporal relations. Whitehead's method of *extensive abstraction*, as he called it, was, in fact, originally devised for the definition of points in terms of perceptible objects. Its application to the definition of momentary instants was first studied by Norbert Wiener (1914).

Whitehead's method depends on a subtle device which, in view of its many and diverse fruitful applications, must be regarded as one of the most powerful methodological innovations of modern times. It was first exemplified in pure mathematics when the limit of an infinite sequence was actually identified with the sequence itself. Thus, the irrational number $\sqrt{2}$, for example, was identified with the set of all rational numbers whose squares are less than 2. This definition was found to satisfy all the formal requirements expected of $\sqrt{2}$ and so, despite its initial strangeness, came to be generally accepted. Then, Frege in Germany (in 1883) and independently Russell in England (in 1901) arrived at a definition of cardinal number which involved the same basic idea; for example, the number 2 was defined as the class, or set, of all couples, and so on. Similarly, Whitehead defined a point, on the Chinese boxes analogy, as the set of all volumes which enclose the point.† To make this idea fully satisfactory and logically rigorous was by no means easy, but the underlying general principle is no more difficult to accept here than in the previous cases. One particular requirement of the method must be mentioned. It has to be proved that it is not circular, and that in defining a point in this way we are not tacitly appealing to the idea of point in constructing the set of volumes in question, namely all those that in fact enclose the point. Fortunately, it can be shown that a set of volumes converging to a point can be defined in terms of certain relations holding between members of the set without reference to the point. Nevertheless, a continuous space of points cannot be generated in this way solely from sense data, since it is necessary to suppose that there is no lower limit to the size of the volumes considered, although sense data cannot be arbitrarily small.

At first sight we might expect that the definition of durationless instants (which we shall from now on call simply 'instants') would be a simpler problem than the definition of spatial points, because time has only one

† Any such set he called an 'abstractive set'. This is the origin of the term 'extensive abstraction'.

dimension whereas space has three. Nevertheless, progress has been slow. The ultimate objective is to derive from non-durationless events the continuum of instants which is postulated by mathematical physicists and hence to justify this hypothesis, for it is not obvious that the temporal order of physics must inevitably be of this type. This continuum is ordinally linear, or similar to the continuum of real numbers, but ordered sets of much more complicated type are known to exist in pure mathematics. Consequently, the derivation of this linear continuum of time from an acceptable set of axioms relating to perceptible events is not just an abstract academic exercise in logic.

In his Lowell Lectures, Russell (1914) described two different ways of tackling the problems. Instants might be constructed from events (of non-zero duration either by means of temporal enclosure in a manner similar to Whitehead's definition of points in terms of spatial enclosure, or by consideration of temporal overlap. To generate a continuous series of instants by the former method, however, appeal must be made to events of arbitrarily short duration (just as, in the case of space, arbitrarily small volumes had to be introduced), although there is no reason to suppose that such events actually occur. Because it corresponds more closely to our experience—compare the suggestion by William James (referred to on pp. 74–5) that at each moment we experience brain processes which overlap each other—we shall confine attention to the method of overlap.†

Our actual experience of time may be analysed in terms of two fundamental relations: simultaneity and temporal order (or precedence). In respect of these, any event is judged to be either *simultaneous with* or else either *earlier than* or *later than* any other event. Two events which are co-present for some time, however short compared with their respective durations, are said to be simultaneous or to overlap. Hence, of two events which do not overlap one must be earlier than, or precede, the other, and this relation is transitive, namely if one event precedes another and this precedes a third then the first precedes the third. If we begin by considering two simultaneous events, any other event which is simultaneous with both must exist during (but not necessarily only during) the time when all three overlap. Russell therefore *defined* an instant as a set of events, any two of which are simultaneous and such that there is no other event (i.e. one not contained in the set) which is simultaneous with them all. The *existence* of instants so defined was assumed.

An event is said to be 'at' an instant when it is a member of the set defining that instant. Temporal order of instants is then defined by

† In using this method we shall assume that the resulting set of instants is everywhere dense.

stipulating that one is earlier than another if there is some event at the former which is earlier than some event at the latter. If neither instant is earlier than the other, then they are simultaneous (identical).

Having defined instants by this method, we now face the crucial question: does this definition yield the temporal continuum of instants postulated by physicists? This continuum, which will be denoted by the symbol T, has the following formal properties†.

(i) *T is a simply ordered set.* By this we mean that, if p and q are any two instants, then either p is simultaneous with q, or p precedes q, or q precedes p, and these three relations are mutually exclusive. Furthermore, if p precedes q and q precedes another instant r, then p precedes r, and q is said to be between p and r.

(ii) *T is a dense set.* This means that, if p precedes r, there is at least one q which is between p and r.

(iii) *T satisfies Dedekind's postulate*, namely if T_1 and T_2 are any two non-empty parts of T such that every instant of T belongs either to T_1 or T_2 and every instant of T_1 precedes every instant of T_2, then there is at least one instant t such that any instant earlier than t belongs to T_1 and any instant later than t belongs to T_2.

(iv) *T contains a linear framework F which is a denumerable subset such that between any two instants of T there is at least one instant which belongs to F.*

In his 1914 paper, Wiener obtained the necessary conditions under which (i) is satisfied and also considered the conditions leading to (ii). Shortly afterwards, Russell investigated the conditions under which an event is 'at' at least one instant, in particular a first instant. In a later paper (Russell 1936), he showed that for (ii) to be satisfied it is sufficient that (a) no event lasts only for an instant and (b) any two overlapping events have at least one instant in common.

In this later paper, Russell was primarily concerned with the problem of the *existence* of instants. He showed that this could be *deduced* if certain special assumptions were made about events. However, as he pointed out, there is 'no reason, either logical or empirical, for supposing these assumptions to be true'. For example, one of these assumptions was that the whole set of events can be 'well ordered', i.e. every subset has a first member. The others concern the existence of certain kinds of well-ordered series of events. 'But in the absence of such possibilities, I do not know of any way of proving the existence of instants anywhere if it is possible that all the events existing at the beginning of some event (or at the end) continue during a period when others begin and cease (or

† Similar properties characterize the linear continuum of real numbers and of points on a continuous line in geometry.

have previously existed during such a period)'. And he concluded that, if the assumptions stated were not true, then 'instants are only a logical ideal' to which it is possible to approximate indefinitely, but which cannot be reached.

Ten years later the problem of defining instants in terms of durations was re-examined by A. G. Walker (1947). He was primarily concerned with the logical analysis of a continuum of fundamental particles forming the framework of a theoretical world model, but he was led to develop a theory of temporal order which was independent of this particular application. He showed how temporal instants can be defined in terms of durations by means of the idea of section in a *partially* ordered set. Properties of such sets had previously been studied by H. M. Macneille (1937). In Walker's view, the ordering concept of *precedence* associated with the set of events, or durations, should be considered as *partial*, for, given any two members of the set, it may or may not happen that one precedes the other. Walker assumed that if a, b, c, and d are any four durations such that a precedes b, b overlaps c, and c precedes d, then a precedes d. This will be called *Walker's postulate.* It includes the property that if a precedes b, and b precedes d, then a precedes d. An instant is then defined as an ordered set of three classes of durations (A, B, C) constructed in the following way. A is the class of all durations a, and B the class of all durations b such that every a precedes every b. The class C is the set of all durations which belong neither to A nor to B. *These three classes define an instant if any member of C is overlapped by some member of A and by some member of B.* The instant (A, B, C) so defined is said to precede the instant (A', B', C') if the class A' *includes* the class A, where the term *includes* signifies that every member of A also belongs to A' but there are members of A' which are not members of A. If the classes A and A' are identical, then it follows that B' coincides with B and C' with C. In this case, we say that the instant (A', B', C') is simultaneous with the instant (A, B, C).

It can be proved that the set of instants so defined is simply ordered, so that (1) for any two instants p and q either p precedes q, or q precedes p, or p and q are simultaneous and (2) for any three instants p, q, and r, if p precedes q and q precedes r then p precedes r. To establish these results, we first prove the preliminary theorem that, if every duration a of the class A precedes every duration b of the class B and if every duration a' of the class A' precedes every duration b' of the class B', then either A and A' are identical or one includes the other.

Walker established this preliminary theorem by *reductio ad absurdum* of its contrary. For, if the theorem were false, then it would follow that there exists a duration a (belonging to A) which is not a member of A'

and also a duration a' (belonging to A') which is not a member of A. We shall show that if this were so then a would necessarily overlap a'. This follows because it can readily be shown that neither can precede the other. For, if a preceded a' then, since a' precedes every b', it would follow by the basic postulate that a must precede every b'; consequently, a would be a member of A', which contradicts the hypothesis that a is not a member of A'. Similarly, we can show that a' cannot precede a. Hence, a and a' must overlap. We next prove that there is a member b of B such that b and a' overlap. For, since a' is not in A there exists a duration b of class B which either precedes or overlaps a'. The former alternative, however, is impossible, for since a precedes b it would follow from the basic postulate that a must precede a', and we have already seen that this is impossible. Consequently, b must overlap a', and similarly we can prove that a must overlap b'. We thus find that a precedes b which overlaps a' which precedes b'. Hence, by the basic postulate, a must precede b'. However, this contradicts our previous result that a must overlap b'. Consequently, the contrary of the preliminary theorem is false, and so we find that either A and A' are identical or else one includes the other.

We are now able to prove that two instants p and q are either simultaneous or else one precedes the other. For, if $p=(A, B, C)$ and $q=(A', B', C')$, then from the preliminary theorem it follows that either A is identical with A', or else A' includes A, or A includes A'. Consequently, p is either simultaneous with q, or p precedes q, or q precedes p. Finally, if p precedes q and q precedes r, where $r=(A'', B'', C'')$, then A' includes A and A'' includes A', whence A'' includes A; therefore p precedes r.

The same method can be used to prove that the set of instants constructed in this way is *closed*, in the sense that every bounded monotonic sequence in the set has a limit which is a member of the set. For, consider an infinite sequence of instants

$$p_1, p_2, p_3, \ldots, p_n \ldots$$

where $p_n=(A_n, B_n, C_n)$, such that p_1 precedes p_2, p_2 precedes p_3, and so on, and suppose that every member of the sequence precedes some instant $q=(A^*, B^*, C^*)$. Then the limit of the sequence can be shown to be the instant $p=(A, B, C)$, where A is the class of all durations which are members of at least one of

$$A_1, A_2, A_3, \ldots, A_n, \ldots,$$

B is the class of all durations which are common to all of

$$B_1, B_2, B_3, \ldots, B_n, \ldots,$$

and C is the class of all durations which belong neither to A nor to B. The instant p exists, since the classes defining it exist; for example, B exists since it includes B^*. Moreover, if $p' = (A', B', C')$ is any instant preceding p, then A includes A', and, since there is always a finite value of n such that A_n includes A', it follows that p' precedes p_n which precedes p. Consequently, p is the limit of the sequence in the sense that for every p' preceding p there is a member of the sequence p_n such that p_n is intermediate between p' and p. A similar result can be established for an infinite sequence in which p_1 is preceded by p_2, p_2 is preceded by p_3, and so on.

It can also be shown that the original durations correspond to intervals of the simply ordered set of instants constructed from them. An interval is defined as the set of all instants which either are preceded by a given instant p, or precede a given instant q, or both.† A duration c which belongs to the class C is said to contain the instant t, where $t = (A, B, C)$. Suppose that c is preceded by some durations a and itself precedes some durations b. We can then define instants p and q such that if t is contained by c then p precedes t and t precedes q, and conversely if t is preceded by p and precedes q it follows that c contains t. These instants are defined as follows: $p = (A_1, B_1, C_1)$, where A_1 is the class of durations which precede c, and B_1, C_1 are the corresponding B and C classes; $q = (A_2, B_2, C_2)$, where B_2 is the class of durations which are preceded by c, and A_2, C_2 are the corresponding A and C classes. Now, if t is contained by c, it follows that c is a member of C and consequently c is not a member of B. However, c is a member of B_1, and therefore B_1 must include B, whence it follows that p must precede t. Similarly, we can prove that t must precede q.

Conversely, if t is preceded by p and precedes q, then A_2 includes A, A includes A_1, B_1 includes B, and B includes B_2. To prove that c contains t, we must show that c is a member of C. This follows if we can show that c belongs neither to A nor to B. Now, if c were a member of A, then c would precede every duration x contained in B. Hence, x would be a member of B_2, and therefore B_2 would include B, which contradicts our previous condition that B includes B_2. Therefore, c cannot belong to A. Similarly, we can prove that c cannot belong to B. Hence, c must belong to C and consequently c contains t.

Although this method enables us to derive a simply ordered set of instants from a partially ordered set of durations in such a way that the original durations correspond to intervals of the set of instants, it is incapable of yielding the temporal continuum postulated by physicists.

† This definition implies that, in general, intervals are open sets of instants, and does not imply that they are all non-null.

Indeed, the analysis has yielded only conditions (i) and (iii) above, and is compatible with the hypothesis of the chronon, i.e. with the hypothesis that every finite duration contains a finite integral number of instants. This method does show, however, that, *if it is assumed that durations obey Walker's postulate, then they can be regarded as compounded of instants which form a unidimensional sequence satisfying Dedekind's postulate.*

To construct the temporal continuum, further conditions must therefore be imposed on the simply ordered set of instants T obtained from the observer's experience of durations. We shall show† that, if this set has property (ii) above, i.e. is everywhere dense, then it also has property (iv) and hence is isomorphic with the continuum of real numbers, provided we impose a further condition concerning a dense set of clocks.

Let t_0 be any given instant of T which precedes some other instant t_1. Then by (ii) it follows that we can choose an instant t_2 which precedes t_1 and is preceded by t_0. We shall denote this by the symbolism $t_0 < t_2 < t_1$. Similarly, we can choose an instant t_3 such that $t_0 < t_3 < t_2$, and in general an instant t_n which is such that $t_0 < t_n < t_{n-1}$ for all positive integers n. By appealing to our previous result that the set of instants is closed, or alternatively by invoking condition (iii), we can show that any definite sequence t_n of this type must tend to a unique *limit* τ in the set T, in the sense that any instant preceding τ also precedes every t_n and any instant preceded by τ is also preceded by some t_n. For, we can divide T into two non-empty sets T_1 and T_2 according to the following criterion: any instant t of T is a member of T_1 if it precedes every t_n, and it is a member of T_2 if there is an instant t_n which precedes it. It is clear that every instant belongs either to T_1 or to T_2 and that neither of these sets is empty, for t_0 is a member of T_1 and t_1 is a member of T_2. Moreover, since the set T is simply ordered, it immediately follows that every instant of T_1 precedes every instant of T_2. Consequently, by Dedekind's postulate, there is at least one instant τ such that any instant earlier than τ belongs to T_1 and any instant later than τ belongs to T_2. However, since the set T is everywhere dense, it follows that τ must be unique. For, if there were two distinct instants of this type, there would exist an intermediate instant which would belong both to T_1 and to T_2, and this is impossible.

We now introduce the concept of *a monotonically ordered dense set of 'clocks'*. First, by a 'clock' we shall mean a hypothetical 'mechanism' which, if 'set'‡ at *any* given instant x, will 'strike' at a unique later instant y. The functional relationship of these two instants will be denoted by $y = \theta(x)$. We shall also stipulate that if $x_1 < x_2$, then $y_1 < y_2$, where y_1, y_2

† The analysis which follows was suggested to the author by further work of A. G. Walker (1948) and A. A. Robb (1936).

‡ This term is used as an abbreviation for 'wound up.'

are the y instants corresponding to the x instants x_1 and x_2, respectively. Beginning with any instant t, a *temporal chain* of instants can be constructed so that, if the clock is set at t, it strikes at instant $\theta(t)$, if set at $\theta(t)$ it strikes at $\theta\{\theta(t)\}$, and in general if set at $\theta^p(t)$ it strikes at $\theta^{p+1}(t)$, where $\theta^{p+1}(t)=\theta\{\theta^p(t)\}$, for all positive integers p. This chain can be extrapolated backwards: if the clock strikes at t it was set at the instant $\theta^{-1}(t)$, and, in general, if it strikes at $\theta^{-q}(t)$, it was set at $\theta^{-q-1}(t)=\theta^{-1}\{\theta^{-q}(t)\}$, for all positive integers q. With this definition of 'clock', we shall postulate that the chain of instants (of striking) constructed with the aid of a given clock from any given instant t covers† all the instants of T, in the sense that any other instant t^* of T is either an instant of the chain or else is such that an integer p (positive, negative, or zero) can be found so that

$$\theta^{p-1}(t)<t^*<\theta^p(t).$$

A set of clocks is said to be *monotonically ordered* if the order of striking of any pair when set at the same instant x is independent of x, i.e. is always the same. We shall assume that there exists a *dense set* of such clocks, so that given any two instants α and β, where $\alpha<\beta$, there is a member of the set which when set at α strikes at an instant which precedes β. Together, these definitions and postulates define a monotonically ordered continuous set of clocks each of which covers T. Ideally, no assumption concerning 'mechanism' is required, other than the hypothesis that rules exist for isolating subsets of instants forming temporal chains with the properties specified above.

We shall now show that T contains a linear framework F which is a denumerable subset of instants such that between any two instants α and β of T there is at least one instant which belongs to F. We begin by choosing a definite sequence of instants $t_1, t_2, \ldots, t_r, \ldots$, where t_{r+1} precedes t_r, for $r=1, 2, \ldots$, converging to some instant τ. We also select from the set of clocks introduced above a subset, associated with functions θ_r, such that the instant of striking of clock θ_r when set at τ is $\theta_r(\tau)$, where

$$\tau<\theta_r(\tau)<t_r,$$

and clock θ_{r+1} strikes before clock θ_r whenever both are set at the same instant. We also choose a further clock, associated with function θ, such

† This postulate can be regarded as the analogue of the *axiom of Archimedes* in geometry (cf. also Euclid, V, Def. 4, namely, 'Magnitudes are said to have a ratio to one another if they are capable when multiplied of exceeding one another'). Unlike the axiom of Archimedes, however, it contains no reference to measurement and congruence. It merely asserts that the clock concerned keeps going all the time.

that if set at instant α it will strike at $\theta(\alpha)$, where

$$\alpha < \theta(\alpha) < \beta.$$

Since the sequence of instants t_r converges to τ, we can find a particular member of the sequence, say t_n, which precedes $\theta(\tau)$. Consequently, since $\theta_n(\tau) < t_n$, it follows that $\theta_n(\tau) < \theta(\tau)$, and hence $\theta_n(t) < \theta(t)$, for any instant t. It follows that $\theta_n(\alpha) < \theta(\alpha)$, and, since $\theta(\alpha) < \beta$, we deduce that $\theta_n(\alpha) < \beta$. Because the clock θ_n covers all the instants of T, it follows that there must exist an integer p (positive, negative, or zero) such that

$$\theta_n^{p-1}(\tau) \leqslant \alpha < \theta_n^p(\tau),$$

where $\leqslant$ denotes precedence or identity. Since

$$\theta_n^p(\tau) = \theta_n\{(\theta_n^{p-1}(\tau)\} \leqslant \theta_n(\alpha) < \beta,$$

we deduce that

$$\alpha < \theta_n^p(\tau) < \beta.$$

The subset of instants $\theta_n^p(\tau)$ is denumerable, since p and n are both integers and τ is fixed. We have therefore constructed a linear framework of instants F such that between any two instants of T there exists at least one member of F, and F is a denumerable subset of T.

The mathematical physicist's abstract unidimensional continuum of instants which is isomorphic with the mathematician's continuum of real numbers, and hence with the geometer's continuum of points on a line, is thus seen to be a logical construct from the observer's experience of finite overlapping durations, provided it is assumed:

(i) that, if duration a precedes b, b overlaps c and c precedes d, then a precedes d;

(ii) that the set of instants T obtained by selecting appropriate classes of durations, as previously explained, is everywhere dense;

(iii) that subsets of instants can be selected so as to yield a monotonically ordered dense set of 'clocks' covering T.

In constructing this linear continuum of mathematical time we have used only ordinal definitions and postulates and it has not been necessary to appeal to any metrical concepts. Although the concept of the 'temporal chain' has been introduced, no metrical concept of periodicity was associated with it. Consequently, the association of specific instants with particular real numbers remains arbitrary.

To sum up, our object has not been to establish the 'reality' of point-like instants but instead to analyse the principles underlying their theoretical construction from the empirical data of consciousness, and hence to show why we obtain the same arithmetical continuum for mathematical time as we do for the set of points forming a geometrical line. It is in this sense that we have sought to 'justify' Galileo's geometrization of time, although the continuum of durationless instants thus obtained is essentially a mathematical abstraction.

4.7. The measurement of time

The reduction of the overlapping durations of human time to a continuous series of durationless instants isomorphic with the mathematical continuum of real numbers does not immediately lead to any system of measuring time. Because the number of instants in any finite duration is infinite and has the power of the continuum, there is no purely numerical measure of time in terms of the respective numbers of instants in different durations. In the words of Whitehead (1920), a duration has 'temporal thickness' and 'retains within itself the passage of nature', whereas the instant is devoid of temporal extension and is conceived as in itself without transition, temporal transition being the succession of instants. A similar problem arises in the measurement of lengths along a continuous line made up of extensionless points.

The problem of measurement was discussed in detail in the Middle Ages, notably by the Oxford school of natural philosophers beginning with Grosseteste. They realized that, since the Pythagorean attempt to analyse all lengths into a finite number of minimal units had broken down,† any line must be regarded as made up of an infinite number of extensionless points and that to overcome the resultant difficulty of measurement it was necessary to introduce conventional units. According to Walter Burley, 'to this state of incertitude I say that since the *continuum* is divisible to infinity, therefore in a *continuum* there is no primary and unique measure according to Nature, but only according to the institution of men' (Crombie 1953). Commenting on Aristotle's definition of time as the 'number of movement in respect of before and after', Grosseteste maintained that associated with all measurement there was an inevitable inaccuracy which sprang from the nature of things and made all human measurement conventional.‡

The linear ordering of instants implies that we can assign numbers to specific instants so that the relations 'before', 'after', and 'simultaneous

† Because of the discovery of the incommensurability of the ratio of the diagonal and side of a square.

‡ Measurements of time necessarily depend on the movements of 'instruments', for example, mechanical clocks, planets, etc., and every physical object and our observations on it are subject not only to imperfections but, at best, to random statistical fluctuations. The harmonic analysis of the 'simplest model' of the statistical fluctuations of time has been investigated by Wiener and Wintner (1958).

with' are indicated by the numerical relations 'less than', 'greater than', and 'equal to'. However, even when we adhere to these rules, the assignment of particular numbers to particular instants is otherwise arbitrary. For, if integers n, $n+1$ are assigned to instants α and β, where α precedes β, then in principle any number p in the range $n<p<n+1$ can be assigned to any definite instant τ which is later than α and precedes β. Similarly, any number q in the range $n<q<p$ can be assigned to any given instant which is later than α and earlier than τ, and so on. The numerical labels assigned in this way merely indicate relative position in a linearly ordered series. A good illustration of this type of procedure is provided by Mohs' scale used by mineralogists. 'Harder than' is, like temporal precedence, a transitive asymmetrical relation. One mineral is said to be harder than another if it will scratch it. Mohs' scale is based on the following hypotheses: if A will scratch B and B will scratch C, then A will scratch C; if A will scratch B, B will not scratch A; any body which will neither scratch A nor be scratched by A will scratch all the bodies which A scratches and will be scratched by all the bodies which scratch A. Because of these properties, a finite number of minerals can be arranged in order of hardness: to the softest the number 1 can be assigned, to the next softest the number 2, and so on. Thus, the hardness of diamond is represented by 10, and the hardness of ruby by 9. This scale is arbitrary, in the sense that if A is harder than B and B is represented by 9, say, then A could equally well be represented by, say, 11 or 100 or a million as by 10, provided the relative order of numbering of all minerals is maintained.

Consequently, Mohs' scale is purely ordinal and is not a scale of measurement. No numerical operations on the numbers of this scale have any significance. Thus, the difference between the numbers assigned to diamond and to ruby tells us nothing about the 'extent' to which the former is harder than the latter. Similarly, the method described above for assigning ordinal numbers to instants tells us nothing about the durations between different instants, i.e. the extent to which one precedes or is later than another. It is purely a method of dating and not of measuring time. Like Mohs' scale it is qualitative and not quantitative.

When we turn to the problem of measuring time we might expect that a basic principle would be that the measure assigned to a duration compounded of any two successive durations must be equal to the arithmetic sum $x+y$ of the respective measures x and y of the two component durations. In practice, this principle is obeyed, but as it cannot be regarded as automatically applicable to *all* forms of measurement (Einstein's addition law for parallel velocities in the theory of relativity being a well-known exceptional case), it follows that in a fundamental theoretical analysis we ought to treat the question more generally.

We shall therefore begin by assuming that, if we are to employ numbers usefully to measure durations, temporal addition must be both commutative and associative. In other words, we shall assume that the 'sum' of successive durations x and y is the same as that of y and x, and that any duration which is compounded of three successive durations x, y, and z has the same measure irrespective of whether z is 'added' to the temporal 'sum' of x and y or the temporal 'sum' of y and z is added to x. Denoting the temporal 'sum' of x and y by the single-valued function $f(x, y)$, we therefore require that $f\{f(x, y), z\}$ should be symmetrical in x, y, and z. Writing $\theta_y(x)$ for $f(x, y)$ and $\theta_z\{\theta_y(x)\}$ for $f\{f(x, y), z\}$, we deduce that

$$\theta_z\{\theta_y(x)\} = \theta_y\{\theta_z(x)\},$$

i.e. the functional operators θ_y and θ_z are commutative. Since x and y can take all values in the continuum, it can be shown† that, if the addition function is differentiable, it must be expressible in the form

$$f(x, y) = \theta_y(x) = \phi^{-1}\{\phi(x) + \alpha(y)\},$$

where ϕ is a monotonic functional operator independent of x and y. Since $f(x, y)$ is a symmetrical function, we must have $\alpha(y) = \phi(y)$ and hence

$$f(x, y) = \phi^{-1}\{\phi(x) + \phi(y)\}.$$

Consequently, if w is the measure of a duration which is the temporal sum of two durations measured by x and y respectively,

$$\phi(w) = \phi(x) + \phi(y). \tag{4.1}$$

The general conditions that temporal addition should be both commutative and associative therefore imply that, for some monotonic function ϕ, the measures x and y of any two successive durations into which a duration of measure w can be decomposed must obey equation (4.1). This has a very important consequence, for it follows that, even if the originally chosen scale of measurement is not arithmetically additive, it can be 'mapped' on to another scale which is. All that is required is a new scale of time measurement represented symbolically by $x \to X$, where $X = \phi(x)$. Then, if Y and W denote the new measures assigned to the durations y and w according to the former scale, it is clear that $W = X + Y$. Hence, any method of assigning measurements to durations which obeys the commutative and associative laws of addition can, in principle, ultimately be made to yield quantities which obey the ordinary law of

† For this solution of the problem of commutative functional operators, see Whitrow (1935). Further papers by A. G. Walker and others will be found in the same journal for 1946 *et seq.*

arithmetic addition. Moreover, since the only continuous function f satisfying the equation $f(X+Y)=f(X)+f(Y)$ is $f(X)=\lambda X$, where λ is independent of X, it follows that the new scale of measurement is *unique*, except for an arbitrary multiplicative constant.

These results are illustrated by the following example. Suppose we wish to construct a scale of time *ab initio* by counting the numbers of atoms of a radioactive element disintegrating in different intervals of time. Let us suppose that at some instant we know† the total number of these atoms in a given source which contains no other radioactive elements. Let us also suppose that we can detect the disintegration of each of these atoms, so that we can determine the total number of atoms remaining at any instant and the number disintegrating in any time interval. If at the beginning of a particular duration the total number of atoms of the original element is n_0 and the number disintegrating in that duration is δn_0, we could take the proportion $\delta n_0/n_0$ which disintegrate as the measure x of that duration. If in an immediately succeeding duration the number disintegrating is δn_1, then by the same rule we should assign to that duration a measure $y=\delta n_1/n_1$, where $n_1=n_0-\delta n_0$. However, to the total duration formed by combining these two we should have to assign the measure $w=(\delta n_0+\delta n_1)/n_0$, and it is clear that w would be less than the arithmetical sum of x and y. Indeed, it immediately follows by simple algebra that

$$w=x+y-xy. \tag{4.2}$$

This provides a law of addition which is both commutative and associative, the 'sum' of three durations x, y, and z being given by

$$x+y+z-xy-yz-zx+xyz,$$

and so on. Law (4.2) is readily converted into form (4.1) by observing that $1-w=(1-x)(1-y)$, and hence that

$$\log\left(\frac{1}{1-w}\right)=\log\left(\frac{1}{1-x}\right)+\log\left(\frac{1}{1-y}\right).$$

If we choose a new scale of measurement given by‡

$$X=\log\frac{1}{1-x}, \tag{4.3}$$

we arrive at the addition law $W=X+Y$.

From (4.3) we see that $X=\log(n_0/n_1)$. Consequently, if t denotes the new scale of time, and t is taken to be zero when n, the number of atoms

† We are concerned here with a purely theoretical analysis and not with the practical feasibility of the method discussed.

‡ We choose the reciprocals of $1-x$, etc. to give monotonic increasing X, etc.

of the original element, was equal to n_0, it follows that $n = n_0 \exp(-t)$. More generally, in accordance with our previous remark that the final time scale is undetermined to within a multiplicative constant, we write

$$n = n_0 \exp(-\lambda t),$$

and hence deduce that

$$\frac{dn}{dt} = -\lambda n.$$

This equation is identical in form with the well-known Rutherford–Soddy law of radioactive decay. The empirical content of this law is therefore that:

(i) the scale t defined by it coincides, within the limits of the experimental accuracy, with the uniform time of physics as otherwise determined, for example, by astronomical observations;

(ii) with a given choice of temporal unit, the value of λ is the same for all quantities of a given radioactive element, irrespective of temperature, pressure, etc.

Our preference for a law of simple arithmetical addition of time intervals is influenced by the following criterion. As a general rule, physical laws are formulated so as to be independent of the particular times of occurrence of events to which they apply, although our analysis above has been deliberately developed on a more general basis. Therefore, only the differences of the times of events and not the times themselves are regarded as significant. The measurement of time depends on the idea of a standard interval of time, or period, analogous to the idea of standard unit of distance, different units being chosen in practice according to the magnitude of the temporal intervals considered. These are measured in terms of numerical multiples of a unit period, and therefore automatically obey the law of arithmetic addition.

In concluding our account of these theoretical considerations, we should take note of their practical significance. This is clearly illustrated by the part played in the history of time measurement by the mechanical clock. The crucial importance of this invention was due not so much to its accuracy, great as this eventually became, as to its dependence on a periodic instead of a continuous process, in contrast to the sun-dials, water-clocks, and sand-glasses of antiquity, for this dependence on a mechanical motion which repeats itself over and over again led to a more precise concept of a unit of time analogous to a unit of length. Modern chronometry stems from Galileo's discovery of a natural periodic process—the swinging pendulum—which could be conveniently combined with clockwork to record mechanically the number of swings. The first successful pendulum clock was devised by Christiaan Huygens in 1656. It was the first satisfactory machine for the uniform† division of physical time, since it could achieve an accuracy of about ten seconds a day.

The accuracy of any mechanical clock not only depends on its construction but has to be checked by constant reference to some natural clock. Throughout history the ultimate natural standard of time has been derived from astronomical observations. The hour, minute, and second were for long defined as fractions of the period of one rotation of the earth on its axis, but because of unpredictable small variations in the earth's rate of rotation astrononers decided to introduce in 1956 a more accurate unit of time based on the period of the earth's revolution around the sun. Nevertheless, it was still felt desirable to have a more fundamental natural unit than any which could be derived from astronomical observations. Such a unit is given by the frequency of a particular atomic spectral line. Optical lines are not suitable, because although we can measure their wavelengths we have no means of measuring their frequencies directly, but the discovery of spectral lines in the new radio frequency band led Dr. L. Essen of the National Physical Laboratory to develop, between 1955 and 1957, a new method of time measurement of remarkable accuracy (one part in 10^{10}), the caesium atomic clock or frequency standard. In this clock the magnetic field produced by an oscillating electric current was synchronized with certain specific vibrations of caesium-133 atoms. These atoms have a single electron in their outermost electronic shells. The electron can spin in the same sense as the nucleus or in the opposite sense. By supplying a little energy the electron can be induced to change the sense in which it is spinning. When it changes back again the energy given is released and produces a sharp line in the radio range with a frequency of about 9200 MHz, corresponding to a wavelength of about 3 cm. In this way a fundamental time scale wholly independent of, and more accurate than, any astronomical determination of time was obtained. As a result, in 1967 a new definition of the second was introduced; it was defined as the duration of 9 192 631 770 periods of the radiation corresponding to the transition between the two hyperfine levels of the ground state of the caesium-133 atom. In recent years the accuracy of the caesium atomic clock has been further improved to one

† Dependence on a non-uniform natural basis of time measurement long delayed the development of satisfactory clocks. The 'temporal hour' of antiquity was one-twelfth part of the day from sunrise to sunset and so varied throughout the year. The need to show temporal hours was responsible for the great complexity of ancient water-clocks. Despite attempts by astronomy in antiquity to introduce the hour of constant length, it was not generally adopted until the middle of the fourteenth century, with the advent of the striking clock.

part in 10^{12}, corresponding to a clock error of only one second in 30 000 years† (Essen 1973).

References

AITKEN, A. C. (1959). *Listener* **62,** 885.

BAILEY, J. *et al.* (1977). *Nature (Lond.)* **268,** 304.

BARROW, I. (1735). *Lectiones geometricae* (transl. E. Stone). Lect. 1., London.

BERGSON, H. L. (1910). *Time and free will* (transl. F. C. Pogson), p. 112. Allen and Unwin, London.

—— (1911). *Matter and memory* (transl. N. M. Paul and W. Scott Palmer), p. 252. Allen and Unwin, London.

BOYER, C. B. (1949). *The concepts of the calculus.* Hafner, New York.

BROAD, C. D. (1913). *Mind* **22,** 318.

BROUWER, L. E. J. (1913). Intuitionism and formalism. *Bull. am. math. Soc.* **20,** 85.

CAJORI, F. (1919). *A history of mathematics*, p. 27. The Macmillan Company, New York.

—— (1934). *Sir Isaac Newton's mathematical principles*, pp. 38–9. University of California Press, Berkeley, Cal.

CANTOR, G. (1883). *Grundlagen einer allgemeinen Mannigfaltichkeitslehre*, p. 29. Leipzig.

ČAPEK, M. (1972). The fiction of instants. In *The study of time* (ed. J. T. Fraser, F. C. Haber, and G. H. Müller), pp. 332–44. Springer-Verlag, Berlin, Heidelberg, New York.

CLAGETT, M. (1956) *The Liber de Motu* of Gerard of Brussels and the origins of kinematics in the West. *Osiris* **12,** 73–175.

CORNFORD, F. M. (1941). *The Republic of Plato*, p. 238. Clarendon Press, Oxford.

CROMBIE, A. C. (1953). *Robert Grosseteste and the origins of experimental science*, p. 103. Clarendon Press, Oxford.

—— (1961) *Augustine to Galileo*, vol. 2, p. 62. Heinemann, London.

CROSBY, H. L. (1955). *Thomas of Bradwardine, his tractatus de proportionibus.* University of Wisconsin Press, Madison, Wisc.

DESCARTES, R. (1644). *Principia philosophiae*, LVII. Amsterdam.

—— (1641) Meditations on the first philosophy. In *A discourse on method etc.* (transl. A. D. Lindsay, 1912), pp. 107–8. Dent, London.

DUHEM, P. (1909). *Etudes sur Léonard de Vinci*, vol. III, pp. 314–6. Hermann, Paris.

ESSEN, L. (1973). *The measurements of frequency and time interval.* Her Majesty's Stationary Office, London.

FRAENKEL, A. A. (1953). *Abstract set theory*, p. 227. North-Holland, Amsterdam.

GRÜNBAUM, A. (1955). *Sci. Mon.* **81,** 238.

—— (1968). *Modern science and Zeno's paradoxes.* Allen and Unwin, London.

† Since 1973 the accuracy has become even greater and is now equivalent to a clock error of one second in about 100 000 years.

HAMBLIN, C. L. (1972). Instants and intervals. In *The study of time* (ed. J. T. Fraser, F. C. Haber and G. H. Müller), pp. 324–31. Springer-Verlag, Berlin, Heidelberg, New York.

HAMILTON, W. R. (1837). Theory of conjugate functions or algebraic couples, with a preliminary and elementary essay on algebra as the science of pure time. *Trans. roy. Irish Acad.* **17,** 293–422.

HOBBES, T. (1651). *Leviathan*, I, p. 3.

—— (1656). *The English works of Thomas Hobbes of Malmesbury* (ed. W. Molesworth, 1839), vol. VII, p. 283. London.

JACOBI, H. (1909). Atomic theory (Indian). In *The encyclopaedia of religion and ethics* (ed. J. Hastings), vol. 2, p. 202. Clark, Edinburgh.

JAMES, W. (1909). *A pluralistic universe*, p. 229. Longmans, Green, New York.

KANT, I (1783). *Prolegomena* (transl. E. Belfort Bax, 1891), p. 30. George Bell, London.

KOEHLER, O. (1951). *Bull. anim. Behav.* **9,** 41–5.

LEE, H. P. D. (1936). *Zeno of Elea.* Cambridge Univ. Press, Cambridge.

MACH, E. (1883). *The science of mechanics* (transl. T. J. McCormack), 6th edn (1960), p. 174. Open Court, La Salle, Ill.

MACNEILLE, H. M. (1937). *Trans. am. math. Soc.* **90,** 416–60.

MAIER, Anneliese (1939). *Das Problem der Intensiven Grösse in der Scholastik*, Teubner, Leipzig.

MAIMONIDES, M. (1190). *The guide for the perplexed* (transl. M. Friedlander, 1904), p. 121. George Routledge, London.

PEIRCE, C. S. (1934). *Collected papers* (ed. C. S. Hartshorne and P. Weiss), vol. 5. p. 334. Harvard University Press, Cambridge, Mass.

ROBB, A. A. (1936). *Geometry of time and space*, pp. 103–5. Cambridge.

RUSSELL, B. (1914). *Our knowledge of the external world*, Lecture IV. Allen and Unwin, London.

—— (1936). *Proc. Camb. phil. Soc.* **32,** 216–28.

—— (1937). *The principles of mathematics*, p. 347. Allen and Unwin, London.

—— (1946). *History of western philosophy*, p. 833. Allen and Unwin, London.

SAMBURSKY, S. (1959). *Physics of the stoics*, p. 103. Routledge and Kegan Paul, London.

TANNERY, P. (1922). *Mémoires scientifiques* (ed. J. L. Heiberg and H. G. Zeuthen), vol. 5, pp. 346–7. Gauthier-Villars, Paris.

WALKER, A. G. (1947). *Rev. Sci.* No. 3266, 131–4.

—— (1948). *Proc. roy. Soc. Edinburgh* **62,** 319–35.

WEINBERG, J. R. (1948). *Nicolaus of Autrecourt*, p. 168. Princeton University Press, Princeton, N.J.

WESTFALL, R. S. (1971). *Force in Newton's physics*, p. 132. Macdonald, London.

WHITEHEAD, A. N. (1920). *The concept of nature*, p. 56. Cambridge University Press, Cambridge.

—— (1929). *Process and reality*, p. 95. Cambridge University Press, Cambridge.

—— (1938) *Modes of thought*, p. 207. Cambridge University Press, Cambridge.

WHITROW, G. J. (1935). *Q. J. Math.* (*Oxford*) **6,** 249–60.

Wieleitner, H. (1914). Über den Funktionsbegriff und die graphische Darstellung bei Oresme. *Bibl. Math.* **14,** 193–243.
Wiener, N. (1914). *Proc. Camb. phil. Soc.* **17,** 441–9.
—— and Wintner, A. (1958). *Nature* (*Lond.*) **181,** 561–2.
Wilson, C. (1956). *William Heytesbury: medieval logic and the rise of mathematical physics*, p. 21. University of Wisconsin Press, Madison, Wisc.

5

RELATIVISTIC TIME

5.1. Time experienced and time inferred

The linear continuum of instants constructed from overlapping durations by the method discussed in the previous chapter was shown to possess the important property that the initial durations correspond to intervals of this continuum. These durations were associated with the individual time of the sentient observer. On the other hand, the intervals or durations of clock-time depend, as we have seen, on specific external phenomena and so have long been regarded as measures of the universal time of physics. The correlation between these two different kinds of time was based on the principle of simultaneity. According to this hypothesis, to each instant of the abstract linear continuum of time constructed from our individual experience of overlapping durations corresponds a definite state of the physical universe. Depending on the abstract concept of ideally precise observation, there arose the idea that time is a 'moving knife-edge', not restricted to any particular place but covering all places simultaneously.

Despite its general appeal, due perhaps to a deep-rooted natural tendency to correlate the microcosm (oneself) with the macrocosm (the universe), this idea of a single world-wide time order is, nevertheless, a highly sophisticated concept. It is an intellectual hypothesis which far transcends our perception of phenomena, for the order of succession of our perceptions cannot be regarded as identical with the order of succession of the external events which initiate the respective chains of phenomena resulting in these perceptions. First we see the flash of lightning and then we hear the thunder, but we believe that both are manifestations of the same electric discharge. Sometimes we are even led to postulate a complete reversal of the temporal order of external events as compared with the temporal order of their perception. Thus, to quote an example from Sigwart's penetrating discussion of the 'Determination of Time' in his *Logic*, 'when a spectator watches a battalion exercising from a distance he sees the men suddenly moving in concert before he hears the word of command or bugle-call, but from his knowledge of causal connections he is aware that the movements are the result of the command, hence that objectively the latter must have preceded the former' (Sigwart 1895).

It might therefore be thought, as Kant argued, that a universal, or objective, determination of time necessarily depends on the principle of causality and that only in so far as I know that one event is the cause of another can I say with certainty that it precedes the other. However, our belief in any particular causal relation is itself based on the observation that one type of event is regularly observed to precede the other. Hence, our knowledge of causality has its roots in the time sequences observed by us.† Consequently, as Sigwart pointed out, we are confronted by a circular argument, for before we can establish a causal relation 'we must be able to affirm with objective validity that B has followed A; but before we can affirm this with objective validity we must have recognized a causal connection between A and B'.

Instead of accepting Kant's contention, we must recognize that our whole procedure is hypothetical. We begin by *assuming* that the objective order of events is identical with the subjective sequence of our corresponding perceptions. We maintain this hypothesis until it conflicts with the general body of accepted knowledge. Whenever such a conflict arises we make further assumptions concerning the time relations between events and our perceptions, and we accept these assumptions not in response to some *a priori* Kantian principle of causality, nor for purely empirical reasons, but because of the resulting coherence of all the consequences of these hypotheses with knowledge as a whole. In practice, we appeal to the hypothesis that the objective order of two given associated events must be coherent with the previously known order of similar associated events, and to the further hypothesis that any difference between this order and the perceived order can be correlated with some difference in the connections between the respective objects and percepts.

We are therefore faced with two problems: (i) how can we choose a standard pair of events with which to compare the particular events in question; (ii) to what differences of connection are we to ascribe any differences between the time relations of our percepts of these particular events and of the standard pair?

The solution of the first problem is to concentrate on events for which the order of our percepts is automatically the same as that of the events themselves. Such events are those that are 'simultaneous' with the percepts to which they give rise. Ideally, they occur 'at' ourselves, but in practice they occur sufficiently near for the time of transmission of the relevant signal, for example light or sound, to be negligible. Thus, in the case mentioned by Sigwart, the spectator compares the movements of the troops and the bugle call with similarly associated events which he has observed close at hand. Turning to the second problem, we find that the

† For further discussion of this point, see Chapter 7, §2.

discrepancy between his observations of distant troops and of those near by can be explained by attributing a slower speed to sound than to light.

Fundamental in this analysis is the concept of simultaneity in the individual time directly associated with the observer's percepts, the correlation of these percepts with epochs in 'one line of time' being presupposed. As we have previously remarked, this perceptual simultaneity must be distinguished from the precise point-like concept of the instant in mathematical time, but it can be refined and made more precise with instrumental aids. On the other hand, the concept of simultaneity in universal, or world-wide, time is an inferred concept, depending on the relative position of the external event and the mode of connection between it and the observer's perception of it. If the distance of an external event is known and also the velocity of the 'signal' connecting it and the resulting percept, the observer can *calculate* the epoch at which the event occurred and can correlate this with some previous instant of his individual time. However, this calculation will obviously be a distinct operation for each observer, and there is no prior guarantee that it will result in a single objective time sequence of events the same for all observers. Nevertheless, when Sigwart wrote, it was universally accepted by philosophers and scientists as an intuitive, or self-evident, truth that, when we have found the rules according to which the time of perception is determined by the time of the event, all perceived events can be brought into a single time sequence. Despite the clarity of his analysis, Sigwart assumed without question that this method led automatically from the subjective time of the perception to the objective time of the event.

The first to query this conclusion, and to appreciate fully the consequences of rejecting it, was Albert Einstein in his famous paper *On the electrodynamics of moving bodies* (Einstein 1905). Einstein realized that the above method led only to the assignment of a *subjective*, not an objective, time to an external event. Moreover, he not only saw that it was a *hypothesis* to assume that, if they calculate correctly, *all* observers must ascribe the same time to any given event, but he produced cogent reasons why, in general, this hypothesis should be rejected.†

Einstein's theory was based on the assumption that there are *no* instantaneous connections between external events and the observer. In

† Having already come to the conclusion that the measurement of time depends on the idea of simultaneity, Einstein suddenly realized one morning in May 1905 that there was a great gap in the classical treatment of time and that it is not obvious that two events in different places which are simultaneous for one observer must necessarily be simultaneous for another.

view of the crucial importance of this hypothesis, let us review the historical evidence which bears upon it. The fact that sound travels with a finite speed is easily deduced from the time lag of echoes, but the transmission of light was for long believed to be instantaneous. An early exception to this general belief was the Greek philosopher Empedocles (*c.* 490–435 B.C.) who, according to Aristotle, 'was wrong in speaking of light as "travelling" or being at a given moment between the earth and its envelope, its movement being unobservable by us' (Smith 1931). The Greeks tended to believe that vision was due to an emanation from the eye and not from the visible object, and on this hypothesis the famous Hellenistic technologist and inventor of steam engines Hero of Alexandria produced the following empirical demonstration that the velocity of light must be infinite. He pointed out that if you turn your head to the heavens at night keeping your eyes closed and then suddenly open them you will see the stars immediately. Therefore, since no time elapses between the instant of opening the eyes and the instant at which the stars are first seen, light (or vision) must travel instantaneously. The famous Islamic scientist Avicenna (980–1073) believed instead that light is due to the emission of certain particles by the luminous source, and he therefore concluded that its speed must be finite. A similar conclusion was reached by Alhazen (*c.* 965–*c.* 1039) who, in his treatise on optics, maintained that light is a movement and is therefore at one instant in one place and at another instant in another place. Consequently, since it is not in both places at the same time, there must be a lapse of time between the two and hence the transmission cannot be instantaneous (Cohen 1944, p. 9).

Nevertheless, centuries later we find Kepler in his *Dioptrics*, published in 1611, reverting to Aristotle's view and arguing that since light is immaterial it can offer no resistance to the moving force and therefore has an infinite velocity. On the other hand, Galileo in his famous dialogues on mechanics, published in 1638, discussed the question from a far more modern point of view (Drake 1974). The experiment which he proposed was that two persons each equipped with a lighted lantern and a shutter should take up positions several miles apart. As soon as one sees the uncovered light of the other he is to shine his own light. As in the case of his investigation of falling bodies, Galileo's theoretical analysis of this problem far outstripped his experimental technique and performance, and he admitted that he had in fact only tried the experiment at a distance of less than a mile.

The first successful empirical demonstration that the speed of light is finite resulted from the discovery that the satellites of Jupiter—which, incidentally, were first observed by Galileo with his telescope—provided the appropriate regularly covered and uncovered lanterns required to make Galileo's experiment work. The earliest tables of motion of these satellites which were accepted as reasonably reliable by other astronomers were published by G. D. Cassini in 1668. Among those who subsequently investigated the irregularities in the times of the eclipses of Juptier's satellites was the young Danish astronomer Ole Christensen Roemer (1644–1710), who worked on this problem at the Paris Observatory in 1675. In September 1676 he announced to the members of the Académie des Sciences that the eclipse of the innermost satellite expected on 9th November would occur ten minutes later than the time computed from observations of previous eclipses. He explained that this delay arose from the fact that light was not propagated instantaneously, as the astronomers had in effect assumed, but 'gradually', and that the observed time of eclipse depended on the distance of the earth from Jupiter, the total variation of which is equal to the diameter of the earth's orbit about the sun. Roemer's prediction was more or less confirmed and on 21st November he read another paper to the Académie in which he stated that the time required for light to traverse the earth's orbit was about 22 minutes (Cohen 1944, p. 26). The correct value is about $5\frac{1}{2}$ minutes less than this. Indeed, Roemer's numerical result was not confirmed by other investigators. Consequently, his *interpretation* of the observed irregularities was not generally accepted in France, although it was in England, except by Robert Hooke. However, this did not mean that all English astronomers accepted Roemer's value for the speed of light. For example, Newton, in the *Opticks*, published first in 1704, stated in Proposition XI, Book II, Part III, that 'Light is propagated from luminous Bodies in time, and spends about seven or eight minutes of an Hour in passing from the Sun to the Earth'. Whether this revised figure resulted from his own calculations or from those of Halley or Flamsteed we do not know.

Roemer's hypothesis only came to be universally accepted after its independent confirmation by the English astronomer James Bradley in 1728, as a consequence of his attempt to determine the parallaxes (and hence the distances) of the stars by utilizing the motion of the earth in its orbit. He chose a particular star for which he found an annual displacement in a small ellipse, but to his surprise the observed displacements were 90° out of phase from those expected. Observations of other stars revealed similar results. It is well known that Bradley found the explanation for this phenomenon when sailing on the Thames. While observing the combined action of the motions of the wind and ship on the ship's flag, it occurred to him to explain his stellar observations by a combination of the motion of light with that of the earth. In his classic paper, communicated to the Royal Society, he wrote, 'At last I conjectured that all the phenomena hitherto mentioned proceeded from the progressive motion of light and the earth's annual motion in its orbit. For I perceived that, if light propagated in time, the apparent place of a fixed object would not be the same when the eye is at rest as when it is moving in any other direction than that of the line passing through the eye and the object; and that when the eye is moving in different directions, the apparent place of the object would be different' (Bradley 1728). The direction in which a telescope must be pointed at a given star will therefore be determined by combining the velocity vector of the earth with that of the light from the star, and this will vary as the earth's velocity vector changes in direction throughout the year. He deduced from his value of the 'constant of aberration' (which he stated would not occur if the velocity of light were infinite) that the time required for light to travel from the earth to the sun should be 8 minutes 12 seconds, a value much closer to Newton's value than to Roemer's.

Bradley's work led to the final acceptance of the hypothesis of the finite velocity of propagation of light, but not until 1849 was Galileo's experiment at last performed on a purely terrestrial scale by the French physicist Fizeau. In his experiment a beam of light was allowed to pass between the teeth of a rotating toothed wheel and after travelling 8.6 kilometres it was reflected back along the same path. By rotating the wheel sufficiently fast, it could be arranged that the returning light did not pass between the teeth. The technique was improved by Cornu who, in 1874, obtained the value 300 330 kilometres per second for the velocity of light in air. For the velocity of light *in vacuo* the current standard value, recommended by Comité des Poids et Mesures in October 1973, is 299 792 458 metres per second. This is considered to be accurate to within 4 parts in 10^9. Previously the most accurate determination was $299\,792\,500 \pm 100$ metres per second, obtained by K. D. Froome (1958).†

In 1873 Clerk Maxwell published his great treatise on *Electricity and Magnetism* in which he explained that light was a form of electromagnetic radiation, and in 1887 Hertz first succeeded in generating the type of electromagnetic waves which we now call radio waves. In recent years waves belonging to this and other parts of the spectrum have been detected coming from the sun, the Milky Way, and remote objects in extragalactic space. All the available evidence points to the fact that there are no connections between external events and ourselves which are more

† For a useful account of modern work on the determination of this fundamental constant see Sanders (1965).

rapid in their operation than the electromagnetic.† Consequently, all our observations of distant events are associated with some time lag. This means that *the world-as-observed at a given instant of individual time cannot be identified with the world-as-it-is at a definite instant of universal time*, for the more distant an object the greater the time lag between it and the observer. Instead of seeing a succession of spatial states of the universe, we observe a succession of spatial–temporal cross-sections. As Eddington once remarked, 'Time, as we now understand it, was discovered by Roemer', for, as we have slowly come to realize over the intervening centuries, his investigations first revealed a fundamental physical difference between local time and time at a distance.

5.2. The determination of time at a distance (i)

Since the determination of the times of occurrence of distant events is not a simple process of clock reading, we are faced with the following problem: given an observer with a temporal experience and a clock which measures time intervals in this experience, how can he determine the times of distant events? I shall refer to this as *Einstein's problem*, because it was the starting-point of his theory. Clearly, the solution must depend on the nature of the connections between distant events and the observer. If the observer knows the distance of the place where an event occurred and the velocity with which the signal from the event travelled towards him, then from the observed time of reception of the signal he can calculate the time of occurrence of the event. However, velocity is measured in terms of space and time, and so it would seem that we must already know how to measure time at all the places traversed by the signal before an epoch can be correctly assigned to distant events. We therefore appear to be in danger of invoking a *petitio principii*.

This difficulty can be avoided if we *assume* that the velocity of the signal is a universal constant. Provided then that a signal is transmitted from an event to him, the observer only needs to know the distance of the event in order to determine its time of occurrence. In the case of electromagnetic waves, Maxwell's theory yielded a constant which he

† We can, of course, consider higher velocities than c (the speed of electromagnetic waves in free space), for example, in media with refractive index less than unity, or when we take a long rod inclined at a very small angle α with a given straight line and move the rod at right angles to the straight line so that the velocity of the point of intersection exceeds c, although the velocity of the rod does not. *In neither case, however, can we signal with speeds exceeding c.* In the former case, the velocity concerned is the wave velocity, but signals can only be given by pulses, or groups of waves, which travel with the *group velocity*, and Sommerfeld has shown, in a classic investigation, that this is less than c for all media. In the latter case, the velocity concerned can never act as a connection between cause and effect. Hence, *c is the maximum speed of transmission of information.*

identified with the velocity of light in free space. Einstein, in his paper of 1905, assumed that this velocity was a universal constant of nature, the same for all observers associated with inertial frames of reference.† This hypothesis went far beyond the experimental data, for, although the methods of determining the velocity of light discussed above yield concordant results, the empirical information which they provide is limited: Fizeau's method yields a value for this velocity within a comparatively short range on the earth's surface, Bradley's method involves only the velocity of light as it enters the telescope, and Roemer's refers to its average value between Jupiter's orbit and the earth. Einstein was fully aware of the *conventional* nature of time at a distance. Indeed, by his explicit recognition of this fundamental point he opened a new chapter in the history of physics.

In the Special Theory of Relativity, which forms the subject matter of his 1905 paper, it is assumed that light is propagated uniformly in straight lines in Euclidean space. In the General Theory of Relativity, developed a decade later, this condition was restricted to situations where gravitational attraction could be neglected. In the presence of gravitating bodies, the velocity of light is *not* strictly uniform, although the local velocity of light as determined by congruent clocks and measuring rods is everywhere the same. In view of this later development, it becomes clear that Einstein's original hypothesis of 1905 involved three distinct principles:

(i) the *local* speed of light, as measured in his immediate neighbourhood by any observer associated with an inertial frame and equipped with a standard clock and measuring rod, is a universal constant c;

(ii) in so far as the effect of gravitation can be neglected, light is propagated in Euclidean straight lines;

(iii) according to each observer associated with an inertial frame, the speed of light *non-locally* is also the same universal constant c, if gravitation is negligible.

Of these principles, (i) and (ii) are open to empirical test, provided that we have *independently* defined the clock and measuring rod employed, but (iii) functions as a disguised definition enabling the observer to assign epochs to non-local events.

A famous experiment that tests (i) was performed by Michelson and Morley in 1887. With the aid of the interferometer invented by the former, it was found that the average to-and-fro velocity of light over a distance of 22 metres in the laboratory was effectively the same in all

† These are the frames for which Newton's laws of motion are valid. They are in uniform relative motion, including relative rest.

directions.† As for (ii), evidence for the rectilinear propagation of light has long been known, and it was not until the solar eclipse expeditions of 1919 that any astronomical evidence conflicting with this principle was obtained. We now believe that light rays from a distant star suffer a small angular deflection when passing near a centre of gravitational attraction, such as the sun.

Turning to (iii), we are confronted by a proposition of a different character. As a means of defining time at a distance it might be assumed that it is essentially an arbitrary convention chosen solely for its mathematical simplicity, but the mere fact that in General Relativity we are led to relax it shows that more complex issues are involved. In that theory space and time are themselves influenced by gravitation, and the velocity of light (according to a given observer) is not uniform throughout the whole gravitational field of a body. When the localized effects of particular bodies are neglected, theoretical cosmologists consider the universe as a whole to be approximately homogeneous—on a sufficiently grand scale—but observational evidence suggests that the universe may change systematically with time. Hence, it follows that the time taken by light to pass from one given galaxy‡ to another varies according to some

† During the short period of time involved in each experiment it can be assumed that the observer concerned can be identified with a definite inertial frame. The experimental result supports the hypothesis that the velocity of light *in vacuo* is the same with respect to all inertial frames. Michelson and Morley themselves were hoping that the experiment would enable them to determine the earth's motion with respect to the stationary luminiferous ether. In their joint paper (Michelson and Morley 1887) they promised to repeat the experiment at intervals of three months (for a year) to ensure against the possibility that at any one time of the year when the experiment might be performed the earth's orbital motion about the sun might happen to be cancelled out by the sun's motion with respect to the ether. A detailed critical discussion of their experiment and of their failure to repeat it throughout the year has been given by Swenson (1972). The experiment has been frequently repeated since 1887, and the accuracy of the null result for the 'ether drift' was greatly improved in the autumn of 1958 by a different experiment using the maser device for amplifying radiation that had been recently developed by C. H. Townes and his colleagues at Columbia University, New York. The maser has a cavity into which is directed a beam of excited ammonia molecules accelerated to high speed. The molecules give off radio waves, and the measurement of their frequencies provides a precise measurement of the passage of time. The frequencies of two beams oppositely directed east and west and then reversed were compared throughout the day, and the experiment was repeated at intervals throughout a year. Apart from a small magnetic effect due to the earth's magnetism and nearby electrical equipment, there was at no time a frequency shift of more than one-fiftieth of a cycle per second. If the earth's orbital motion had affected the frequencies, there should have been a maximum difference of about 20 cycles per second. This puts an upper limit of some 30 metres per second on the earth's velocity through the ether or about one-thousandth of the earth's orbital speed. This is equivalent to an accuracy for the isotropy of the velocity of light of one part in 10^7, and has been regarded as totally discrediting the fixed ether hypothesis (Cedarholm and Townes 1959).

‡ Strictly speaking, we should speak of clusters of galaxies rather than individual galaxies in this context.

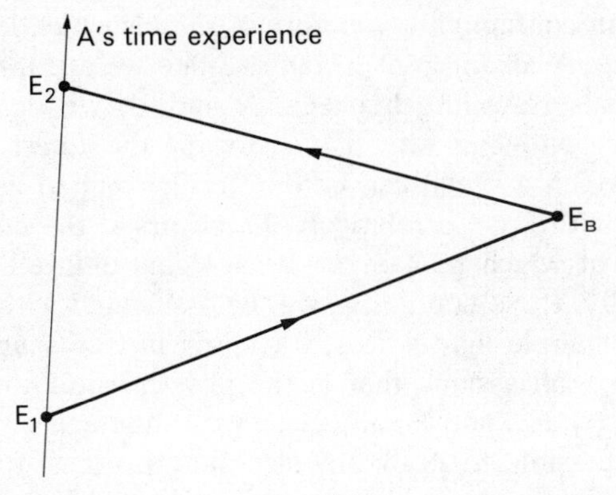

FIG. 5.1

general law. This phenomenon can be expressed either in terms of a continual change of distance between each pair of galaxies, or in terms of a secular variation in the velocity of light.† Therefore, instead of assuming without more ado that time at a distance can be determined, at least in principle, by postulating that the velocity of light at all times and places is a universal constant, I propose to consider the whole question from a more general point of view. We shall find that our analysis will lead to alternative possibilities, with important consequences when we come to study the universe as a whole.

First, we define the clock used by an observer A to record local time as some physical mechanism of the type considered in §7 of Chapter 4. We shall assume that, in principle, there is a precise instant of occurrence of the emission, or reception, of any signal by A, so that he can assign to each such event a definite number in the continuum of real numbers.

We consider next an event E_B which occurs, in general, externally to A. We assume, for the purposes of theoretical analysis, that it occurs at some mechanism B which can reflect instantaneously‡ signals received from A. We shall associate with the distant event E_B two distinct events E_1, E_2 occurring at A, E_1 being the emission of a signal by A which arrives at B at the event E_B, and E_2 being the reception by A of a signal which is emitted by B at the event E_B. We can contemplate a variety of signals which leave A at different instants and yet all arrive at B simultaneously.

† Incidentally, a continual secular variation in the velocity of light as it traverses intergalactic space will imply a corresponding modification of condition (i).

‡ The analysis which follows is based on the postulate that A assigns the same epoch to the departure from B of the return signal as he does to the arrival at B of the outgoing signal. (Later we shall assume that these two events will be simultaneous in the actual experience of an observer located at B.)

Similarly, we can contemplate a variety of signals which leave B together and yet arrive at A at different instants. Since we are interested only in the fastest signals travelling between A and B, we define E_1 and E_2 uniquely in the following way. E_1 occurs at the *latest* time t_1 in A's experience at which a signal can be emitted by him so as to arrive at B simultaneously with E_B. Correlatively, E_2 occurs at the *earliest* time t_2 in A's experience at which he can receive a signal emitted by B when E_B occurs. Physically, these conditions will be associated with electromagnetic waves, for example light waves, travelling in free space.

Although we shall assume that in the physical situation under discussion the events E_1, E_B, and E_2 all occur, we can imagine events E_B which are totally inaccessible to A by any signalling process; for example, any event occurring in a system which is receding from A so fast that no signal from A can overtake it. Again, we can imagine a situation in which E_1 and E_B exist but no finite event E_2 occurs. For the present we shall ignore these possibilities.†

Einstein's problem will now be analysed on the basis of the following axioms:‡

Axiom I. Postulate of causality: $t_2 > t_1$, *unless* E_B *occurs at* A, *in which case* $t_2 = t_1$.

This means that we rule out the possibility that the event E_2 could occur in A's experience *before* the event E_1. In accordance with current practice, we shall call the epochs t_1 and t_2, respectively, the *retarded time* and the *advanced time*, according to the observer A, of the event E_B.

Axiom II. Postulate of spatial isotropy: the epoch t_B *theoretically assigned by A to the event* E_B *is determined by a relation of the form* $t_B = f(t_2, t_1)$, *where f is a single-valued function of* t_2 *and* t_1.

This axiom signifies that the function f, which it is our object to determine, is independent of the spatial orientation of E_B relative to A; for example, if the paths of the signals concerned (which, for convenience, we shall refer to from now on as *light signals*) are rectilinear, then f is independent of the directions of these paths. This may be regarded as a theoretical generalization of the Michelson–Morley result that the average local to-and-fro velocity of light is the same in all directions.

Before formulating the remaining axioms, we introduce the notion of a simply ordered series of events lying on the same *light path*, i.e. the path of a given light-signal.

† The same symbol > can be used to denote the numerical relation of the numbers t_2 and t_1(t_2 greater than t_1) and also the temporal relation of the instants to which they refer (t_2 later than t_1).

‡ See §6 of Chapter 6.

Axiom III. The light paths joining E_1 with E_B and E_B with E_2 are, in general, unique.

We shall say that events E_1, E_B, and E_C 'occur in this order' on the path of a light signal leaving A at event E_1, if a light signal emitted by A at event E_1 can be received by B at event E_B and be instantaneously re-emitted so as to arrive at another mechanism C at the event E_C, coincident† with the arrival at C of a light signal emitted by A at event E_1. A similar definition can be formulated for events on a light path terminating at A at the event E_2. If the advanced time, according to A, of E_C is $t_3(>t_2)$, then in accordance with our rule above (Axiom II) the epoch theoretically assigned by A to E_C will be $t_C = f(t_3, t_1)$.

With regard to these theoretically assigned epochs, we impose the following condition:

Axiom IV. Postulate of time order at a distance: if E_1, E_B, E_C occur in this order on the path of a light signal leaving A at E_1, then $t_C > t_B$; similarly, if E_C, E_B, and E_2 occur in this order on the path of a light signal received by A at E_2, then $t_B > t_C$.

We now introduce three axioms concerning the spatial distance, according to A, described by a light signal between the hypothetical events E_B and E_C on its path:

Axiom V. Postulate of spatial homogeneity: the distance, according to A, described by a light signal between any two events E_B and E_C on its path to which the respective epochs t_B and t_C are theoretically assigned by A is given by $r(E_C, E_B) = \psi(t_C, t_E)$, where ψ is a positive single-valued function of t_C and t_B and is of the same form irrespective of whether the signal is emitted from, or received by, A.

Axiom VI. The addition law for contiguous segments of the same light path:

$$r(E_C, E_B) + r(E_B, E_1) = r(E_C, E_1),$$

where E_1, E_B, E_C occur in this order on a path beginning at E_1, and

$$r(E_2, E_F) + r(E_F, E_D) = r(E_2, E_D),$$

where E_D, E_F, E_2 occur in this order on a path terminating at E_2.

Axiom VII. According to A, the distance described by the light signal emitted by A at E_1 and instantaneously reflected at E_B is equal to the distance described by the signal on its return path terminating at A at E_2.

Axiom V is called 'the postulate of spatial homogeneity', because it signifies that the distance described by a light signal between any two given epochs (theoretically assigned by A) is the same in whatever part of space the light is travelling. Later we shall consider an extension of this

† In accordance with our restriction of attention to 'fastest' signals.

axiom to light paths which neither emanate from nor terminate at the observer.

Axiom VII is equivalent to postulating that *the observer A regards himself as being at rest.* If A were to regard himself as being in motion, this axiom would, in general, be inappropriate, for during the interval of time between E_1 and E_2 he would regard himself as having moved in relation to the spatial position of the event E_B.

From axiom VI it follows that

$$\psi(t_C, t_B) + \psi(t_B, t_1) = \psi(t_C, t_1),$$

where $t_C > t_B > t_1$. Writing $x = t_C$, $y = t_B$, and $a = t_1$, we obtain

$$\psi(x, y) = \psi(x, a) - \psi(y, a).$$

Hence, on substituting $\xi(x)$ for $\psi(x, a)$ and $\xi(y)$ for $\psi(y, a)$, we deduce that ψ, which must be positive, is of the form

$$\psi(x, y) = \xi(x) - \xi(y),$$

where $x > y$, and ξ is any single-valued monotonically increasing function† of its argument. Combining this result with axiom VII, it follows that

$$r(E_B, E_1) = \xi(t_B) - \xi(t_1) = \xi(t_2) - \xi(t_B). \tag{5.1}$$

Consequently, writing r for $r(E_B, E_1)$ and t for t_B, we deduce that

$$r = \tfrac{1}{2}\{\xi(t_2) - \xi(t_1)\} \tag{5.2}$$

and

$$\xi(t) = \tfrac{1}{2}\{\xi(t_2) + \xi(t_1)\}. \tag{5.3}$$

Since $\xi(t)$ is a monotonically increasing function of t, it possesses a unique inverse $\xi^{-1}(t)$, and so we can rewrite equation (5.3) in the form

$$t = \xi^{-1}[\tfrac{1}{2}\{\xi(t_2) + \xi(t_1)\}]. \tag{5.3'}$$

Formulae (5.2) and (5.3′) provide general criteria, in accordance with our axioms, for assigning an epoch t and a distance r to any event in terms of its retarded time t_1 and advanced time t_2.

It has often been remarked by writers on the foundations of relativity that, in our notation, the time t theoretically assigned to a distant event ought to be such that $t_1 < t < t_2$, so that

$$t = t_1 + \epsilon(t_2 - t_1), \tag{5.4}$$

where $0 < \epsilon < 1$, but usually no more is said than that Einstein's condition $\epsilon = \frac{1}{2}$ is the simplest rule to adopt (Reichenbach 1958, Törnebohm 1952).

† We shall assume that this function is differentiable.

Since $\xi(t)$ is monotonically increasing, it is immediately seen that the t defined in terms of t_1, t_2 by equation (5.3) automatically satisfies the condition $t_1 < t < t_2$, irrespective of the particular form of ξ. Equation (5.3), however, is much more informative than equation (5.4). It has an interesting interpretation in terms of the theory of means and convex functions, for it defines a general mean between t_1 and t_2 which includes the familiar arithmetic, geometric, and harmonic means as particular examples arising when $\xi(t)$ is t, $\log t$, and $-1/t$ respectively (Hardy *et al.* 1934).

By inspection, we see that the addition of an arbitrary constant μ to the function ξ makes no difference to equations (5.2) and (5.3). More generally, we can show that the general form of ξ which yields the same value of t for any given values of t_1, t_2 is expressible by the relation

$$\xi = \lambda\xi_0 + \mu, \tag{5.5}$$

where ξ_0 is any particular form of ξ yielding this value of t and λ is an arbitrary positive constant. For, if ξ and ξ_0 both satisfy (5.3) for the same t_1, t_2, and t, it follows, on writing $F = \xi\xi_0^{-1}$, $t_1 = \xi_0^{-1}(x)$ and $t_2 = \xi_0^{-1}(y)$, that

$$F\left(\frac{x+y}{2}\right) = \tfrac{1}{2}\{F(x) + F(y)\}.$$

Since ξ and ξ_0 are continuous, F must also be continuous. It is easily seen, for example graphically, that F must be a linear function, and so equation (5.5) follows. (Corresponding to the change from ξ_0 to ξ, there is a change of r to λr, but all *ratios* of distances remain the same.)

The particular form $\xi(t) = t$ gives the arithmetic mean $t_a = \frac{1}{2}(t_2 + t_1)$. In the general case it is natural to compare t with t_a. If $t > t_a$ for all t_1, t_2, then

$$\xi\left(\frac{t_2+t_1}{2}\right) < \tfrac{1}{2}\{\xi(t_2) + \xi(t_1)\},$$

and so the function $\xi(t)$ must be convex.† Similarly, if $t < t_a$ for all t_1, t_2, the inverse function $\xi^{-1}(t)$ must be convex. More generally, the necessary and sufficient condition that the value of t given by a function ξ should always exceed (be later than) that given by another function ξ^* for the same t_1, t_2 is that the function $\xi\xi^{*-1}$ should be convex.

The function $\xi(t)$ has a simple physical interpretation. From equations (5.1), on replacing t_B by t, we deduce that the velocity of a light signal according to A is given by $\pm\xi'(t)$, where $\xi'(t)$ denotes the derivative of

† It is well known that, if a function possesses a second derivative, a necessary and sufficient condition for the function to be convex is that its first derivative should be a non-decreasing function of its argument.

$\xi(t)$, the plus sign referring to an outgoing signal and the minus sign to an incoming signal. We observe that the speed of a light signal is the same at all points at the same theoretical epoch t. So far we have confined our attention to signals travelling radially according to A, but we can extend axiom V to any two events which do not necessarily lie on the same light path through A, and hence it follows that A will assign the same speed $\xi'(t)$ to any light signal passing any point B in any direction at epoch t, according to A. The particular case in which $\xi'(t)$ is a constant, corresponding to $t=\frac{1}{2}(t_2+t_1)$, was that originally adopted by Einstein.

5.3 The determination of time at a distance (ii)

So far we have derived a more general† solution of Einstein's problem than that assumed by Einstein himself in developing the Special Theory of Relativity. We shall return to this more general solution in Chapter 6, but for the present we shall consider the effect of imposing the following additional axiom:

Axiom VIII. The time interval (t_C-t_B) *between the epochs* t_B, t_C, *theoretically assigned by A to any two events* E_B, E_C, *is independent of the choice of time-zero on A's clock.*

This signifies that when the clock kept by A is readjusted by change of zero so that $t_1 \to t+\alpha$, $t_2 \to t_2+\alpha$, then t given by equation (5.3′), namely,

$$t=\xi^{-1}[\tfrac{1}{2}\{\xi(t_2)+\xi(t_1)\}],$$

is also subject to the transformation $t \to t+\alpha$. Hence, for all α,

$$\xi(t+\alpha)=\tfrac{1}{2}\{\xi(t_2+\alpha)+\xi(t_1+\alpha)\}. \tag{5.6}$$

On comparing equations (5.3) and (5.6), it follows from condition (5.5) that

$$\xi(t+\alpha)=\lambda(\alpha)\xi(t)+\mu(\alpha), \tag{5.7}$$

where $\lambda(\alpha)$ and $\mu(\alpha)$ are independent of t, and $\lambda(\alpha)\neq 0$.

We have already seen that there is no change in (5.2) or (5.3) following the addition of an arbitrary constant to $\xi(t)$. We are therefore free to take

† Still more general solutions arise if we relax the homogeneity postulate (Axiom V). It is interesting to note, however, that equation (5.3) for time at a distance stills holds when the radial signal velocity is given by a function of the form $\xi'(t)/\phi'(r)$, although in equation (5.2) the symbol r is now replaced by $\phi(r)$, the integral with respect to r of $\phi'(r)$. The particular case arising when $\xi'(t)=c$ and $\phi'(r)=(1-2Gm/c^2r)^{-1}$ gives the radial velocity of light corresponding to the well-known Schwarzschild metric of General Relativity, G being the constant of gravitation, m the central mass, and c the limiting velocity of light at infinity. (In General Relativity, c is the limiting (maximum) velocity of light at places 'infinitely' distant from the sources of gravitational fields.)

$\xi(0)=0$, if we wish. With this condition we find that equation (5.7) gives $\xi(\alpha)=\mu(\alpha)$, and hence this equation can be replaced by

$$\xi(t+\alpha)=\lambda(\alpha)\xi(t)+\xi(\alpha). \tag{5.8}$$

On interchanging t and a and subtracting, we see that

$$\frac{\lambda(t)-1}{\xi(t)}=\frac{\lambda(\alpha)-1}{\xi(\alpha)}.$$

Consequently, since t and α are independent, it follows that

$$\lambda(t)=1+a\xi(t), \tag{5.9}$$

where a is independent of t.

If $a=0$, then equation (5.8) becomes

$$\xi(t+\alpha)=\xi(t)+\xi(\alpha).$$

This is Cauchy's functional equation, the only continuous solution of which is

$$\xi(t)=ct, \tag{5.10}$$

where c is a constant. If, however, $a\neq 0$, then on substituting (5.9) in (5.8) we find that

$$\lambda(t+\alpha)=\lambda(t)\lambda(\alpha).$$

This equation is readily reduced to Cauchy form and has the solution

$$\lambda(t)=\exp(kt),$$

where k is a constant. Consequently, we deduce from (5.9) that

$$\xi(t)=\frac{\exp(kt)-1}{a}, \tag{5.11}$$

which gives (5.10) when $a=k/c$ and $k\to 0$.

Omitting the additive constant in (5.11), which in general is irrelevant, and replacing $1/a$ by r_0 and k by $1/t_0$ we obtain $\xi(t)=r_0\exp(t/t_0)$, whence (5.2) and (5.3) give

$$r=\tfrac{1}{2}r_0\{\exp(t_2/t_0)-\exp(t_1/t_0)\} \tag{5.12}$$

and

$$\exp(t/t_0)=\tfrac{1}{2}\{\exp(t_2/t_0)+\exp(t_1/t_0)\}. \tag{5.13}$$

We can immediately verify that (5.13) obeys axiom VIII. On the other hand, a change in the choice of time zero has the effect of multiplying all distances r, given by (5.12), by the same scale factor. All *ratios* of such distances, however, are unaltered by a change in the zero of time.†

Next, we consider the effect of changing the unit of time measurement. We introduce

Axiom IX. All times theoretically assigned by A to distant events are multiplied by the same scale factor when the unit of time of A's clock is changed arbitrarily.

This axiom will be interpreted as signifying that *the formula expressing t as a function of t_1 and t_2 contains no dimensional constant.*‡ Since $t \to Kt$ whenever $t_1 \to Kt_1$ and $t_2 \to Kt_2$, we deduce by an argument similar to the one following axiom VIII that, for all $K>0$,

$$\xi(Kt) = \lambda(K)\xi(t) + \mu(K),$$

where $\lambda(K)$, $\mu(K)$ are functions of K and $\lambda(K) \neq 0$. By taking $\xi(t)$ as a function of $\log t$, we find that the general form of ξ is

$$\xi(t) = \frac{\exp(k \log t) - 1}{a} = \frac{t^k - 1}{a}, \tag{5.14}$$

to within an arbitrary additive constant. Incidentally, we note that when $a = k/b$ and $k \to 0$ we have $\xi(t)$ proportional to $\log t$. Writing $a = 1/\alpha$, formula (5.14) gives§

$$r = \tfrac{1}{2}\alpha(t_2^k - t_1^k) \tag{5.15}$$

and

$$t^k = \tfrac{1}{2}(t_2^k + t_1^k). \tag{5.16}$$

The singular case, $\xi(t)$ proportional to $\log t$, gives

$$r = \tfrac{1}{2}r_0 \log(t_2/t_1) \tag{5.17}$$

† It is not difficult to prove that equation (5.11) gives the only form of $\xi(t)$ for which this is true, for ξ must satisfy a functional equation of the type

$$\xi(t_2+\alpha) - \xi(t_1+\alpha) = K(\alpha)\{\xi(t_2) - \xi(t_1)\}$$

for all admissible t_1, t_2, and α, where $K(\alpha)>0$ and is independent of t_1 and t_2. Since

$$\xi(t_2+\alpha) - K(\alpha)\xi(t_2) = \xi(t_1+\alpha) - K(\alpha)\xi(t_1),$$

it follows that, for all t,

$$\xi(t+\alpha) - K(\alpha)\xi(t) = L(\alpha),$$

where $L(\alpha)$ is independent of t. On comparison with (5.7) we see that $\xi(t)$ must be of the form (5.11).

‡ The associated formulae for distance and light velocity will *necessarily* involve some dimensional constant.

§ We observe that formulae (5.16) and (5.18), unlike formula (5.13), contain no constant of temporal dimension.

and

$$t = \surd(t_2 t_1), \tag{5.18}$$

the velocity of light being inversely proportional to t. In this case all epochs must have the same sign. The epoch $t = 0$ is singular, and at this instant the velocity of light would be infinite.

We find that the only form of $\xi(t)$ compatible with both axiom VIII and axiom IX is that given by (5.10). We therefore conclude that the only solution† of Einstein's problem compatible with all the axioms above is the one which Einstein himself adopted in formulating his Special Theory of Relativity, namely

$$t = \tfrac{1}{2}(t_2 + t_1). \tag{5.19}$$

This particular rule for assigning times to distant events is correlated with the distance law

$$r = \tfrac{1}{2}c(t_2 - t_1) \tag{5.20}$$

associated with a constant velocity of light c.

It will have been observed that no appeal has been made in this analysis to the notion of the ideal rigid body presupposed by the classical theory of spatial measurement. Nevertheless, we have not only obtained a rule for assigning times to distant events but also a rule of spatial measurement (particularly if we invoke the extended form of axiom V to determine the lengths of all light paths in free space). In his paper of 1905, Einstein explicitly retained the rigid-body concept, but in a paper on the dynamics of the electron, published in the following year, Poincaré (1906) pointed out that if we postulated the existence of a finite invariant light-signal speed c we could dispense with the concept of the rigid body as the foundation of spatial measurement. Instead, all spatial measurements could be defined in terms of appropriate time readings, the distance travelled by a light signal according to an observer A between two epochs being defined as the product of c and the numerical difference of the epochs. Consequently, distances in free space would be equal if traversed by light (or other electromagnetic waves) in equal times.

Although on the intermediate scale of classical laboratory physics the rigid-body concept plays an important role (at least implicitly, for example in measurements made with a graduated ruler), on the atomic and sub-atomic scale and also on the astronomical and cosmological scale this

† The solution (5.13) associated with a velocity of light proportional to $\exp(kt)$ applies in the case of the de Sitter universe (see p. 285).

concept has no direct relevance. On these scales we are obliged to appeal to the properties of electromagnetic waves rather than to those of rigid bodies. Nevertheless, it is sometimes argued that length measurement by means of a ruler is necessarily fundamental in physics, because it is the only measurement which does not involve reference to other types of physical magnitude. Despite its superficial plausibility this argument is invalid, for the measurement of length involves at least two instants of time: the instant at which the observer notes that one mark of the scale coincides with one end of the distance to be measured and the other instant at which he notes that another mark on the scale coincides with the other end point. Moreover, not only is the precise control of a standard metre rod or yardstick a difficult problem in practice, involving the most delicate regulation of factors such as temperature, but the underlying theoretical concept is by no means simple either. The naïve classical idea of absolute rigidity must be abandoned, since it implies that a disturbance can be propagated with infinite velocity through a body, and this is incompatible with our principle that there is a finite upper limit to the speed of transmission of a signal. Thus the apparent primacy of spatial measurement becomes less obvious the more closely it is examined.†

On the other hand, the measurement of local time, although in practice often recorded spatially by means of a pointer revolving on a dial, is not necessarily dependent on spatial measurement. As we have already pointed out, the earliest clocks recorded time solely by means of a striking mechanism; the latest and most accurate types of clock depend on the natural frequencies of atomic and molecular systems, and epochs are recorded by numerical counters. Indeed, whereas any measurement of distance necessarily involves some reference to time and is based on *two* distinct simultaneity judgments, the assignment of an epoch to an event in the observer's immediate experience depends only on *one* simultaneity judgment, for example the coincidence of the event with a particular stroke of the clock. Simultaneity judgments at the observer are primary data of physical measurement. In practice, this idea has been combined with electromagnetic signalling in the techniques of interferometry and radar to determine distances by means of reflected signals. In recent years

† Indeed, as Bondi (1959) has pointed out the size of our measuring rods is determined by atomic interactions which are fully characterized by atomic *frequencies*, in accordance with the fundamental rule $E = h\nu$, where E denotes energy, ν frequency, and h Planck's constant. Thus, only time standards are true primary standards, and units of length are determined from them by use of c. For example, the wavelength of light (and other forms of electromagnetic radiation) of given frequency ν is the distance c/ν travelled in one cycle at speed c.

the radar method has been employed in geodesy, notably by E. Bergstrand (see Sanders 1965), and has even been used to measure astronomical distances.†

Acceptance of the principle that time readings can be treated as fundamental and spatial measurements as subsidiary does not, however, necessarily imply universal acceptance of Einstein's rule for assigning epochs to events at a distance. As we have seen, other rules can be formulated and may be more appropriate in certain contexts. Nevertheless, Einstein's is the simplest rule of its kind. It is independent of spatial location and orientation, and does not depend on any particular choice of time origin nor involve any constant of temporal dimension. However, before considering its application to more than one observer and the problem of correlating clocks in different places, we must refer to A. A. Robb's categorical rejection of the basic idea that an observer can assign any epoch to a distant event.

Although Robb (1921) agreed with Einstein that 'the present instant, properly speaking, does not extend beyond here', so that 'the only really simultaneous events are events which occur at the same place', he was more ruthless in jettisoning the classical conception of world-wide simultaneity *in toto*, since in his view 'there is no identity of instants at different places at all'. He based his theory of time and space on the concept of one instant being *after* another, and maintained that the abstract power of anyone or anything at a particular instant at A to produce an effect at some other particular instant at B is not merely a sufficient but also a *necessary* condition that the instant at B be after the instant at A. Similarly, for the instant at B to be *before* some instant at A it is essential that some influence at the instant at B can produce an effect at the instant at A. Consequently, referring to the situation illustrated by Fig. 5.1, since no physical influence or entity which leaves A *after* E_1 can arrive at B at the event E_B nor can any influence or entity which leaves B at E_B arrive at A *before* E_2, Robb argued that the range of instants at A between E_1

† The determination of the respective time lags between outgoing and returning radar signals to Venus and Mercury has led to remarkable precision in the calculation of the 'astronomical unit' or mean earth–sun distance. The analysis of 5 years of radar data has led to the value of this constant being given in light seconds as $\tau = 499{\cdot}004780 \pm 0{\cdot}000001$, which is more than 30 000 times as precise as the best previous value based on visual observations. It corresponds to an accuracy within 300 metres of a distance that is very nearly 150 million kilometres! (Ash *et al.* 1971).

Incidentally, it may be mentioned that the radar method for determining distance seems to be used by bats. It has been found by Galambos and Griffin, in America, that bats emit supersonic vibrations in short pulses, the time taken for a pulse of vibrations to return to the bat providing an estimate of the distance of the reflecting surface. Echo location by sound, or *sonar*, is also used by birds, fish, and other animals, and by man for underwater location, etc. (Griffin 1958).

and E_2 can be in no temporal relationship whatsoever with E_B. Thus, in Fizeau's experiment, any instant at the sending apparatus which is after the instant of departure of a flash of light and before the instant of its return is neither before nor after the instant of reflection at the distant mirror. In particular, there could at most be one instant at A which could be correlated chronologically with E_B, but 'we have no means of saying which instant it is'.

To take account of this lack of any correlation between E_B and events at A between E_1 and E_2, Robb introduced the idea that 'elements of time', i.e. events, form a system in what he called 'conical order'. This was defined in a purely formal axiomatic way as an extension of the simple linear order of events actually experienced by an observer. It can be illustrated by means of ordinary geometric cones, as shown in Fig. 5.2.

In relation to any given event E, all other events can be depicted in a four-dimensional diagram as lying inside, on, or outside a double-sheeted hypercone (α and β) with vertex at E. The generators of the cone represent the light paths through E. Events in β are *before* E, those in α are *after E*, and events outside both α and β have no temporal relation to E. No event which is not depicted as coincident with E can be simultaneous with it.†

Robb was careful to point out that he was attacking not Einstein's mathematics but his philosophy. He complained that, in effect, Einstein was using the term 'simultaneity' in two distinct senses. Observed simultaneity, i.e. the perception by A that one event in his experience is simultaneous with another, is an inescapable fact, unlike the definition of simultaneity for a distant event and one in A's experience. According to Robb (1936) in the one case the word 'is employed correctly to describe something absolute while in the other it would be used to describe a mere convention', and, moreover, a convention dependent on the assumption that the observer can regard himself as being at rest.

Granted, however, that events at A after E_1 and before E_2 are in an

† Robb neatly epitomized this principle in the following sequence of quotations:

> The Bird of Time has but a little way
> To fly—and Lo! the Bird is on the Wing.
>
> *Omar Khayyam*
>
> I could not have been in two places at once
> unless I were a bird.
>
> *Sir Boyle Roche*
>
> Contrary to the view so generally held; not
> even 'the Bird of Time' can be in two places
> at once.
>
> *A. A. Robb*

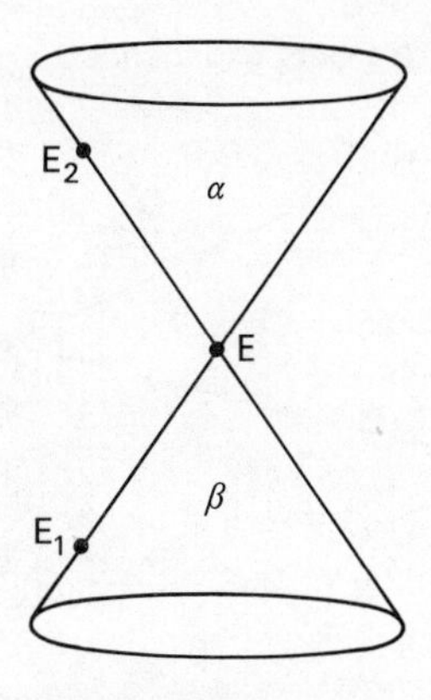

FIG. 5.2

empirically undetermined order with respect to event E_B at B, must we accept Robb's contention that Einstein was mistaken in allowing A to assign a *theoretical* epoch to E_B? In other words, if we reject the classical doctrine of time which stipulates that there *must* be a unique event at A which is absolutely simultaneous with E_B, does it follow that Einstein ought not to have ascribed a definite conventional system of time relations (earlier than, simultaneous with, and later than) between E_B and all events at A? The function of convention in the construction of theories is descriptive simplicity, and it must be admitted that Einstein's Special Theory of Relativity is simpler than Robb's alternative.† However, that is not all. As we have seen, Einstein's conventional rule by which A assigns a theoretical epoch to E_B is not a 'mere' convention in the sense of being wholly arbitrary, for, although it is a convention in so far as it is freely chosen and not imposed upon us, it can be isolated uniquely from other admissible rules by means of the axioms stated above. With all due respect to Robb, the essential question is not the conceptual legitimacy of Einstein's convention but its practical scope, i.e. the range of physical contexts to which it can be most usefully applied.

5.4. The correlation of time perspectives

So far we have considered only a single observer A. Unlike Frank and Rothe (1911), Whitehead (1919, pp. 147–60), and others who sought to

† This is no reflection on the magnificent rigour of Robb's analysis. Indeed, he can be regarded as having done for the theory of time relations what Euclid did long ago for the theory of space relations. A more recent and no less thorough axiomatic development of special relativity, which, unlike Robb's, is based on the kinematics of light signals and inertial particles, has been given by J. W. Schutz (1973). It owes much to previous work by A. G. Walker (1948, 1959).

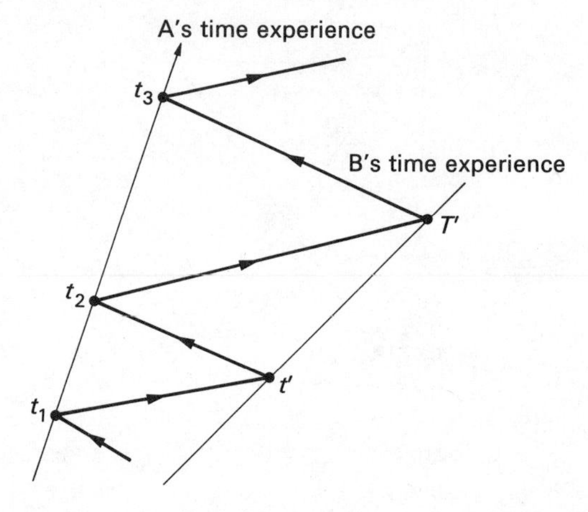

FIG. 5.3

deduce the existence of a finite universal velocity from more primitive postulates, we have not found it necessary to consider the correlation of the space and time coordinates assigned to a distant event by different observers. Although this presented no special difficulty for the classical Newtonian physicist who believed in an absolute world-wide simultaneity and an absolute physical space governed by the laws of Euclidean geometry, as soon as these assumptions were abandoned the problem had to be re-examined. It is now generally recognized that the most satisfactory method of solution is to consider first the correlation of two observers' clocks by means of the same experiment in light signalling that we introduced above (pp. 233).

There we considered the assignment by A of times to events occurring at B. As we have seen, Einstein's solution was based on his postulate that the velocity of light according to A is a universal constant, independent of position and direction of propagation. We must now consider the correlation of this theoretical time assigned by A to an event at B with the empirical epoch t' which would actually be recorded on a clock placed at B. To make the problem precise we postulate that B is now an observer 'similar' to A. In particular, this implies that B carries a clock 'similar' to the one carried by A. For example, if A carries a particular type of atomic or molecular clock, we assume that B carries another clock of identical construction.† With the aid of this clock, B can partake in A's light-signalling experiment, the signals being instantaneously reflected back to either observer on arrival at the other, as indicated in Fig. 5.3.

† If we assume that all natural clocks carried by a given observer keep the same time, then we need only stipulate that B graduates his clock in the same way as A graduates his.

In the Special Theory of Relativity it is assumed that A and B are associated with inertial frames of reference. Consequently, they are either at relative rest or in uniform relative motion. The Principle of Relativity on which the theory is based was formulated by Poincaré (1904) in a lecture at Saint Louis, U.S.A. in September 1904. According to his statement, 'the laws of physical phenomena must be the same for a "fixed" observer as for an observer who has a uniform motion of translation relative to him: so that we have not, and cannot possibly have, any means of discerning whether we are, or are not, carried along in such a motion'. Shortly afterwards, and independently,† the principle was enunciated *in a much more explicit form* by Einstein: 'the same laws of electrodynamics and optics will be valid for all frames of reference for which the equations of mechanics hold good'. This principle presupposes that the observers associated with such frames of reference employ similar measuring instruments, for example clocks, and adopt the same metrical rules and definitions. Therefore, if A assigns a universal value c to the speed of light, then B must do the same.

It is customary when considering the correlation of the clocks and time perspectives of A and B in Einstein's Special Theory to concentrate on the case in which they are in uniform relative motion. Instead, in view of its importance for establishing one of the main results in the following

† In recent years there has been considerable discussion on Einstein's role in the formulation of the Special Theory of Relativity. In his well-known history of modern physics, Vol. II of his *History of the Theories of Aether and Electricity*, published in 1953, Sir Edmund Whittaker gave to the chapter on this theory the provocative title 'The Relativity Theory of Poincaré and Lorentz'. The strength of Einstein's position compared with that of Poincaré and Lorentz was that the investigations of the latter were based on the full theory of electrodynamics (including the concept of the ether) and were essentially restricted to phenomena associated with it, whereas Einstein developed his theory from elementary considerations of light signalling. Subsequent developments have revealed the importance of this distinction, for Einstein's theory is in no way restricted to electrodynamics and is quite independent of our views on the ultimate nature of the interaction between elementary particles. Poincaré's role has been criticized by the French historian of science Taton (1957). According to Taton, although Poincaré knew what was required, 'he did not dare to explain his thoughts, and derive all the consequences, thus missing the decisive step separating him from the real discovery of the principle of relativity'. In support, Taton quoted the following judgment by de Broglie (1951): 'Why did Poincaré fail to advance to the limits of his thought? No doubt this was due to his somewhat hypercritical turn of mind, or perhaps to the fact that he was a pure mathematician. He had a somewhat sceptical attitude towards physical theories, and thought there was generally an infinity of different viewpoints and different ideas, all logically equivalent, from which the scientist only chooses for reasons of convenience. It appears that this nominalism caused him sometimes to misunderstand the fact that, amongst possible logical theories, there are, nevertheless, some which are closer to physical reality or, in any case, better adapted to the physicist's intuition, and therefore more apt to aid his efforts'. A full-scale critical analysis of Poincaré's views has been given by Arthur Miller (1973). See also Holton (1973), who draws attention to Poincaré's doubts about the validity of the principle of relativity following experimental results obtained by W. Kaufmann in 1906. As for Lorentz, he did not believe in relativity at all!

chapter, I shall begin by considering the case in which they are at relative rest. If A and B have similarly graduated clocks, then, apart from the possible adjustment of an additive constant depending on the choice of zero-time on each clock, the principle of relativity can be reduced, as far as kinematics is concerned, to the following:

Axiom X. Principle of kinematic symmetry: t_2 is the same function of t' as t' is of t_1.

Hence, there must be functional relations of the form

$$t_2 = \theta(t'), \qquad t' = \theta(t_1). \tag{5.21}$$

Consequently, the function θ, which we shall call the *signal function* correlating A and B, must be such that

$$t_2 = \theta\theta(t_1). \tag{5.22}$$

However, since B is at a fixed distance from A and the light signals travel with constant speed, it follows that $(t_2 - t_1)$ must be a constant. Hence, θ must be such that

$$\theta\theta(t_1) = t_1 + 2a, \tag{5.23}$$

for all values of t_1 and some constant a. If we drop the subscript, an obvious solution of this functional equation is given by $\theta(t) = t + a$.

More generally, by operating on each side of (5.23) with θ we deduce that

$$\theta(t+2a) = \theta(t) + 2a,$$

whence it immediately follows that $\theta(t)$ must be of the form

$$\theta(t) = t + \omega(t),$$

where $\omega(t)$ is of period $2a$. To reduce this to the particular form $\theta(t) = t + a$, we must consider other similar stationary observers. Thus, if A, B, and C are collinear, with B lying between A and C, and ϕ is the signal function correlating B and C, then A and C will be related by the signal function ψ given by $\psi = \theta\phi = \phi\theta$. Consequently, θ and ϕ must be commutative functions. Since C is at a fixed distance from B, ϕ must satisfy a functional equation of the form

$$\phi\phi(t) = t + 2b, \tag{5.24}$$

where b is some constant. It is then easily proved that

$$\psi\psi(t) = t + 2(a+b),$$

and so we deduce that A and C are at a fixed distance apart equal to the sum of the respective distances of A and B and of B and C. By operating on both sides of (5.24) with the function θ and appealing to the commutative property of θ and ϕ, we deduce that

$$\theta(t+2b)=\theta\phi\phi(t)=\phi\phi\theta(t)=\theta(t)+2b,$$

whence it follows that

$$\theta(t)=t+\omega(t),$$

where $\omega(t)$ is of period $2b$. Hence, $\omega(t)$ must admit both $2a$ and $2b$ as periods. If A, B, and C are any three members of a continuum of relatively stationary observers, then $2a$ and $2b$ will, in general, be incommensurable. Consequently, by a well-known theorem the only continuous form for the function $\omega(t)$ is a constant, and so from equation (5.23) it follows that $\theta(t)=t+a$.

With this solution for $\theta(t)$, equations (5.21) give

$$t'=\tfrac{1}{2}(t_2+t_1).$$

By comparison with equation (5.19), we deduce that $t'=t$, i.e. the time recorded on B's clock when any event occurs at B is the same as the time theoretically assigned to that event by A on the basis of the uniform velocity of light. Therefore, all relatively stationary observers assign the same time to any given event, and this time agrees with that actually recorded on the clock kept by the observer at the point where the event occurs. In this conventional sense, there is world-wide simultaneity of events, and therefore universal time, for all relatively stationary observers.

The above analysis was based on the 'kinematic symmetry' of relatively stationary observers with similarly graduated clocks who assign the same constant value to the speed of light signals passing between them in free space. In his Special Theory of Relativity, Einstein† showed how the same principle of kinematic symmetry in light-signalling experiments could be extended to observers in uniform relative motion, although the consequences are not entirely the same as for relatively stationary observers. In particular, there is no longer world-wide simultaneity, and hence no universal common time, for the aggregate of uniformly moving observers. Consequently, although the theory is based on the hypothesis that the *general laws* governing physical formulae are of the same form for an observer associated with any inertial frame in uniform relative motion as for an observer associated with any inertial frame at relative rest, there

† In his General Theory of Relativity, however, observers associated with general frames of reference do not assign the same *universal* value to the speed of light and the simple type of light-signalling analysis characteristic of the Special Theory no longer applies.

are important differences regarding the epochs assigned to particular events.

To see this most simply, we again consider light-signalling from A to B and from B to A, as in Fig. 5.3 but this time we stipulate that the two observers concerned move away from coincidence with each other at a particular epoch with uniform velocity in a radial direction.† We also postulate that the two similar clocks were synchronized to read time zero at the original instant of coincidence. As before, we consider a signal emitted by A at time t_1, recorded on A's clock. We suppose that this signal is instantaneously reflected on arrival at B at time t', according to B's clock, returning to A at time t_2, according to A. From the principle of kinematic symmetry it follows that, if $t' = \psi(t_1)$, then $t_2 = \psi(t')$. Therefore,

$$t_2 = \psi\psi(t_1). \tag{5.25}$$

However,

$$t_2 = t + r/c, \qquad t_1 = t - r/c,$$

where r is the distance of B from A, according to A, at the instant of reflection, and t is the epoch theoretically assigned by A to this event. Since B is moving away radially from coincidence with A at time zero, it follows that

$$r = Vt,$$

where V is the relative speed of B. Hence,

$$t_2 = \alpha^2 t_1, \tag{5.26}$$

where

$$\alpha^2 = \frac{1 + V/c}{1 - V/c}.$$

Consequently, on comparing (5.25) and (5.26) we see that the function ψ must be such that for all values of the variable t

$$\psi\psi(t) = \alpha^2 t. \tag{5.27}$$

By operating on each side of this equation with ψ, we deduce that

$$\psi(\alpha^2 t) = \alpha^2 \psi(t),$$

whence

$$\psi'(\alpha^2 t) = \psi'(t), \tag{5.28}$$

† The following method of obtaining the basic kinematic formulae of Special Relativity was first developed by Whitrow (1933, 1935). It was afterwards used by Bondi (1964), who wrote k for α and called the technique the 'k-calculus', whereas Whitrow called it the 'signal-function method'.

the prime symbol denoting the derivative. The only solution of equation (5.28) which is continuous as $t \to 0$ (positively) is $\psi'(t) = k$, where k is a constant.† Since $t' = 0$ when $t_1 = 0$, it follows that $\psi(0) = 0$, and hence we must have $\psi(t) = kt$. Comparison with (5.27) yields $k^2 = \alpha^2$. In order to obtain the unique solution $k = \alpha$, and hence

$$\psi(t) = \alpha t, \tag{5.29}$$

where α is positive, we must invoke a further axiom:

Axiom XI. The order of reception of light signals by B, according to B, corresponds to the order of emission of these signals by A, according to A.

We have seen that, according to A, there is at any point at a given (theoretically assigned) epoch a *unique* value for the speed of light in free space. It follows that the order, according to A, of arrival of light signals at B must be the same as the order of their emission from A, for, if a signal emitted by A at some epoch were to arrive at B, according to A, before an earlier signal emitted from A, then, assuming continuity, there would be some event occurring in between A and B at which the second signal would overtake the first and pass it. At such an event there would be, according to A, *two* values for the speed of light in free space. Axiom XI can therefore be regarded as asserting that the theoretically assigned time order of events at B, according to A, agrees with the time order of these events as actually experienced by B. In this sense, we can speak of the time order of these events according to A being in the same sense as the time order of the same events according to B. By the principle of relativity, A and B are interchangeable in axiom XI.

Since $t_2 = \alpha t'$, $t' = \alpha t_1$, and $t = \frac{1}{2}(t_2 + t_1)$, where t is the time theoretically assigned by A to the arrival (and reflection) of the signal at B, it follows that

$$t = \tfrac{1}{2}\left(\alpha + \frac{1}{\alpha}\right) t' = \frac{t'}{\surd(1 - V^2/c^2)}. \tag{5.30}$$

Hence, we deduce that, although A and B agree on the time order of events at B, they will assign different measures to the time interval between any two instants at B.

More generally, we consider next a light signal which is emitted by A at time t_1 according to his clock, passes B at time t_2', according to B, is reflected instantaneously at some event E in line with A and B, passes B again at t_3' according to B, and returns to A at t_4. Then, if (t, r) are the theoretically assigned epoch and distance of E according to A and (t', r')

† From (5.28) it follows that $\psi'(t) = \psi'(\alpha^{-2n}t)$, and $\alpha^{-2n}t \to 0$ as $n \to \infty$, since $\alpha^2 > 1$, because $0 < V < c$.

are the theoretical epoch and distance of E according to B, it follows that

$$\left.\begin{aligned} t_4 &= t + r/c, & t_1 &= t - r/c, \\ t_3' &= t' + r'/c, & t_2' &= t' - r'/c. \end{aligned}\right\} \tag{5.31}$$

Since

$$t_4 = \alpha t_3', \qquad t_1 = t_2'/\alpha, \tag{5.32}$$

we find, on substituting (5.31) in (5.32) and solving, that

$$t' = \frac{t - Vr/c^2}{\surd(1 - V^2/c^2)}, \qquad r' = \frac{r - Vt}{\surd(1 - V^2/c^2)}. \tag{5.33}$$

These are the celebrated Lorentz formulae† for an event in line with the two observers A and B. In the case of an event E which occurs *anywhere*, provided that it is acessible to A and B by means of light signalling (i.e. E must lie on light paths emanating from A and B and also on light paths terminating at these observers), it can be proved (see Appendix) that, if the x co-ordinate of E is measured by A in the direction of B and the x' co-ordinate of E is measured by B in the direction opposite to A, (x, y, x) and (x', y', z') being orthogonal triads (Cartesian axes) which coincide at time zero, then the Lorentz formulae correlating the respective co-ordinates and epochs of E according to A and B are expressible in the form,

$$x' = \frac{x - Vt}{\surd(1 - V^2/c^2)}, \qquad y' = y, \qquad z' = z, \qquad t' = \frac{t - Vx/c^2}{\surd(1 - V^2/c^2)}. \tag{5.34}$$

These formulae can also be expressed in the reciprocal form,

$$x = \frac{x' + Vt'}{\surd(1 - V^2/c^2)}, \qquad y = y', \qquad z = z', \qquad t = \frac{t' + Vx'/c^2}{\surd(1 - V^2/c^2)} \tag{5.35}$$

so that apart from the change in the sign of V (due to the unsymmetrical choice of the directions of the x, x' axes) we find that the Lorentz formulae are *self-reciprocal*, in accordance with the principle of relativity on which the theory is based.‡

† The term Lorentz transformation is due to Poincaré (1905).

‡ From formulae (5.35) we can immediately deduce Einstein's velocity addition formulae

$$u = \frac{u' + V}{1 + u'V/c^2}, \qquad v = \frac{v'\surd(1 - V^2/c^2)}{1 + u'V/c^2}, \qquad w = \frac{w'\surd(1 - V^2/c^2)}{1 + u'V/c^2}$$

for the velocity components (u, v, w) relative to A of a particle moving with velocity components (u', v', w') relative to B. A remarkable feature of these formulae, in particular the formula for u, is that velocities are not additive. Instead, in the case of velocities in the same straight line, for example u and V, we find that $\tanh^{-1} u/c = \tanh^{-1} u'/c + \tanh^{-1} V/c$. The function $\tanh^{-1} u/c$ has been called by Robb the *rapidity* corresponding to velocity u. In the case of non-collinear velocities it can be proved that the law of composition of the corresponding rapidities is given by the triangle law of hyperbolic (Lobatchewskian) trigonometry. Incidentally, we observe that if $V = c$ then $u = c$, irrespective of the value of u'.

We are primarily interested here in the formulae for t' and t. These replace the classical formula $t'=t$, expressing the universal nature of Newtonian time and simultaneity. The appearance of the spatial coordinates x and x' in the respective expressions for t and t' makes it inevitable that events at a distance which are (conventionally) simultaneous for one observer are not (conventionally) simultaneous, in general, for the other. Thus, although the Special Theory of Relativity is compatible with the universal simultaneity of events at the same place, it denies the universal simultaneity of spatially separated events.† Consequently, the simultaneity of events throughout the universe becomes an indeterminate concept until a frame of reference (or observer) has been specified. Just as observers in different places have different spatial perspectives of the universe, so we now find that observers with different velocities have different temporal perspectives.

5.5. Time dilatation

Although, when Einstein's theory of physical time first became widely known, most philosophers concentrated their attention on his rejection of the classical concept of world-wide simultaneity, the really novel feature of his theory was his contention that the measurement of time intervals varies from one observer to another, depending on their relative motion. From formula (5.30) correlating the epoch assigned by A to an event at B with the epoch actually recorded by B, we immediately deduce that, if $\delta t'$ denotes the time interval recorded on B's clock between any two events at B, the duration assigned to this interval by A must be δt, where

$$\delta t = \frac{\delta t'}{\sqrt{(1-V^2/c^2)}}. \tag{5.36}$$

Consequently, $\delta t=\delta t'$ only if $V=0$, and if V is not zero we must have $\delta t>\delta t'$, although the difference will not be significant when V is a small fraction of c (as it normally is in everyday life), since it depends on the

† In a well-known but curious passage, Whitehead criticized Einstein's 'signal-theory' definition of simultaneity in a way which confuses the essential distinction between these two kinds of simultaneity. Arguing that the signal theory exaggerates the importance of light signals in our lives, Whitehead wrote: 'The very meaning of simultaneity is made to depend on them. There are blind people and dark cloudy nights, and neither blind people nor people in the dark are deficient in a sense of simultaneity. They know quite well what it means to bark both their shins at the same instant' (Whitehead 1919, p. 53). This misses Einstein's essential point that, whereas the simultaneity of two events at the same place can be directly experienced and is universal, the simultaneity of distant events is 'fictitious' and relative. As Eddington pointed out, Einstein's definition 'is simply an announcement of the rule by which we propose to extend fictitious time-partitions through the world' (Eddington 1922). In the case of distant events, the discovery by Roemer of the finite nature of the velocity of light had already demolished the commonsense view that things exist simultaneously with their perception.

square of V/c. For values of V close to c we find that δt becomes arbitrarily large. In the limit, when we consider B moving with speed c, the time interval δt becomes infinite. For speeds exceeding c, no correlation of A's clock and B's clock is possible. We call $\delta t'$ the *proper time*† of the interval between the two events at B, and the result that any observer in uniform relative motion will assign a greater measure than $\delta t'$ to this interval of time is called the phenomenon of *time dilatation*. It is independent of the sign of V, and so it does not matter whether the two observers are receding from or approaching each other. The effect depends solely on their relative speed. Einstein's conclusion that the time interval δt assigned by A exceeds the time interval $\delta t'$ recorded on B's clock was *partially* anticipated by Lorentz and also by Larmor. In his book *Aether and matter*, published in 1900, Larmor argued that a clock moving *relative to the ether* with speed V must run slower than a stationary clock, the ratio of the rates being approximately $\surd(1-V^2/c^2)$ to unity.‡ He arrived at this result after studying the Michelson–Morley experiment. The *original object* of this experiment was to determine optically the 'absolute' motion of the earth with respect to the luminiferous ether by utilizing Michelson's interferometer to compare the to-and-fro transmission of light along equal arms at right angles to each other. The null result obtained was interpreted by Einstein as confirmatory evidence for his hypothesis that the velocity of light is the same with respect to all frames of reference in uniform relative motion. Previously, however, the Irish mathematical physicist Fitzgerald had suggested that the result of this experiment could be explained if it were assumed that the length, measured in the direction of motion, of a body moving with velocity V relative to the ether is automatically diminished, in virtue of this motion, in the ratio $\surd(1-V^2/c^2)$ to unity.§ This hypothesis—which was later suggested independently by Lorentz—was simply an *ad hoc* assumption that motion produces an actual physical contraction, although this could not be detected by an observer moving with the body since all his measuring instruments would be affected in the same way. This point of view led Lorentz to consider the effect of elastic forces on the electronic and atomic constitution of matter with the object of accounting for the

† *Proper time* is another name for *local time*, i.e. time which either is, or could be, recorded directly by a clock.

‡ Larmor only stated the second-order approximation in V/c, although Whittaker (1953) attributed to him the factor $\surd(1-V^2/c^2)$. A careful criticism of Larmor's point of view and failure to anticipate Einstein's relativistic concept of time dilatation has been made by Rindler (1970).

§ Strictly speaking, in his first published reference to this contraction, a letter to *Science*, Fitzgerald (1889) referred to the contraction of bodies moving through the ether only as 'depending on the square of the ratios of their velocities to that of light'.

occurrence of the same contraction phenomenon in all forms of matter. In a similar way, he also tried to introduce and account for the analogous effect on the rate of a moving clock, but it was much more difficult for physicists to accept the idea that relative motion could influence the temporal working of a physical clock than it could cause a 'rigid body' to contract, for it was possible to imagine some mechanistic explanation of the latter effect (due to compression by the ether) but not of the former.

It was one of the great merits of Einstein's approach to these questions that he circumvented the problem of the structure of matter and turned his attention to the theory of measurement. Instead of assuming that there are *real*, i.e. structural, changes in length and duration owing to motion, Einstein's theory involves only *apparent* changes, and these are independent of the microscopic constitution and hidden mechanisms controlling the structure of matter. Moreover, unlike the previously postulated real changes, these apparent phenomena are *reciprocal*: just as B's measuring rod seems to A to have contracted in the direction of motion, so equally A's measuring rod seems to B to be shortened in exactly the same way, and, just as B's clock seems to A to run slow, so conversely A's clock seems to B to lag behind his own. Because of this reciprocity, or relativity, of A and B, Einstein discarded the idea of the luminiferous ether as a preferential frame of reference.

An important experiment was performed by Kennedy and Thorndike (1932) to discriminate between the older view that the Fitzgerald contraction is a real effect and Einstein's view that it is only apparent. Their experiment was a modification of the Michelson–Morley experiment with arms which were unequal, although sufficiently close to each other for good interference fringes to be observed. Allowing for the Fitzgerald contraction, but assuming the existence of a luminiferous ether, the time difference for the transmission of light along the two arms should be a function of the diurnally and annually varying velocity of the apparatus. The observed absence of any such effect was a powerful empirical argument for Einstein's contention that the velocity of light is the same for different observers in motion relative to each other.

Nevertheless, there was no direct experimental evidence in support of the time dilatation effect until 1938, when Ives and Stilwell (1938, 1941) confirmed formula (5.36) to the second order of the ratio V/c. The natural clock used in their experiment was a rapidly moving positively charged hydrogen atom (a canal ray), the rate of this 'atomic clock' being measured by the frequency of the light emitted by it. According to Einstein's theory, the apparent rate of such a clock when in motion should contribute to the Doppler effect, as can be shown in the following way which is somewhat simpler than Einstein's original derivation.

We consider a source of radiation at B emitting light (or other electromagnetic waves) of proper frequency ν' at proper time† t' in the direction of the observer A who receives it at time t_2 with apparent frequency ν_2, according to his clock. In a short interval of proper time $\mathrm{d}t'$ the number of waves emitted by B is $\nu'\,\mathrm{d}t'$. If these arrive at A in an interval $\mathrm{d}t_2$, according to A, then

$$\nu'\,\mathrm{d}t' = \nu_2\,\mathrm{d}t_2.$$

If t denotes the time assigned by A to epoch t' at B, and r is the distance of B according to A at this epoch, then $t_2 = t + r/c$, and hence

$$\frac{\nu'}{\nu_2} = \frac{\mathrm{d}t_2}{\mathrm{d}t'} = \left(1 + \frac{V_r}{c}\right)\frac{\mathrm{d}t}{\mathrm{d}t'},$$

where V_r denotes the radial velocity of B with respect to A, being positive if B is receding and negative if B is approaching. (If B moves only radially, then V_r is the same as the relative velocity V.) Provided B is moving uniformly with respect to A, the ratio $\mathrm{d}t/\mathrm{d}t'$ is given by equation (5.36), and hence

$$\frac{\nu'}{\nu_2} = \frac{1 + V_r/c}{\surd(1 - V^2/c^2)}. \tag{5.37}$$

Since wavelength is inversely proportional to frequency (for light travelling with constant speed), it follows that

$$\frac{\lambda + \delta\lambda}{\lambda} = \frac{\text{apparent wavelength}}{\text{proper wavelength}} = \frac{1 + V_r/c}{\surd(1 - V^2/c^2)}. \tag{5.38}$$

This is Einstein's formula for the Doppler effect‡ on light received from a uniformly moving source. If the phenomenon of time dilatation were neglected, as in Newtonian physics where $t = t'$, then equation (5.38) would give the classical formula for this effect,

$$\frac{\delta\lambda}{\lambda} = \frac{V_r}{c}. \tag{5.39}$$

When $V_r = 0$, there is no classical Doppler effect but there is a second-order relativistic effect

$$\frac{\delta\lambda}{\lambda} = \frac{1}{\surd(1 - V^2/c^2)} - 1 = \frac{1}{2}\frac{V^2}{c^2} \tag{5.40}$$

† This signifies the epoch which would, in principle, be recorded on a clock travelling with the source. *Proper frequency* means frequency with respect to this clock.

‡ Nowadays, astronomers and cosmologists usually denote the shift ratio $\delta\lambda/\lambda$ by the symbol z.

if terms of order higher than the second power in V/c are neglected. This is known as the transverse Doppler effect. It is due solely to time dilatation. Furthermore, we note that, if the relative motion of the source is purely radial so that $V_r = \pm V$, time dilatation yields a positive second-order correction $\frac{1}{2}(\delta\lambda/\lambda)^2$ to the classical Doppler effect shift ratio $\delta\lambda/\lambda = \pm V/c$.

Following a paper on canal rays by Stark in 1906, Einstein (1907) suggested that they might be used to observe the transverse Doppler effect which he had predicted. Stark's technique was too primitive for this purpose as it did not yield sufficiently sharp lines in the spectroscope, and 30 years elapsed until the experiment could be performed satisfactorily by Ives and Stilwell, following work by A. J. Dempster. In their apparatus, canal rays of high and uniform velocity were produced and the insertion of a mirror led to spectroscopic pictures containing lines corresponding to stationary atoms and to atoms moving towards and away from the observer. In the absence of time dilatation, the mean of the lines due to atoms moving away from and towards the observer with the same speed, as given by (5.39), would coincide with the central line due to the stationary atoms. According to Einstein's formula, however, the mean would be displaced slightly to the red by the amount $\frac{1}{2}(\delta\lambda/\lambda)^2$ approximately. The experiment gave this result and so confirmed within the expected limits of accuracy Einstein's quantitative formula for the time-dilatation effect.†

Since then extremely cogent, although somewhat less precise, data derived from cosmic-ray phenomena provided further confirmatory evidence which caught the imagination of physicists. The elementary particles known as muons, found in cosmic-ray showers, disintegrate spontaneously, their proper lifetime (i.e. time from production to disintegration according to an observer travelling with a muon) being about 2 microseconds, more precisely $(2{\cdot}09 \pm 0{\cdot}03) \times 10^{-6}$ seconds. These particles are mainly produced at heights of the order of 10 kilometres above the earth's surface. Consequently, those observed in the laboratory on photographic plates must have travelled that far. However, in 2 microseconds a particle moving with the speed of light travels less than a kilometre, and according to the theory of relativity all material particles travel with speeds *less* than that of light. The time-dilatation factor $(1 - V^2/c^2)^{-\frac{1}{2}}$ is equal to the ratio E/m_0c^2, where E is the energy of a particle (according to the observer in the laboratory) and m_0 is its rest mass (i.e. its mass

† It is a curious fact that Ives and Stilwell themselves did not accept Einstein's theory but believed that their results confirmed that a moving clock runs slow in the absolute sense, as suggested by Larmor and Lorentz, owing to its motion through the ether.

according to an observer moving with it).† For the muons in cosmic-ray showers this ratio is of the order of 10, and so their speed V is almost that of light, being about 0·995c. The time-dilatation factor is thus of the order required to explain why it is that to the observer in the laboratory these particles appear to travel about ten times as far as they could in the absence of this effect (Rossi and Hall 1941, Frisch and Smith 1963).

A more precise check on the time-dilatation factor of Special Relativity has since been made with the aid of the caesium atomic clock (see pp. 220). In October 1971 four of these clocks were flown on commercial jet planes, travelling at about 600 miles per hour, on two trips circumnavigating the Earth, one eastward and the other westward (Hafele and Keating 1972, Schlegel 1974). Compared with a similar clock τ_0 in a non-rotating inertial frame (e.g. at the North or South Pole), a clock τ moving eastwards with speed v (small compared with the velocity of light) relative to the earth's surface in the equatorial plane will run slow, approximately according to the formula

$$\tau - \tau_0 = -\frac{(R\omega + v)^2}{2c^2}\,\tau_0$$

where R is the Earth's radius and ω its diurnal angular velocity. For a clock moving westwards in the equatorial plane a similar result follows except that v is replaced by $-v$. A clock at rest on the earth's surface will run slow, approximately according to the formula

$$\tau - \tau_0 = -\frac{(R\omega)^2}{2c^2}\,\tau_0.$$

Hence the two circumnavigating clocks will run slow relative to an earthbound clock according to the formulae

$$\tau - \tau_0 = \left(\mp\frac{R\omega v}{c^2} - \frac{v^2}{2c^2}\right)\tau_0 \tag{5.41}$$

respectively. Since for the value of v concerned the magnitude of the second term in the bracket is less than that of the first term, the westward moving clock actually runs ahead of the earthbound clock. There is also an additional time-gain factor due to the earth's gravity (gravitational red shift, see pp. 279) to take into account. It is given by $(gh/c^2)\tau_0$, where g is

† The equality of the time-dilatation factor and the ratio E/m_0c^2 can be used to estimate the mean proper lifetimes of transient elementary particles, some of which are extremely short lived compared with the muon. (The shortest known is of the order 10^{-22} seconds.) For, if such a particle travels in its proper lifetime t_0 a distance l, relative to the observer, at a speed V close to c, then l must be approximately ct, where $t/t_0 = E/m_0c^2$. Hence, $t_0 = m_0cl/E$, and so t_0 can be calculated if E, m_0, and l have been determined.

the gravitational acceleration and h the flying altitude, which is very small compared with R. Hence formula (5.41) is replaced by

$$\tau-\tau_0=\left\{\frac{gh}{c^2}\mp\frac{R\omega v}{c^2}-\frac{v^2}{2c^2}\right\}\tau_0. \tag{5.42}$$

In the actual experiment the aircraft did not fly in the equatorial plane nor did they maintain constant altitude, ground speed, or latitude. At latitude λ, R in formula (5.42) is replaced by $R\cos\lambda$, and if θ is the angle between the direction of v and the circles of latitude it is easily seen that (5.42) is replaced by

$$\tau-\tau_0=\int\left\{\frac{gh}{c^2}-\frac{R\omega v\cos\lambda\cos\theta}{c^2}-\frac{v^2}{2c^2}\right\}\mathrm{d}\tau \tag{5.43}$$

where the integral is calculated along the actual flight path. In nanoseconds (10^{-9} seconds) the mean time differences for the eastward and westward travelling clocks compared with the earthbound clock in the Hafele–Keating experiment are given by Table 5.1. The agreement between theory and experiment was good. It is the only *direct* test so far of the time-dilatation formula of Special Relativity.

This experiment involved clock velocities that were small compared with the velocity of light. Recently, a still more precise check of Einstein's time-dilatation formula has been made by Bailey *et al.* (1977) for velocities very close to the velocity of light. This team, using the Muon Storage Ring at CERN, obtained values for the time-dilatation effect on muons in circular orbit with average velocities of 99.94 per cent of the speed of light, involving a time-dilatation factor of about 29.327. The accuracy of Einstein's time-dilatation formula was thereby confirmed to within about 0.1 per cent. This is much the most precise check of this effect so far made.†

TABLE 5.1

	Eastward	*Westward*
Observed	-59 ± 10	273 ± 7
Predicted	-40 ± 23	275 ± 21

† The most accurate high-velocity test of Special Relativity that has yet been made concerns the 'Thomas precession'. By precise measurement of the frequency of rotation of electrons in a magnetic field, this relativistic kinematic effect has been confirmed to within 5 parts in 10^9 (Newman *et al.* 1978).

5.6. The clock paradox

'If we placed a living organism in a box ... one could arrange that the organism, after any arbitrary lengthy flight, could be returned to its original spot in a scarcely altered condition, while corresponding organisms which had remained in their original positions had already long since given way to new generations. For the moving organism the lengthy time of the journey was a mere instant, provided the motion took place with approximately the speed of light'.†

This startling prediction was made by Einstein (1911) on the basis of the Special Theory of Relativity. Already in 1905 in his first paper on that theory this prediction was foreshadowed by his contention that, if a clock B of identical construction and operation to another clock A and initially coincident with it were moved with constant speed V along a closed curve until it returned to A, the journey lasting t seconds according to A, then according to B the journey would last only $\surd(1-V^2/c^2)t$ seconds, and so B's clock would appear to A to be $\{1-\surd(1-V^2/c^2)\}t$ seconds slow. This retardation appeared to Einstein to be an immediate consequence of the phenomenon of time dilatation associated with a standard clock in *uniform relative motion.* He argued that this retardation still occurs if the clock B is no longer restricted to move in a straight line but is allowed to move in any polygonal line, and hence in the limit in any curved line, and also if the end points of this line coincide.‡

An enormous literature has sprung up in the last 70 years around this famous 'clock paradox', almost rivalling that associated with the paradoxes of Zeno.‡ Since 1957 there has been a fresh burst of interest and lively debates have been conducted in *Nature* and other scientific journals by H. Dingle, W. H. McCrea, and others. Moreover, at the seventh International Astronautical Congress held in Rome in 1956 the chief German delegate, in a paper on 'The possibility of reaching the fixed stars', talked confidently of nuclear-powered space-ships which could approach the speed of light and maintained that, according to Einstein's theory, members of the crew would find on return to Earth that their children would have grown old during a voyage which to them had

† Einstein (1905) concluded that a balance clock at the equator must go more slowly, by a very small amount (due to the Earth's diurnal rotation), than a precisely similar clock situated at one of the poles under otherwise identical conditions. With his usual scientific 'tact', although he had not yet developed his theory of gravitation (general relativity), he was careful to exclude the use of a pendulum clock in this context.

‡ In France the 'clock paradox' is usually called 'Langevin's Paradox', because the famous French physicist wrote a classic early paper on the subject (Langevin 1911). Arzeliès (1966) reviews the controversy that resulted, particularly in the 1920's, and gives over 100 references to the clock paradox literature. The most thorough discussion in English is that given by Marder (1971), who lists nearly 250 references.

seemingly lasted only a few days! Thus, whereas in the Red Queen's country Alice found that it was necessary to run as fast as she could to remain in the same place, in the physical universe it would seem that by travelling sufficiently fast we could remain at effectively the same epoch 'all the time'.

It is not always made clear that Einstein's clock paradox has two distinct aspects: (1) it appears to be contrary to common sense that two individuals can part and then meet again to find that one has lived longer than the other between the same two events; (2) it also appears that a logical antinomy is involved. The former, however puzzling it may seem,† is not the aspect that has given rise to the main controversy. Indeed, once we accept the idea of time dilatation because there is confirmatory evidence for it, the purely commonsense objection to Einstein's argument loses its force. Moreover, as we saw when discussing biological time in Chapter 3, similarly constructed natural clocks do not 'tick' at the same rate in all circumstances. In the case of cold-blooded animals, physiological time is affected by external temperature; for example, for a lizard physical events which we regard as uniform in time probably appear to be non-uniform, so that as the Sun rises its rate of rising must seem to decrease and as it sets its rate of setting must seem to increase. Even in the case of homeothermic man, it has been suggested that life can be prolonged by artificial refrigeration‡ so that, quite independently of relativistic time dilatation, it could happen that a space traveller who left the earth for a trip around one of the nearer stars and passed most of the journey in a state of suspended animation might arrive back on earth to find that several hundred years had elapsed whereas he himself had aged hardly at all! In view of this alternative (hypothetical) method of arriving at the same result, we cannot summarily dismiss Einstein's argument merely because it conflicts with our intuitive prejudices concerning time, for these are based merely on the implicit assumption§ that time is 'absolute', existing in its own right.

The point is sometimes made that, strictly speaking, Einstein's argument should apply only to purely physical, or inorganic, clocks and that we ought not to assume that it applies automatically to metabolic and other biological clocks. If, however, relative motion causes a physical

† The normal commonsense reaction is that Einstein's contention is no less a fable than the story of the monk who went into the woods, heard a bird break into song, listened enraptured for a trill or two, and then returned to find himself a stranger at the gates of his monastery for he had been absent fifty years and of all his comrades only one had survived to recognize him! (R. L. Stevenson, 'The lantern bearers', in *Across the plains*).

‡ See Chapter 3, pp. 126–7.

§ cf. Barrow's remark, quoted on p. 187, that he did not believe that 'there is anyone but allows that things existed equal times which rose and perished together'.

clock to appear to run slow, we should expect a biological clock to show the same effect, for, if not, then biological processes in an organism *at rest* with respect to a moving physical clock would be speeded up relative to purely physical processes, and this would imply a profound difference between the physics of organic processes and of the inorganic entities involved in them, for which there is no evidence.

The second objection to Einstein's argument is far more serious, for it appears to lead to a genuine paradox. According to the principle of relativity on which Einstein founded his theory, it should be possible to regard *either* of the two originally coincident and synchronous clocks as cruising through space with the same uniform relative speed V, in which case each clock must lag behind the other on their reunion. However, this is a logical contradiction, and therefore impossible. Supporters of Einstein have argued that *in the circumstances envisaged* the two clocks are not interchangeable and that the argument leading to the logical paradox is therefore invalid. They claim that the time-dilatation formula (5.36) applies only to the case in which A and B are associated respectively with two definite inertial frames of reference in uniform relative motion. Consequently, if A and B meet once they cannot meet a second time, although in the circumstances envisaged A and B necessarily meet twice. Therefore, if one observer is associated throughout with a unique inertial frame, the other cannot be, and during certain intervals—when he moves over from one inertial frame to another—he must be accelerated. Unfortunately, this argument appears to give the *coup de grâce* not only to those who reject Einstein's conclusion that on reunion B's clock will lag behind A's but also to the train of thought by which Einstein was led to this conclusion, for his appeal to formula (5.36) would no longer be legitimate, since this formula was derived on the assumption that *each* clock is associated with a unique inertial frame throughout. Therefore, the crucial argument of those who support Einstein automatically undermines Einstein's own position, as well as that of his opponents! Consequently, it is not surprising that the fog of confusion has descended on the battleground. As the editor of *Discovery* remarked in his introductory comments to a correspondence between Professor McCrea and Sir Ronald Fisher published in that journal in February 1957, the result of previous correspondence on this topic was 'to stimulate uncertainty about this question, not to quell it'.

It has long been realized by many relativists that Einstein's Special Theory of Relativity is not adequate for a complete discussion of the clock paradox if accelerations are involved. Appeal has therefore been made from time to time to Einstein's General Theory. Unfortunately, this has tended to cloud the issue still further. Until recently the General

Theory played a very minor role in modern physics compared with the Special Theory and still cannot be regarded as so well established, despite the fact that no rival theory of gravitation has attracted anything like the same amount of support. Moreover, even if we agree to accept this theory unreservedly, we find that those who have appealed to it for the elucidation of the present problem have been neither clear nor cogent. For example, R. C. Tolman's well-known solution (Tolman 1934) has been aptly criticized by G. Builder (1957). 'In effect', he writes, 'the "paradox" was resolved by denying the applicability of the restricted theory to the problem and then using instead conclusions that had been derived from the restricted theory by means of the principle of equivalence. This tortuous procedure succeeded in hiding the paradox rather than in resolving it'.

In the appeal to General Relativity, for example by C. Møller (1972), it is usually assumed, on the authority of Einstein, that the *acceleration* of a clock relative to an inertial system has no influence on its rate. In other words, the time dilatation associated with a clock when moving relative to the observer with speed V is given by the same formula (5.36) irrespective of whether the clock is moving uniformly or is accelerated, although in the latter case δt and $\delta t'$ must now be restricted to infinitesimal values, since V changes with time. So far, however, all that can be said about this hypothesis is that (i) as the result of a detailed analysis based on General Relativity,† Møller (1955) concluded that acceleration has no influence

† Alternatively, we may consider in a manner similar to that in which we studied two equivalent clocks in uniform relative motion, the case in which they are in uniform relativistic acceleration $\mathrm{d}W/\mathrm{d}t = f$, where $W = V(1-V^2/c^2)^{-1/2}$, from initial coincidence at relative rest at epoch zero. If each assigns the constant value c to the speed of light, it can be shown (Milne and Whitrow 1938) that, using the notation of Fig. 5.3, and replacing f by $2c/k$.

$$1/t_1 = 1/t' + 1/k, \qquad 1/t_2 = 1/t' - 1/k,$$

whence, writing $t_1 = t - r/c$, $t_2 = t + r/c$, and $V = \mathrm{d}r/\mathrm{d}t$, it follows that

$$\sqrt{\left(\frac{1-V/c}{1+V/c}\right)} = \frac{t_1}{t_2} = \frac{1-t'/k}{1+t'/k}$$

and therefore

$$\sqrt{\left(1-\frac{V^2}{c^2}\right)} = \frac{1-t'^2/k^2}{1+t'^2/k^2}.$$

The time t assigned by A to the event at B at proper time t' is given by

$$t = \tfrac{1}{2}(t_1+t_2) = \frac{t'}{1-t'^2/k^2}$$

and hence

$$\frac{\mathrm{d}t}{\mathrm{d}t'} = \frac{1+t'^2/k^2}{(1-t'^2/k^2)^2}.$$

Footnote cont'd. on page 264

on the rate of certain idealized clocks and (ii) some experiments have been interpreted as evidence that, in the circumstances concerned, clock-rate was unaffected by acceleration (Sherwin 1960).

In view of the additional complications and ambiguities involved in considering the possible effects† of accelerated motion, I suggested to E. A. Milne some years ago that fresh insight into the alleged paradox might be obtained by formulating it in a context which involves no reference whatsoever to accelerations. Let us therefore now suppose that B moves relative to A with uniform velocity V in a finite universe of constant positive curvature (the three-dimensional non-Euclidean analogue of the two-dimensional surface of a Euclidean sphere). As before, we stipulate that A and B are initially together at zero epoch (according to each). After a certain time-interval $t = cl/V$, according to A, where l is the time taken by light, according to A, to circumnavigate the universe, the two observers will again meet. At this event the time t' recorded by B will be $\sqrt{(1-V^2/c^2)}t$.

We shall now verify this result *de novo*. The particular epoch t_1, according to A, at which a light signal must leave A in order to return to A at epoch t (after circumnavigating the universe) is given by

$$t_1 = t - l.$$

The epoch t', according to B, at which this signal arrives at B will be, by formula (5.29),

$$t' = \alpha t_1,$$

where

$$\alpha = \sqrt{\left(\frac{1+V/c}{1-V/c}\right)}.$$

Hence

$$t' = \alpha(t-l) = \alpha t(1-V/c) = \sqrt{(1-V^2/c^2)}t. \tag{5.44}$$

Footnote cont'd. from Page 263.

Consequently,

$$\frac{dt'}{dt} = \frac{2(1-V^2/c^2)}{1+\sqrt{(1-V^2/c^2)}} < \sqrt{(1-V^2/c^2)}.$$

Except initially when $V = 0$, we see that the time dilatation associated with relative velocity V is greater than in the corresponding case of uniform motion. Hence, we find that the uniform acceleration of B does affect the apparent rate according to A of the clock carried by B. This result is, of course, not based on General Relativity. Nevertheless, in that theory a uniform acceleration (in the classical sense, i.e. dV/dt = constant) is considered to be indistinguishable from a uniform gravitational field (Einstein's Principle of Equivalence), and the latter does affect the rate of a clock (gravitational red shift)!

† In any case, these are really side issues.

It should be noted that *all* the epochs mentioned in this proof, t_1, t', and t, are actual clock readings and that none is an epoch assigned to a distant event according to some theoretical rule. Consequently, in the discussion of the present problem no question of the arbitrariness, or otherwise, of such a rule can arise. The only assumptions needed to obtain (5.44) are as follows:

(i) A and B carry identical clocks which run in the same sense;

(ii) the transmission of light signals between A and B is governed by the principle of relativity, in the sense that if the two clocks are synchronized at the instant when A and B first coincide then the local time of reception of a signal by one observer is, in each case, the same function of the local time of emission by the other;

(iii) A and B assign the same uniform speed c to the velocity of light;

(iv) A regards B as moving radially with uniform velocity V;

(v) A (but not B) is at rest relative to the local background in a finite universe which can be circumnavigated by light in a constant time interval, according to A's clock.

It is clear that, although A's and B's clocks synchronize when they first coincide, B's clock will lag behind A's when they meet again. Moreover, however strange this result may seem from the point of view of our commonsense experience of the world, no logical antinomy or paradox is involved. The time lag of B's clock compared with A's clock is *absolute* and not relative, and there is no conflict with the principle of relativity (which governs the transformation of co-ordinates from one observer to the other), because there is an absolute difference between A and B *in their respective relations to the universe as a whole.*

As the point at issue is important but is easily misunderstood, it may help to put it another way in terms of the Fitzgerald contraction. If A and B each carry equivalent measuring rods of the same proper length† (in the direction of their relative motion), then A will regard B's rod as shorter than his own, and B will regard A's as similarly contracted. Now, in principle, A can regard the whole circuit of the universe from A back to A as a rod R. Similarly, B can regard the circuit from B back to B as another rod R′. *However, these will not be equivalent rods,* for B will assign a shorter length to R′ than A does to R. In fact, A will assign a length cl to R and B a length $cl\surd(1-V^2/c^2)$ to R′. The time it will take B to move through the whole length of A's rod will be cl/V, according to A, and the time it will take A to move relative to B through the length of B's rod will be $cl\surd(1-V^2/c^2)/V$, according to B. Consequently, it is in complete

† The length of A's rod according to A is $\frac{1}{2}c$ multiplied by the time according to A for light to traverse it from A back to A with instantaneous reflection at the far end, and similarly, *mutatis mutandis*, for the length of B's rod according to B.

conformity with the principle of special relativity that A's clock and B's clock do not agree when A and B meet the second time, and it would be a breach of that principle if they did.

In this particular thought experiment we have found that a clock which always moves uniformly relative to the material background of a closed static model universe records a shorter interval of time between successive epochs of meeting a similarly constructed stationary clock than is recorded by the latter. In this imaginary experiment each clock is associated throughout with a unique inertial frame, and the question of the possible effects of acceleration does not arise. *The essential difference between the two clocks concerns their relations to the universe as a whole.*

In the usual formulation of Einstein's clock paradox no explicit reference to the universe is normally made, but only one of the clocks is associated throughout with the same inertial frame. Because of this asymmetry, it is not surprising that the two clocks do not synchronize when they meet the second time. The class of inertial frames, however, is presumably determined by the general distribution of matter in the universe. Consequently, in this case too we can regard the two clocks as having different relations to the world as a whole, and it is this, rather than any particular consequence of acceleration as such, which leads us to accept Einstein's conclusion that a rapidly moving organism could return to its point of origin younger than if it had remained there all the time.

To obtain further insight into the relations between time and the universe we must now introduce the concepts of *space–time*† and *cosmic time.*

† For further discussion of the clock paradox, in terms of this concept, see pp. 277–8. In criticism of the argument above, Nerlich (1976), after correctly stating that the difference between two clocks is 'a matter of the length of their world lines (which the clocks measure)', finds it 'not entirely clear how the different orientation of their world lines with regard to those of distant matter', i.e. with respect to the universe as a whole, can be 'a significant feature of the situation'. The answer is that, in the usual formulation of the paradox (relative motion to and fro in a straight line), one clock (A) is associated *throughout* with a unique inertial frame, whereas the other (B) is not. In other words, B eventually 'turns round' and comes back to A, but A itself never 'turns round'. The only way in which this distinction can be meaningful is for there to be background matter to refer to (at least implicitly)—in other words, the rest of the universe. Relative to B, not only A but the whole universe appears to 'turn round' when A and B start approaching each other after their initial mutual recession. Relative to A, however, only B 'turns round'. According to the relativistic length-contraction effect, the spatial path described by B (against the background of the universe) will be longer, according to A, than the spatial path that B will regard A as describing in their to-and-fro relative motion, and, since their relative velocities are the same, clock B will record a shorter time interval between their first and second meetings than will clock A.

References

Arzeliès, H. (1966). *Relativistic kinematics.* Pergamon Press, Oxford.

Ash, M. E., Campbell, D. B., Dyce, R. B., Ingalls, R. P., Penttengill, G. H., and Shapiro, I. I. (1971). Astronomical constants from analysis of inner planet radar data. *Bull. am. astron. Soc.* **3,** 474.

Bailey, J. *et al.* (1977). *Nature* (*Lond.*) **268,** 301–5.

Bondi, H. (1959). *Rep. Prog. Phys.* **22,** 105.

—— (1964). Foundations of special relativity. In *Lectures on general relativity* (eds. A. Trautman, A. E. Pirani and H. Bondi), vol. 1, pp. 386–406. Brandeis Summer Institute in Theoretical Physics. Prentice-Hall, Englewood Cliffs, N.J.

Bradley, J, (1728). *Phil. Trans. Roy. Soc.* **35,** 697.

de Broglie, L. (1951). *Savants et découvertes.* Albin Michel, Paris.

Builder, G. (1957). *Aust. J. Phys.* **10,** 261.

Cedarholm, J. P. and Townes, C. H. (1959). *Nature* (*Lond.*), **184,** 1350–1.

Cohen, I. B. (1944). *Roemer and the first determination of the velocity of light.* The Burndy Library, New York.

Drake, S. (1974). *Galileo Galilei*: *Two new sciences* (transl. S. Drake), pp. 49–51. The University of Wisconsin Press, Madison, Wisc.

Eddington, A. S. (1922). *Mathematical theory of relativity*, p. 29. Cambridge University Press, Cambridge.

Einstein, A. (1905). *Annln. Phys.* **17,** 891–921.

—— (1907). *Annln. Phys.* **23,** 197–8.

—— (1911). *Vierteljahrsschr. Naturforsch. Ges. Zürich*, **56,** 1–14 (see also Kopff 1923).

Fitzgerald, G. F. (1889). *Science* **13,** 390.

Frank, P. and Rothe, H. (1911). *Annln. Phys.* **34,** 825.

Frisch, D. H. and Smith, J. H. (1963). *Am. J. Phys.* **31,** 342–55.

Froome, K. D. (1958). *Proc. Roy. Soc.* A, **247,** 109.

Griffin, D. R. (1958). *Listening in the dark.* Yale University Press, New Haven, Conn.

Hafele, J. C. and Keating, R. E. (1972). *Science*, **177,** 166–70.

Hardy, G. H., Littlewood, J. E., and Polya, G. (1934). *Inequalities,* Chap. III. Cambridge University Press, Cambridge.

Holton, G. (1973). *Thematic origins of scientific thought*: *Kepler to Einstein*, pp. 189–90. Harvard University Press, Cambridge, Mass.

Ives, H. E. and Stilwell, G. R. (1938). *J. opt. Soc. Am.* **28,** 249–60.

—— (1941). *J. opt. Soc. Am.* **31,** 369.

Kennedy, R. J. and Thorndike, E. M. (1932). *Phys. Rev.* **42,** 400–18.

Kopff, A. (1923). *The mathematical theory of relativity* (transl. H. Levy), p. 52. Methuen, London.

Langevin, P. (1911). *Scientia* **10,** 31.

Marder, L. (1971). *Time and the space-traveller.* Allen and Unwin, London.

Michelson, A. A. and Morley, E. W. (1887). *Am. J. Science* (*Ser.* 3) **34,** 333–45; *Phil. Mag.* (*Ser.* 5) **24,** 449–63.

Miller, A. I. (1973). A study of Poincaré's "Sur la dynamique de l'électron". *Arch. Hist. exact Sci.* **10,** 208–328.

Milne, E. A. and Whitrow, G. J. (1938). *Z. Astrophys.* **15,** 344.

Møller, C. (1955). *K. Dan. vidensk. Selsk. mat.-fys. Medd.* **30,** (10), 1–28.

—— (1972). *The theory of relativity* (2nd. edn.), pp. 46–8, 292–8. Clarendon Press, Oxford.

Nerlich, G. (1976). *The shape of space*, pp. 266–7. Cambridge University Press, Cambridge.

Newman, D., Ford, G. W., Rich, A. and Sweetman, E. (1978). *Phys. Rev. Lett.* **40,** 1355–8.

Poincaré, H. (1904). *Bull. Sci. Math.* (Ser. 2) **28,** 302–24. (English transl. by G. B. Halsted, 1929. *The Foundations of Science*, p. 300. The Science Press, New York.)

—— (1905). *C. R. Acad. Sci. Paris* **140,** 1504–8.

—— (1906). *Rend. Circ. mat. Palermo*, **21,** 129–75.

Reichenbach, H. (1958). *The philosophy of space and time* (transl. M. Reichenbach and J. Freund), p. 127. Dover Publications, New York.

Rindler, W. (1970). *Am. J. Phys.* **38,** 1111–5.

Robb, A. A. (1921). *The absolute relations of time and space*, p. 13. Cambridge University Press, Cambridge.

—— (1936). *The geometry of time and space*, p. 12. Cambridge University Press, Cambridge.

Rossi, B. and Hall, D. B. (1941). *Phys. Rev.* **59,** 223.

Sanders, J. H. (1965). *The velocity of light.* Pergamon Press, Oxford.

Schlegel, R. (1974). *Am. J. Phys.* **42,** 183–7.

Schutz, J. W. (1973). *Foundations of special relativity: kinematic axioms for Minkowski space–time. Lecture notes in mathematics, No. 361* (eds. A. Dold and B. Eckmann). Springer-Verlag, Berlin, Heidelberg, New York.

Sherwin, C. W. (1960). *Phys. Rev.* **120,** 17.

Sigwart, C. (1895). *Logic* (transl. H. Dendy), vol. 2, pp. 242–5. Swan Sonnenschein, London.

Smith, J. A. (1931). In *The Works of Aristotle* (ed. W. D. Ross), vol. 3, *De anima*, (transl. J. A. Smith), 418b. Clarendon Press, Oxford.

Swenson, L. S. (1972). *The ethereal aether: a history of the Michelson–Morley–Miller aether-drift experiments, 1880–1930*, chap. 4. University of Texas Press, Austin, London.

Taton, R. (1957). *Reason and chance in scientific discovery* (transl. A. J. Pomerans), p. 135. Hutchinson, London.

Tolman, R. C. (1934). *Relativity, thermodynamics and cosmology*, p. 194. Clarendon Press, Oxford.

Törnebohm, H. (1952). *A logical analysis of the theory of relativity*, p. 20. Almqvist and Wiksell, Stockholm.

Walker, A. G. (1948). *Proc. Roy. Soc. Edinburgh* **62,** 319–35.

—— (1959). Axioms for cosmology. In *The axiomatic method with special reference to geometry and physics* (ed. L. Henkin *et al.*), pp. 308–21. North-Holland, Amsterdam.

WHITEHEAD, A. N. (1919). *An enquiry concerning the principles of natural knowledge.* Cambridge University Press, Cambridge.

WHITROW, G. J. (1933). *Q. J. Math.* (*Oxford*) **4,** 161–72.

—— (1935). *Q. J. Math.* (*Oxford*) **6,** 249–60.

WHITTAKER, E. T. (1953). *History of the theories of aether and electricity: the modern theories, 1900–1926*, p. 32. Nelson, London, Edinburgh.

6

SPACE–TIME AND COSMIC TIME

6.1. The concept of space–time

John Locke in his *Essay concerning Human Understanding*—which has been described by J. M. Keynes (1951) as 'the first modern English book'—after discussing space and time separately devoted a chapter to their joint consideration. At the end of that chapter he wrote, 'To conclude: expansion and duration do mutually embrace and comprehend each other; every part of space being in every part of duration, and every part of duration in every part of expansion'. He then made the prophetic comment: 'Such a combination of two distinct ideas is, I suppose, scarce to be found in all that great variety we do or can conceive, and may afford matter to farther speculation'.

A century later, in the *Critique of Pure Reason*, Kant contended that in order to represent to ourselves the ideas of time and change we are compelled to appeal to the idea of space. In discussing the origin of our intuition of change, which he characterized as a 'combination of contradictorily opposed determinations in the existence of one and the same thing', he argued that we cannot conceive this intuition without an example drawn from 'external', i.e. spatial, perception: 'For in order to make even internal change thinkable we must represent time (the form of inner sense) figuratively as a line, and the internal change by the drawing of this line (motion) and so we are obliged to employ external perception in order to represent the successive existence of ourselves in different states'. Nevertheless, even though J. A. Gunn (1929) exaggerated when he claimed that Locke's suggestion was 'ignored' by Kant, the fact remains that Kant was concerned only with the *a priori* justification of Galileo's geometrical representation of time. No essentially new idea concerning the union of space and time was advanced until the present century.

On 21st September 1908, the mathematician Hermann Minkowski delivered his famous lecture 'Space and Time' to the members of the *Naturforscher-Versammlung* at Cologne. In his address he explained in semi-popular form the ideas on the formal unification of space and time which he had presented mathematically the previous year in his paper 'Die Grundlagen für die elektromagnetischen Vorgänge in bewegten Körpern'. Instead of asserting with Locke that *every* part of space is in *every* part of time and *every* part of time in *every* part of space, Minkowski (following Einstein) pointed out that 'Nobody has ever noticed a place except at a time, or a time except at a place'. A point of space at a point of time he called a *world point*,† and the totality of all conceivable world points he called the *world*. A particle of matter or electricity enduring for an indefinite time will correspond in this representation to a curve which he called a *world line*,‡ the points of which can be labelled by successive values of a parameter t associated with a clock carried by the particle. 'The whole universe', he wrote, 'is seen to resolve itself into similar world lines', and he suggested that 'physical laws might find their most perfect expression as reciprocal relations between these world lines' (Minkowski 1923).

Minkowski's object was to provide a new substitute for the Newtonian absolute space and absolute time discarded by Einstein. In their place he advocated his absolute 'world' which gave different 'projections' in space and in time for different observers (associated with inertial frames of reference). This absolute 'world' was later called *space–time*. Mathematically, its absolute character can be established as a direct consequence of Einstein's postulate of the invariance of the velocity of light (with respect to all inertial frames). Let a light path connect two 'neighbouring' world points (x, y, z, t) and $(x+\mathrm{d}x, y+\mathrm{d}y, z+\mathrm{d}z, t+\mathrm{d}t)$, where the co-ordinates are determined with reference to a particular observer A. Since the spatial distance between (x, y, z) and $(x+\mathrm{d}x, y+\mathrm{d}y, z+\mathrm{d}z)$ is given by $\sqrt{(\mathrm{d}x^2+\mathrm{d}y^2+\mathrm{d}z^2)}$ in Euclidean geometry, and since light travelling with speed c describes this distance in time $\mathrm{d}t$, according to A, it follows that

$$\mathrm{d}t^2-\frac{\mathrm{d}x^2+\mathrm{d}y^2+\mathrm{d}z^2}{c^2}=0. \tag{6.1}$$

Similarly, if primed symbols refer to the co-ordinates assigned to the same world points by another observer B also associated with an inertial frame, it follows that

$$\mathrm{d}t'^2-\frac{\mathrm{d}x'^2+\mathrm{d}y'^2+\mathrm{d}z'^2}{c^2}=0. \tag{6.2}$$

More generally, we can consider *any* two neighbouring world points not necessarily connected by a light path. Because the vanishing of the left-hand side of (6.1) always entails that of the left-hand side of (6.2),

† The terms *point instant* and *event* have since been used.

‡ This concept has affinities with Bertrand Russell's concept of *causal line* (Russell 1948). He contends that, if we abandon the classical philosophical concept of *substance* (still retained, however, by Emile Meyerson in *Identity and Reality*), then identity must be defined as a causal line, being a temporal series of events which indicates the persistence of 'something'—a constancy of structure, or of quality, or of gradient change of both, in short, anything which can be depicted as a world line, long or short, straight or curved.

and vice versa, it follows that

$$dt'^2 - \frac{dx'^2 + dy'^2 + dz'^2}{c^2} = \phi(x, y, z, t)\left\{dt^2 - \frac{dx^2 + dy^2 + dz^2}{c^2}\right\}, \quad (6.3)$$

where ϕ is some function of (x, y, z, t) not involving the differentials (dx, dy, dz, dt). By the principle of relativity, we can interchange primed and unprimed symbols throughout equation (6.3), and hence we deduce that

$$\phi(x, y, z, t)\phi(x', y', z', t') = 1.$$

The simplest possible form† of ϕ is clearly $\phi = 1$. We readily find from the Lorentz transformation that condition (6.3) does hold for this value of the function ϕ. Consequently, we deduce that, for all inertial frames of reference,

$$ds^2 = dt^2 - \frac{dx^2 + dy^2 + dz^2}{c^2} \quad (6.4)$$

is an invariant. We call ds the *space–time interval* between neighbouring world points.

In problems involving only one dimension of space, Minkowski space–time can be simply represented on paper by means of a *Minkowski diagram*, as in Fig. 6.1. The diagram applies to a particular inertial observer A whose own space–time locus, or world line, is given by the t axis. The world lines E_0L and E_0M are the space–time paths of light rays which leave A at any given event E_0 in A's experience in the directions of the positive and negative spatial x-axes, respectively. Similarly, $L'E_0$ and $M'E_0$ are the space–time paths of light rays arriving at A at E_0. More generally, we must imagine the space–time diagram to be drawn four dimensionally, the y axis and the z axis being similar to the x axis and forming with it an orthogonal triad. The aggregate of theoretically possible light rays leaving A at E_0 generate the *forward light cone* at this event, and the aggregate of those arriving at A at E_0 generate the *backward light cone* at E_0. All paths of material particles, or other objects, moving relative to A with speeds less than c will be represented by world lines which are everywhere inclined at smaller angles to the t axis than the generators of the light cones, the slopes relative to the t axis depending on the respective speeds. The diagram provides a simple and beautiful distinction between inertial and accelerated motion, the former corresponding to straight world lines, for example lines 1, 2, and 3, and the latter to curved world lines, for example line 4.

† Other forms of ϕ are possible when non-inertial frames of reference are introduced.

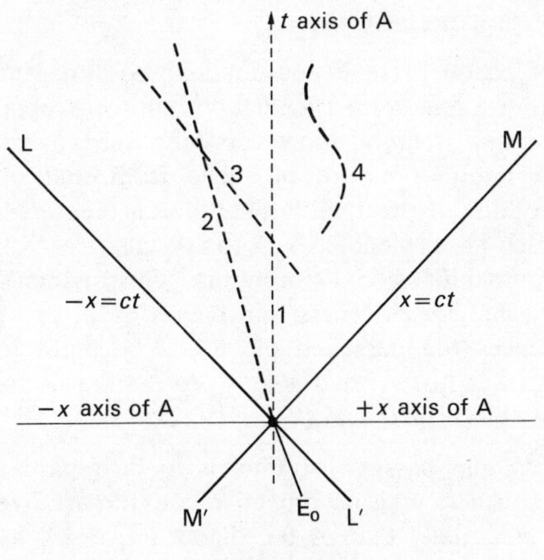

FIG. 6.1

The geometry of the Minkowski diagram differs from that of ordinary Euclidean space based on the positive definite metric $d\sigma^2 = dx^2+dy^2+dz^2$ because the space–time metric ds^2, given by equation (6.4), involves negative signs. Minkowski space–time is, however, similar to Euclidean space in being open in *every direction*. Consequently, no pair of rectilinear world lines (inertial lines) can meet more than once. Nevertheless, this property of the Minkowski diagram does not settle the question of the global structure of the universe which may conceivably be closed in its spatial directions.

We have already noted that ds^2 is an invariant for all observers associated with inertial frames, although neither the time-interval component dt nor the space-interval component $d\sigma$ possesses this property separately. Minkowski attached the greatest importance to this result. He wrote ds^2 in the form $c^2\,dt^2-(dx^2+dy^2+dz^2)$, so that its physical dimensions were those of the square of a *length* in a four-dimensional world with a pseudo†-Euclidean geometry. In his enthusiasm he exclaimed, 'Henceforth space by itself, and time by itself, are doomed to fade away into mere shadows, and only a kind of union of the two will preserve an independent reality' (Minkowski 1923). This famous, but excessive, claim tended, however, to reduce the importance of time much more than that of space. Indeed, Minkowski space–time was envisaged as a new kind of hyperspace in which events do not 'happen' but we 'come across them'.

† Because of the negative signs.

As Hermann Weyl expressed it,

> 'The scene of action of reality is not a three-dimensional Euclidean space, but rather a *four-dimensional world, in which space and time are linked together indissolubly*. However deep the chasm may be that separates the intuitive nature of space from that of time in our experience, nothing of this qualitative difference enters into the objective world which physics endeavours to crystallize out of direct experience. It is a four-dimensional continuum, which is neither "time" nor "space". Only the consciousness that passes on in one portion of this world experiences the detached piece which comes to meet it and passes behind it, as *history*, that is as a process that is going forward in time and takes place in space' (Weyl 1922).

In other words, the *passage* of time is to be regarded merely as a feature of consciousness with no objective counterpart. Weyl's view, like Einstein's, was essentially that of the 'block universe', to use the term coined by William James to denote the hypothesis† that the world is like a film strip: the photographs are already there and are merely being exhibited to us. Although, as Weyl says, the four-dimensional continuum is neither 'time' nor 'space', it is nevertheless more spatial than temporal. Indeed, one is reminded of Wagner's lines in *Parsifal*, Act 1, Scene 1: 'Du siehst, mein Sohn, zum Raum wird hier die Zeit'. ('You see, my son, time changes here to space').

The first philosopher who tried to construct a metaphysical system on the space–time hypothesis was Samuel Alexander whose Gifford Lectures for 1916–18 were published in 1920 under the title *Space, Time and Deity*. He maintained that space and time by themselves are abstractions from space–time, and 'if they are taken to exist in their own right without tacit assumption of the other they are illegitimate abstractions of the sort which Berkeley censured ... The real existence is Space–Time, the continuum of point-instants or pure events' (Alexander 1920, p. 48). Alexander regarded this space–time as a Cartesian *plenum* which is the synthesis of all of its perspectives, a perspective being the relation of space–time to any one of its constituent point instants. He claimed that this conception of the universe, which he had arrived at metaphysically, harmonized with the mathematical-physical hypothesis of Minkowski. According to Alexander, 'all things no matter what their qualities are bits of space-time' (Alexander 1920, p. 223). In particular, empirical things are 'vortices or eddies in the stuff of Space–Time, and universals are the laws of their construction' (Alexander 1920, p. 226).

† For further discussion of this hypothesis, see p. 348.

Alexander's concrete conception of space–time was submitted to penetrating criticism by C. D. Broad who pointed out that the Special Theory of Relativity did not break down the *distinction* of space and time but only their *isolation*. Moreover, having got rid of the absolute theory of space and time we must not introduce it again for space–time. Space–time should not be regarded as a kind of generating matrix, for it no more creates events than the framework or organization of an army creates wars. Moreover, if anyone were led to suppose that such organizations 'were substances that existed side by side with the soldiers he would be talking nonsense; and it would be the same kind of nonsense as is talked by people who imagine Space–Time to be an existent substance which pushes and pulls bits of matter about' (Broad 1923). When we discuss the properties of physical space–time all we are doing is simply analysing the general structure of the spatio-temporal totality which is the universe.

It is true that Alexander was a metaphysician and not a scientist, but his attitude towards space–time was broadly similar to that of many scientists, following the publication in 1915 of Einstein's General Theory of Relativity. In this theory Minkowski's concept was generalized to include gravitational phenomena. We have seen that in Minkowskian space–time the mathematical distinction between straight and curved world lines corresponds precisely to the physical distinction between inertial and accelerated frames of reference. The motion of particles and bodies associated with accelerated frames depends on the action of *forces*, whereas that of inertial particles and bodies is 'free' motion occurring only in the absence of force. We may therefore correlate the action of force on a particle with the curvature of its world line. Now, the uniform motion of a particle free from force can be regarded as a purely kinematic phenomenon because it is independent of inertial mass. Einstein pointed out that Galileo's law of falling bodies signifies that in a uniform gravitational field the accelerated motion of particles can also be regarded kinematically, since all bodies fall according to the same law. This is because

(i) in Newtonian terminology, gravitational mass has been found by experiment to be identical (at least to one part in a hundred million†) with inertial mass;
(ii) *locally*, i.e. within a region small compared with the dimensions of the earth, the gravitational field can be regarded as uniform.

Hence, Einstein argued, within any region small enough for the gravitational field therein to be effectively uniform, acceleration and gravitation

† In recent years this accuracy has been improved to one part in 10^{12}.

are interchangeable concepts (Einstein's *principle of equivalence*). Consequently, gravitation became synonymous with the 'curvature' of space–time, manifesting itself in the bending of light rays and the deviations in the motion of material particles from uniformity, i.e. rectilinearity in Minkowski space–time. The fact that the principle was valid only locally meant that Einstein was obliged to concentrate on the microstructure, or differential geometry, of space–time, whereas Minkowski had been concerned with its structure 'in the large'. However, his analysis was more powerful than Minkowski's, for it automatically took gravitation into account through the field equations connecting the differential geometry of space–time with the energy–momentum tensor of matter and radiation. This intimate association of matter (and energy) with the geometry of space–time led many supporters of Einstein's theory to embrace the view, expressed by Eddington, that 'When we perceive that a region contains matter we are recognising the intrinsic curvature of the world . . . We need not regard matter as a foreign entity causing a disturbance in the gravitational field; the disturbance is matter' (Eddington 1920, p. 190). Just as light does not cause the electromagnetic field to oscillate, for the oscillations constitute the light, and similarly just as heat is the motion of molecules and not something causing that motion, so matter should itself be regarded as 'a symptom and not a cause'. Moreover, whereas it was generally held that by his Special Theory Einstein had consigned the universal ether to oblivion, following the introduction of the General Theory Eddington suggested that the *world*, defined as the aggregate of all point instants, 'might perhaps have been legitimately called the *aether*; at least it is the universal substratum of things which the relativity theory gives us in place of the aether' (Eddington 1920, p. 187).

Eddington drew attention to the fact that Einstein's theory of matter was 'anticipated with marvellous foresight' by the English mathematician W. K. Clifford, who suggested, in an article written in 1875, that 'the theory of space-curvature hints at the possibility of describing matter and motion in terms of extension only' (Clifford 1879). Einstein had, however, an even more remarkable forerunner, René Descartes, for they both aimed at the *geometrization* of physics. The basic philosophy of General Relativity may indeed be fittingly described as neo-Cartesian, because of its emphasis on the extensional, rather than the temporal, aspects of phenomena.

6.2. Space-time and time

We have already seen that in his treatment of Special Relativity, Minkowski wrote ds^2 in a form having the dimensions of the square of a length. If, instead, we write it in the form $dt^2-(dx^2+dy^2+dz^2)/c^2$, then ds has the dimension of time. When ds^2 is positive, so that

$$\left(\frac{dx}{dt}\right)^2+\left(\frac{dy}{dt}\right)^2+\left(\frac{dz}{dt}\right)^2<c^2,$$

we can write the left-hand side of this inequality as V^2, where V denotes the uniform relative velocity, according to the original observer (whom we associate with an inertial frame), of a hypothetical second observer (with whom we can associate another inertial frame) who is at point (x, y, z) at time t and at point $(x+dx, y+dy, z+dz)$ at time $t+dt$. Consequently,

$$ds = dt\sqrt{(1-V^2/c^2)}. \tag{6.5}$$

Since ds is an invariant, it follows that it must be the interval of time which would be recorded by an inertial clock moving, according to the first observer, from (x, y, z) at time t to $(x+dx, y+dy, z+dz)$ at time $t+dt$. Thus, when its square is positive, the physical significance of ds is that it represents proper time, the relativistic substitute for the absolute time of Newtonian physics.

This interpretation provides a clear visual representation of the situation leading to Einstein's clock paradox discussed in the previous chapter. In Fig. 6.2, line 1 denotes the world line according to A of the clock carried by A from event E_0 (when B leaves A) to event E_1 (when B returns to A). Lines 2 and 3 denote two possible world lines according to A of the clock carried by B. Along line 2 the relative velocity of B is

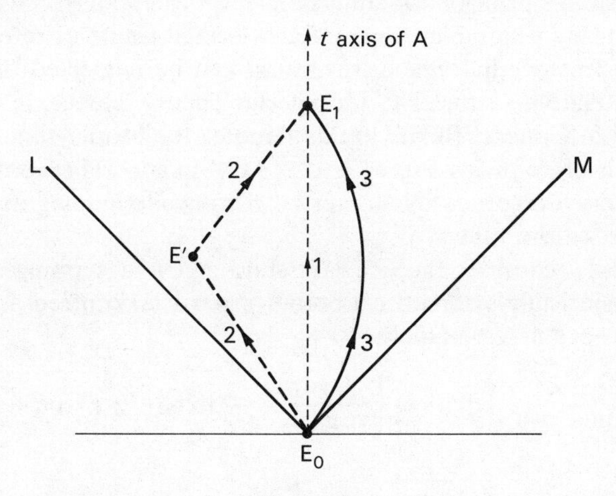

FIG. 6.2

always uniform, but it changes discontinuously at event E′, whereas along line 3 it changes continuously. By virtue of equation (6.5), it follows that world line 1 is the line of maximum 'length' joining E_0 and E_1, i.e. proper time between E_0 and E_1 is greater along this line. If line 2 is divided into a large number of small segments each of which has the same projection δt on the t axis of A, it is clear that the 'length' of each of these segments along line 2 must be less than δt, and hence that their sum must be less than the proper time from E_0 to E_1 according to A. The element of proper time $\mathrm{d}s$ is not a complete differential but depends on the world line followed. In the case of a continuously curved line, such as line 3, the comparison with line 1 is often based on Einstein's 'unprovable assumption' that the rate of the moving clock at any instant depends solely on the relative velocity at that instant and is independent of its acceleration. Nevertheless, this assumption (which we have previously seen to be by no means obviously true) is a side issue as far as the clock paradox is concerned. The main result is already given if B follows line 2, and generally the proper time which elapses between E_0 and E_1 must depend on the particular world line followed. There is no contradiction with the principle of relativity since, in general, only one straight line (inertial line) can be drawn joining the two events. The time lag of B's clock compared with A's clock when they meet at E_1 is due to the fact that the two clocks have followed different *types* of world-line.

In General Relativity the basic invariant $\mathrm{d}s^2$ is taken as the metric or 'distance' element of space–time and is given by a quadratic differential form† $g_{ij}\,\mathrm{d}x^i\,\mathrm{d}x^j$ which reduces *locally*, with appropriate choice of space and time co-ordinates, to the Minkowskian form of Special Relativity. For, by Einstein's principle of equivalence, we can abolish the local effect of gravitation by a suitable choice of accelerated frame of reference, and Einstein postulated that when gravitation can be neglected the General Theory of Relativity reduces to the Special Theory. Hence, in the case of positive $\mathrm{d}s^2$ in General Relativity, $\mathrm{d}s$ denotes the proper time associated with the pair of point-instants x^i and $x^i+\mathrm{d}x^i$ in general co-ordinates, i.e. the time interval recorded by an inertial clock while moving from the one point-instant to the other.

For example, consider the Schwarzschild metric describing space–time outside a spherically symmetrical body (or particle) centred at the origin of spatial co-ordinates, namely

$$\mathrm{d}s^2=\left(1-\frac{2Gm}{c^2r}\right)\mathrm{d}t^2-\frac{1}{c^2}\left\{\frac{\mathrm{d}r^2}{1-2Gm/c^2r}+r^2\,\mathrm{d}\theta^2+r^2\sin^2\theta\,\mathrm{d}\phi^2\right\} \quad (6.6)$$

† Einstein's summation convention operating, for i, $j=1, 2, 3, 4$.

where G is the constant of gravitation, c the speed of light (at great distances from the origin), m the mass of the body and (r, θ, ϕ) are polar co-ordinates. Let A and B denote two points fixed in space (*not* space–time). Consider a light signal which leaves A at time t_A and arrives at B at time t_B. Since the field is static, it follows that a light signal which leaves A at time $t_A + \delta t$ will arrive at B at time $t_B + \delta t$. The proper-time interval between the emission of these signals at A is $\delta s_A = \delta t \sqrt{(1 - 2Gm/c^3 r_A)}$ and the proper-time interval between the reception of these signals at B is $\delta s_B = \delta t \sqrt{(1 - 2Gm/c^2 r_B)}$, where r_A, r_B denote the distances of A and B from the origin. Let the signals be emitted with frequency ν_A at A and be received with frequency ν_B at B. Then, since the number of signals emitted at A is equal to the number received at B, it follows that $\nu_A \delta s_A = \nu_B \delta s_B$. Hence, if B is very remote from the origin compared with A, a standard spectral line (i.e. a definite line due to a specific atomic transition) at A when compared with the corresponding line at B will appear to be redder, the proportional change in frequency being approximately

$$\frac{\delta \nu}{\nu} = \frac{Gm}{c^2 r_A}. \tag{6.7}$$

This way of deriving Einstein's celebrated formula for the gravitational red shift (in a spherically symmetrical gravitational field) is due to McCrea (1956). Incidentally, from (6.6) the time-dilation factor for velocity V in the gravitational field of m is $(1 - 2Gm/c^2 r - V^2/c^2)^{-\frac{1}{2}}$.

The time-gain factor gh/c^2 referred to in the discussion of the Hafele–Keating experiment on pp. 258–9 results from taking Gm/r^2 to be the gravitational acceleration g and the difference between two nearly equal values of r to be h, the ratio $Gm/c^2 r$ being very small when m and r have the values of the mass and radius of the earth respectively. Actually, a very precise experimental test of formula (6.7) is possible with the aid of the effect discovered by Mössbauer (1958) and named after him. Mössbauer showed that some γ-rays are emitted by excited nuclei in the crystal lattice of a solid without individual nuclear recoil, the recoil momentum being delivered to the crystal lattice as a whole. The remarkably sharp spectral line that can be obtained under strict temperature control makes possible extremely accurate determinations of frequency. As a result, Pound and Rebka (1960) were able to measure the gravitational red shift within a terrestrial laboratory. For a vertical height of 22.6 metres in a laboratory tower the value of gh/c^2 is 2.46×10^{-15}. With source and absorber at the top and bottom of the tower, respectively, Pound and Rebka confirmed the Einstein gravitational red shift prediction to within 4 per cent. A repetition of this experiment a few years later by Pound and Snider (1964) improved the accuracy to about 1 per cent. Previously, it had only been possible to test the formula by wavelength measurements of light from the sun and white dwarf stars. The difficulty in these determinations was to eliminate other effects.†

In recent years the development of high-speed electronics and high-power radar has made possible the testing of another time-effect predicted by Einstein's General Theory of Relativity. This concerns the time delay in interplanetary radar echo. It was proposed by Shapiro (1964) that radar signals from earth be bounced off Mercury, preferably when it is at superior conjunction and the signals pass the sun at grazing incidence, precise measurements being made of the time taken. According to Einstein's theory the time taken exceeds what it would be if the sun's gravitation had no effect on the speed of light by an amount‡

$$\frac{1}{c}\left(\frac{4Gm}{c^2}\right)\left(1+\ln\frac{4r_1r_2}{r_0^2}\right),$$

to the first order in Gm/c^2, where m is the mass of the sun, r_0 is its radius, and r_1 and r_2 are the respective distances of the earth and Mercury from the sun. This comes out to be approximately 240 microseconds in a total time interval of approximately 23 minutes. Although there is no difficulty in telling time to within a microsecond over the time interval concerned, the experiment is an extremely difficult one to interpret precisely. Nevertheless, when it was first performed in 1967 good agreement was obtained between observation and theory (Shapiro *et al.* 1968).

Although Einstein's Special Theory of Relativity originated as a new theory of time, it has had profound repercussions on the theory of spatial measurement, in particular on the classical concept of the rigid rod. The concept loses its original simplicity, and we have therefore argued that for this and other reasons the measurement of space should be based on the measurement of time. On the other hand, the General Theory of Relativity as originally developed by Einstein was primarily a quasi-spatial theory (based on Riemannian geometry) in which time played a subsidiary part. We have already advocated a different view, at least in the case of positive ds^2, but the most conclusive argument for regarding time as fundamental in both branches of relativity depends on certain inescapable limitations on the accuracy of measurement of space–time intervals. These limitations are independent of the sign of ds^2.

In 1938, in his Tarner Lectures, Eddington made the point that

† For a history of the solar determinations see Forbes (1961).

‡ A rigorous derivation of this result and careful comparison with observation, taking due account of the identification of the theoretical coordinate system with the spherical polar coordinates of celestial mechanics, has been given by McVittie (1970).

'relativity theory has to go outside its own borders to obtain the definition of length, without which it cannot begin. It is the microscopic structure of matter which introduces a definite scale of things' (Eddington 1939). Nevertheless, as he reminded his listeners, even microscopic theory is not self-sufficient, for its concepts must be related to measurements which the experimenter can actually make and these ultimately depend on 'our own gross sense-organs'. Twenty years later these questions were considered in detail by Salecker and Wigner (1958). They began by stressing that it is not merely possible but *essential* to use clocks to measure both spatial and temporal intervals and to avoid the use of measuring rods, for, in contrast to clocks, rods are necessarily macrophysical objects which during the process of measurement will strongly influence other objects through their gravitational fields. If an accurate micro-clock existed and the recoil of light signals could be disregarded, the space–time interval between two point instants A and B could be measured with arbitrary accuracy by a signalling experiment of the type considered previously and illustrated by Fig. 6.3. For, if A is taken at epoch t_A on the world line of the observer O, and t_1, t_2 are the retarded and advanced times, respectively, of B according to O, and the distances involved are small compared with the radius of curvature of space, then the space–time interval between A and B is given by $\sqrt{(T_1 T_2)}$, where $T_1 = t_1 - t_A$ and $T_2 = t_2 - t_A$. Also, the difference between the epoch assigned to B by the observer O and the time t_A is $\frac{1}{2}(T_2 + T_1)$, and the spatial distance of B from A is $\frac{1}{2}c(T_2 - T_1)$.

The point instants in question are identified as collisions between material objects and photons. Ultimately, of course, as Eddington recognized, all measurements must be recorded on some macroscopic object. If, however, this object were itself part of the clock, there could be no such thing as a micro-clock, even in principle. Salecker and Wigner

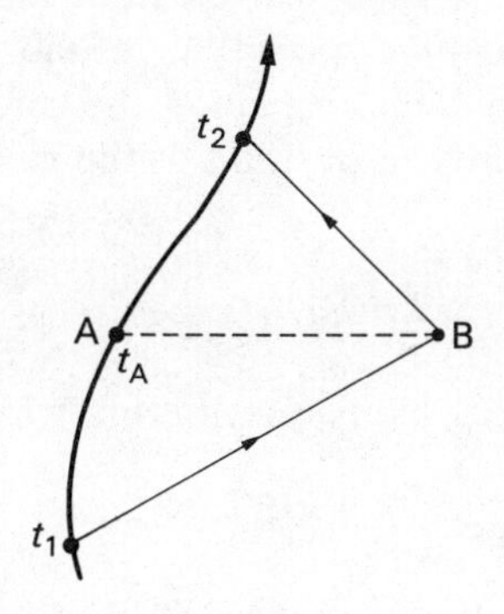

FIG. 6.3

therefore stipulated that the macroscopic recorder should be very distant compared with the displacements of the micro-clock during the process of measurement. However, here a peculiar difficulty is encountered, for a light quantum will reach the recording instrument 'with certainty' only if it does not spread out in all directions. Consequently, in their paper the authors confined attention to a world with only one spatial dimension, in addition to the time-like dimension. The main drawback to this assumption is that in such a world, unlike one of more dimensions, the influence of a body does not necessarily decrease with distance. Therefore, as they admitted, it is not obvious in this case that the disturbing influence of the macroscopic recorder can be reduced merely by placing it far enough away. However, they thought that this difficulty might not be too serious, since the micro-clocks which can be used in practice are 'not wholly microscopic' and the need to focus their signals ought not to modify the results of the investigation 'too much'.

The main problem of this investigation can be regarded as the analogue for time measurement of the classical problem of the ultimate limit of spatial magnification by optical microscopes. Salecker and Wigner found that, although the accuracy of reading of a clock increases with its mass, its gravitational field must not have too disturbing an effect; however, this field also increases with mass. The problem, therefore, is to construct a clock which shall be both as accurate and as light as possible.

Since the clock must not be deflected too much by being read, its mass M must exceed a certain minimum value depending on the accuracy τ with which time is to be measured and the running time T of the clock, i.e. the total time for which it is required to function. When a signal is emitted the velocity of the clock acquires a 'spread' $h/Mc\tau$, where h is Planck's constant.† If we stipulate that the time at which the photon strikes the clock is predetermined to within τ, i.e. that the position of the clock does not introduce a statistical element into the measurement of time, then the corresponding spread λ in the position of the clock throughout T must be less than $c\tau$. Hence, the spread in velocity is of the order $h/M\lambda$. The uncertainty in position of the clock after a time interval T is

$$\lambda + hT/M\lambda < c\tau.$$

The minimum value of this expression occurs when

$$\lambda = \surd(hT/M),$$

† The uncertainty in position of order $c\tau$ yields a corresponding minimum uncertainty of the order $h/c\tau$ in momentum (Heisenberg's uncertainty principle).

and this is less than $c\tau$ if

$$M > \frac{h}{c^2\tau}\left(\frac{T}{\tau}\right). \tag{6.8}$$

The minimum mass uncertainty of the clock is $h/c^2\tau$, since by Heisenberg's principle the minimum energy uncertainty is h/τ. Hence, the minimum mass of the clock exceeds this minimum mass uncertainty by the ratio of the running time and the accuracy. Moreover, if we take the mass m of the proton (order of 10^{-24} gram) for this minimum mass uncertainty, then

$$\tau = \frac{h}{mc^2}, \tag{6.9}$$

which is of the order of 10^{-24} seconds. In the present state of knowledge, we may therefore regard this as *the absolute lower limit to measurable time intervals.*†

6.3. Cosmic time and the expanding universe (i)

In Chapter 1 we saw that it has long been recognized that the concepts of time and the universe stand in a peculiarly intimate relation to each other. The discovery that there is no world-wide simultaneity for observers associated with inertial frames of reference seemed to destroy this relationship and to demolish the ancient concept of universal time. However, just as Einstein in effect resurrected the idea of the ether in 1915 when he formulated his concept of gravitational space–time with an intrinsic structure which was not homogeneous throughout, so in his no less famous paper of 1917 on the application of the General Theory of Relativity to cosmology he re-introduced the idea of *cosmic time.* Associated with this idea was the concept of a homogeneous spatial substratum determined by the distribution of matter in bulk throughout the universe, local irregularities being regarded as 'smoothed-out'. Unfortunately, the precise relationship between the concepts of Einstein's General Relativity and his cosmological theory were not analysed by him systematically and some confusion resulted.

The reason for this confusion was that Einstein's theory of local gravitation was developed first and his theory of world gravitation later. Consequently, the associated theory of cosmic time and space arose as a particular application of his general theory of the fine structure of

† It will be noted that this agrees with the higher of the values obtained for the chronon on p. 204.

space–time. Instead, we ought to take the 'smoothed-out' universe as the ultimate frame of reference and regard General Relativity as primarily a technique for analysing local gravitational fields superimposed on the world-gravitational field. The question then arises whether we are obliged to retain the characteristic features of this essentially local theory when studying the universe as a whole. In fact, when he came to consider the latter, Einstein himself decided to modify his theory by introducing into the field equations a new term involving a new constant of nature, the so-called 'cosmical constant' Λ.

The finite homogeneous world model discovered by Einstein (1917) as a consequence of introducing this term was a static system in spherical (or elliptic†) space. As was soon pointed out by Eddington (1920, p. 163), in the Einstein universe universal space and time are restored for phenomena on a cosmical scale and 'relativity is reduced to a local phenomenon'. Although Eddington 'was inclined to look on the limitation rather grudgingly', he made the important point that the theory of relativity is not concerned to deny the impossibility of cosmic, or world-wide, time 'but to deny that it is concerned in any experimental knowledge yet found'. He therefore argued that we need not be perturbed if the concept of universal time appears in 'a theory of phenomena on a cosmical scale, as to which no experimental knowledge is yet available', and he concluded that just as each limited observer has his own particular separation of space and time, so 'a being coextensive with the world might well have a special separation of space and time natural to him'. Thus, Eddington endeavoured to explain away Einstein's re-introduction of cosmic time by regarding this concept as one of the prerogatives of Newton's ubiquitous Deity, and this is beyond the scope of experimental science.‡

Shortly after Einstein had announced the discovery of his model universe, the Dutch astronomer de Sitter published an important paper (de Sitter 1917) in which he obtained the remarkable and unexpected result that an *empty* universe need not have the metric of Minkowski space–time (the limiting form of the metric of General Relativity far from all gravitating matter). De Sitter's discovery was a direct consequence of Einstein's introduction of the cosmical constant into the equations for the gravitational field, for, if this constant is not zero, the Schwarzschild metric of space–time in the presence of a gravitating particle of mass m at

† The distinction is a topological one (Whitrow 1959).

‡ This argument invites contrast with Newton's own argument, partly based on his actual experiment with the rotating bucket, for the existence of absolute space which, it will be remembered, he regarded as 'the Sensorium of God'!

the origin becomes

$$ds^2 = \left(1 - \frac{2Gm}{c^2 r} - \tfrac{1}{3}\Lambda r^2\right) dt^2 - \frac{1}{c^2}\left\{\frac{dr^2}{1 - 2Gm/c^2 r - \frac{1}{3}\Lambda r^2} + r^2(d\theta^2 + \sin^2\theta\, d\phi^2)\right\} \quad (6.10)$$

and if we put $m = 0$, which is equivalent to supposing that the world is completely empty, we obtain the metric

$$ds^2 = (1 - \tfrac{1}{3}\Lambda r^2)\, dt^2 - \frac{1}{c^2}\left\{\frac{dr^2}{1 - \frac{1}{3}\Lambda r^2} + r^2(d\theta^2 + \sin^2\theta\, d\phi^2)\right\}. \quad (6.11)$$

This reduces to the Minkowskian form if, and only if, $\Lambda = 0$. If $\Lambda \neq 0$, equation (6.11) determines a four-dimensional space–time of constant curvature $\frac{1}{3}\Lambda$. Whereas the Einstein universe given by the metric

$$ds^2 = dt^2 - \frac{1}{c^2}\left\{\frac{dr^2}{1 - \Lambda r^2} + r^2(d\theta^2 + \sin^2\theta\, d\phi^2)\right\} \quad (6.12)$$

is ‘cylindrical’ in space–time (i.e. spherical in its spatial cross-sections given by $t = \text{constant}$, but open and ‘uncurved’ in the time-like direction), the de Sitter universe is a pseudo-hyperspherical space–time. In other words, if we formally replace t by iw, where $i = \sqrt{(-1)}$, we obtain a four-dimensional sphere. Since, however, time is given by t and not by w, it follows that, although sections orthogonal to the time direction are closed spheres, any section which includes the time direction is a hyperboloid of one sheet open at both ends in that direction.

An obvious difference between (6.11) and (6.12) is that the metric given by the latter reduces to dt^2 when dr, $d\theta$, $d\phi$ all vanish, whereas the former reduces to $(1 - \frac{1}{3}\Lambda r^2)\, dt^2$ and so differs from dt^2, except at the origin $r = 0$. However, Lemaître (1925) and also Robertson (1928) found that by introducing a new time variable τ and a new radial distance variable ρ given by the formulae

$$r = \rho e^{\tau/a}, \qquad \sqrt{\left(1 - \frac{r^2}{a^2}\right)} e^{t/a} = e^{\tau/a}, \quad (6.13)$$

where $a = \sqrt{(3/\Lambda)}$, the metric of the de Sitter universe becomes

$$ds^2 = d\tau^2 - \frac{e^{2\tau/a}}{c^2}\{d\rho^2 + \rho^2(d\theta^2 + \sin^2\theta\, d\phi^2)\} \quad (6.14)$$

so that ds reduces to the element of proper time $d\tau$ at all points for which ρ, θ, and ϕ are constants. It is now generally recognized that (6.14) is the physically significant form of the de Sitter metric, i.e. that ρ and τ, rather than r and t, are the appropriate co-ordinates to adopt. In this form the spatial cross-section $d\tau = 0$ gives a three-dimensional space element which increases with time τ, the distance between any two points which are given by fixed values of ρ, θ, and ϕ varying as $e^{\tau/a}$. Consequently, the de Sitter universe is not truly static. Strictly speaking, it is a world model which expands exponentially with time.

The de Sitter universe only appears to be static because of its emptiness. As de Sitter himself discovered, if a free particle is placed in it at distance r from the origin it automatically acquires an outward acceleration $\frac{1}{3}\Lambda c^2 r$. Therefore, once matter is introduced into this model it automatically becomes the scene of systematic motions. This is not surprising. The term involving the cosmical constant introduced by Einstein into the field equations corresponds, in Newtonian terminology, to a force of repulsion. At each point in the Einstein universe cosmical repulsion is exactly balanced by world gravitation. Hence, the Einstein universe is truly static. What de Sitter did was to eliminate world gravitation by taking the mean density of matter in the universe to be vanishingly small, so that only the effects of cosmical repulsion remained. Thus, as Eddington remarked, the changelessness of de Sitter's universe depended on there literally being no matter present, i.e. on 'the simple expedient of omitting to put into it anything that could exhibit change'. As the actual universe is neither perfectly motionless nor perfectly empty, the question arises, 'Shall we put a little motion into Einstein's world of inert matter, or shall we put a little matter into de Sitter's Primum Mobile?' (Eddington 1933, p. 46).

From the point of view of theory, the decisive answer was given by Eddington in 1930 when he discovered that the Einstein model universe is unstable: if its radius increases slightly, cosmological repulsion will exceed gravitational attraction and the radius will continue to increase still further, whereas, if initially the radius shrinks a little, world gravitation will be greater than cosmological repulsion and so the model will continue to contract. In Eddington's opinion, the history of the actual universe is a steady progression from the primeval Einstein state towards the ultimate de Sitter state of zero density.

Eddington's theoretical investigation was prompted by Hubble's announcement in 1929 of the empirical law correlating the red shifts in the spectra of the galaxies with their distances. Assuming that these red shifts were Doppler shifts indicating recessional motion, Hubble obtained the relation, since known as ***Hubble's law***,

$$v = r/T_0, \tag{6.15}$$

where v denotes radial velocity, r denotes distance, and T_0 is the same for all galaxies investigated, and is now called the *Hubble time*. The preponderance of red shifts in extragalactic spectra had been known for some years and had led many astronomers and others to pay particular attention to the de Sitter world model. The discovery of Hubble's law was delayed by the fundamental difficulty of determining a reliable distance scale for extragalactic objects. To determine T_0, it was necessary to advance beyond the so-called Local Group (comprising the Andromeda Nebula, the Milky Way, and some lesser systems such as the Magellanic Clouds) and estimate the distances of some galaxies outside. With the 100-inch telescope on Mount Wilson, Cepheid variables could be detected only in the local galaxies. With their aid, it was possible to estimate how far away these systems are. In some galaxies too remote for the detection of stars of this type Hubble found particular objects which he believed were the brightest constituent stars (other than novae). By comparing their apparent magnitudes with those of the brightest stars in galaxies whose distances had already been determined, he was able, in this step-by-step way, to estimate how far away the galaxies containing them are. When he correlated the distance of these galaxies with their spectral shifts he obtained the relation given by equation (6.15) and found that T_0 appeared to be nearly 2 thousand million years.

Since 1952 Hubble's distance scale for galaxies has been considerably modified, and this has affected his estimate of T_0. In that year Baade showed that the distances of all galaxies, including those in the Local Group, had been underestimated owing to a mistake in the distance scale for Cepheid variables. As a result, it came to be generally accepted that the distances previously assigned to galaxies must be multiplied by a factor of rather more than 2. Then, in 1958, Sandage announced that in galaxies outside the Local Group the objects that Hubble had taken to be bright stars are, in fact, regions of glowing hydrogen gas (known to astronomers as H II regions) of intrinsic luminosity about two magnitudes brighter than that supposed by Hubble on the assumption that they were stars. Consequently, the distances formerly assigned by Hubble to all galaxies outside the Local Group must be multiplied by a factor of between 5 and 10. Although it is generally recognized that a precise revision of Hubble's distance scale is very difficult to make because of the long chain of steps and many uncertainties involved, various lines of investigation have led Sandage and others to conclude that on the available evidence the best estimate of T_0 is about 19.4 thousand million years with an uncertainty of about 1.6 thousand million years (Sandage and Tammann 1976).

Although repeated attempts have been made to find some alternative explanation for the extragalactic red shifts, the recessional interpretation based on the hypothesis that these shifts are due to the Doppler effect has stood its ground well and is accepted by most astronomers.† Moreover, in so far as it has been possible to compare shifts in optical wavelengths with those in other parts of the spectrum, it has been found that the shift ratio $z = \delta\lambda/\lambda$ for lines in the spectrum of a given galaxy is independent of wavelength. This is a necessary, although not a sufficient, condition for the phenomenon to be due to motion. Consequently, the prevailing opinion among astronomers is that the extragalactic clusters are streaming away from the Local Group with velocities proportional to their distances (with possibly some modification of this proportionality at very great distances). This 'expansion of the universe' is generally assumed to be a cosmic phenomenon which would be observed from any cluster. In other words, our Local Group is not regarded as a unique centre of the universe from which the other clusters are receding, but instead we believe that there is a continual increase in the scale factor determining the size of the region occupied by any given set of clusters of galaxies.

The discovery of the expansion of the universe marked a major revolution in man's conception of the cosmos.‡ Here we are concerned primarily with its effect on man's idea of time. In Chapter 1 we raised the problem of a natural 'origin of time' suggested by the hypothesis that the universe has been expanding throughout its history. Before we can examine this question further we must consider the various theoretical possibilities which can arise if we relax the condition that the universe as a whole is static.

6.4. Cosmic time and the expanding universe (ii)

Basically, two different types of relation may be envisaged between matter as a whole and space, depending on whether we adopt the 'relational' or the 'absolute' theory of space. Although the former is usually associated historically with the name of Leibniz, it is in fact much older. According to this view, space is the nexus of spatial relationships of material objects. (A somewhat different, but related, hypothesis was advocated by Descartes who *identified* 'extension', i.e. space, with matter.) On the other hand, Newton believed that space was intrinsically

† In recent years the principle objectors have been those who accept the interpretation of 'discordant' red shifts advocated by Arp (1973).

‡ It is interesting to note that, whereas Newton, like Einstein in 1917, regarded the universe as being in much the same state throughout its history, Descartes assumed that the universe changes in time, beginning with chaos and gradually merging into the order which is now the world.

distinct from matter, that it existed in its own right and was therefore absolute. Although this positive idea of empty space was a cardinal feature of the philosophy of the seventeenth century Cambridge Platonists (who greatly influenced Newton), it was not held by other philosophers such as Locke† until after the publication of the *Principia* in 1687. Newton's theoretical belief in empty space was supported by his interpretation of the rotating bucket experiment‡ as a crucial test that rotational motion is absolute. Absolute space was associated by Newton with the class of inertial frames. This class defined the aggregate of all fixed directions, i.e. the 'compass of inertia', to use Gödel's convenient phrase (Gödel 1949b). Nevertheless, Newton was able to distinguish the class of inertial frames from all other conceivable classes of frames in uniform relative motion with the same compass of inertia only by choosing a particular point as the origin of a particular inertial frame. With this object in view, he arbitrarily identified the centre of mass of the solar system as 'the centre of the world' (Cajori 1934).

Despite his general support of Newton, Locke regarded the question of the relation of the 'place' of the universe to that of infinite space as insoluble, and he made the significant comment that 'when one can find out and frame in his mind clearly and distinctly the place of the universe, he will be able to tell us whether it moves or stands still in the undistinguishable inane of infinite space' (Locke 1690). Although Locke made no mention of it, Newton's rotating bucket experiment establishes

† As Professor James Gibson pointed out in his illuminating account of Locke's theory of space in his book *Locke's Theory of Knowledge and its Historical Relations* (Gibson 1931), Locke's views underwent a significant change between 1678 and 1690. In 1676–8 he published three papers on the metaphysics of space. In the first, he argued that space when separated from matter 'seems to have no more real existence than number has (*sine* enumeration) without anything to be numbered'. In the later papers, he inclined to the view that space is a relation, being in the case of bodies 'nothing but the relation of the distance of the extremities', and in the case of unoccupied space nothing but 'a bare possibility of body to exist'. Hence, 'space, as antecedent to body, or some determinate being, is in effect nothing', and its supposed infinity, though something we are 'apt to conceive', is not a property of any real being. However, in the *Essay concerning Human Understanding*, published in 1690, he argued that space exists in its own right as 'a uniform infinite ocean'. Since 'the distinction between "space itself", as something "uniform and boundless", and the extension of body which is presented to us in sense perception can hardly be regarded as the direct produce of Locke's own principles, it is natural to look for some external influence to account for the doctrine of the Essay'. Locke was a diligent student of Newton's *Principia*. Brewster in his *Memoirs of Sir Isaac Newton*, vol. 1, p. 339, quotes from Desaguliers, who says he was informed by Newton himself, that Locke (at that time an exile in Holland) asked Huygens if all the *mathematical* propositions in the *Principia* were true. When he was assured of this he examined the deductions from them and became a firm follower of Newton.

‡ See pp. 34–5.

only the existence of fundamental directions in the universe. These directions could be determined by the primary distribution of matter and motion and are compatible with world expansion (or world contraction). According to the 'absolute' theory, such expansion would be an expansion of the material universe into outer empty space, like the diffusion of a gas into a surrounding vacuum. According to the 'relational' theory, there is nothing—not even empty space—outside the universe and its expansion is simply a change in the scale relationship of the universe as a whole to the linear dimensions of typical constituents: for example, the diameter of a typical atom, the radius of an electron or proton, or the wavelength of a photon emitted in a specific atomic transition. The 'absolute' directions of the compass of inertia would be automatically determined by the directions of radial recession.

Since the idea of world expansion was first suggested, two different mathematical techniques have been invented for the construction of world models: the technique of an *expanding space* and the *kinematic technique*. It has been customary to regard these merely as two different mathematical methods, and indeed it has been shown that there exists a close relationship between them. Nevertheless, there is a vital philosophical difference, for the expanding-space technique is the natural concomitant of the relational concept of space, whereas the kinematic technique is most naturally associated with the idea of absolute space. Thus, in the one case there is motion *of* space and in the other motion *in* space, i.e. in the former space is the framework of all matter and this framework expands, whereas in the latter attention is concentrated on the type of motion of the fundamental particles† rather than the space structure (Infeld and Schild 1945).

Whichever technique is used, it is customary to follow Weyl (1923) in postulating that in every region of the universe which is sufficiently extensive there is a definite mean motion of matter, the deviations of the actual motions of individual macroscopic bodies in this region from this mean being relatively small (compared with the velocity of light) and unsystematic. A fundamental particle is then defined as a particle having this mean motion and a mass corresponding to the total amount of matter in the region. Associated with this fundamental particle is a space–time frame of reference which can be regarded as a fundamental frame for all macroscopic bodies in the region. The proper time associated with this frame functions as a mean local time for the region. There are two reasons why it is appropriate to call it *cosmic time*. First, in the homogeneous world models usually considered, it is found to function as a

† Idealizations of the principal aggregates of matter (clusters of galaxies).

universal time, like the time co-ordinate in the metrics of the Einstein universe and the de Sitter universe. Also, in each region it may be regarded as the time scale of the basic rhythm of the universe as revealed in local atomic and electromagnetic oscillations, etc. In practice, these natural clocks are subject to minor perturbations due to individual motions, local gravitational fields, and so on, but in principle these can be ironed out by statistical averaging. Consequently, cosmic time is essentially a statistical concept, like the temperature of a gas.

It is also customary to assume that in the space–time frame of each fundamental particle the spatial directions around any point in the three-dimensional space given by the equation $t = \text{constant}$ are indistinguishable. This postulate of spatial isotropy is also essentially a statistical postulate, valid only for the average distribution of matter within a sufficiently large region.

Each fundamental particle is at rest with respect to the local reference system considered. The totality of such particles corresponds to a family of geodesics in space–time each representing the associated proper time. The cross-section $t = \text{constant}$ of this family is a three-dimensional space which changes with lapse of time. According to the relational concept of space, we need not consider the motions of the individual particles but the sequence of changes of the space structure as a whole. It is interesting to find that this idea was anticipated in a general way as long ago as 1885 by Clifford when he remarked that space may have a constant curvature 'but its degree of curvature may change as a whole with time. In this way our geometry based on the sameness of space would still hold good for all parts of space, but the change of curvature might produce in space a succession of apparent physical changes' (Clifford 1885). In 1929, this idea was made precise when Robertson (1929) deduced that the space–time metric of a homogeneous and isotropic world model could be expressed in the form

$$ds^2 = dt^2 - \frac{1}{c^2} R^2(t)\, d\sigma^2, \tag{6.16}$$

where $d\sigma$ is the line element of a space of constant curvature and $R(t)$ is a function of the cosmic time t usually known as the *expansion factor.* This result was obtained by a different method by Walker (1937), and the metric (6.16) is called the *Robertson–Walker metric.*

The alternative kinematic view of the expanding universe was first systematically explored by Milne (1933) who constructed a world model formed by a continuous three-dimensional system of fundamental particles in uniform relative motion from an initial singular state of coincidence at time $t = 0$. From the point of view of an observer associated with any one of these particles all the other particles were assumed to be receding in Euclidean space, the whole system occupying the interior of an expanding sphere in this space. Later Milne found that, if the scale of time were changed from t to τ, where τ varied logarithmically with t, his world model could be depicted as a stationary system in hyperbolic space (constant negative curvature), each fundamental particle being at a fixed point in this space (Milne 1937). It thus appeared that this model could be regarded from both points of view, although when the fundamental particles were identified with particular points of space the associated time scale was not the uniform time of atomic vibrations. Indeed, Milne was led to abandon the hypothesis that there is a *unique* natural scale of time for any observer associated with a fundamental particle.

In the metric (6.16) the co-ordinate t is the proper time not only for a fundamental particle at rest at the origin but also for any other particle at rest with respect to the spatial co-ordinates. The occurrence in the model of this cosmic time is closely related to the fact that the associated three-dimensional spatial cross-section is determined solely by the fundamental particles, i.e. it is a relational space and not an absolute space with an independent existence of its own.† In view of its importance, we shall derive the metric (6.16) in a way which differs from the methods of Robertson and Walker but brings the relationship between cosmic time and world space to the fore.

We begin by adopting the relational hypothesis that the material universe in its large-scale feature can be identified with world space.‡ Moreover, we shall assume that the observer associated with any fundamental particle can choose a scale of length measurement so that, if he wishes, he can regard this space as static. In other words, he can take the universe itself as the basis for a possible scale of length. This point of view is, of course, perfectly compatible with the notion of expansion since all change is relative, for, if the universe is in fact expanding relatively to our

† It may be mentioned that in Milne's model where the fundamental particles move through space uniformly there is no cosmic time; the time assigned to a distant event E by an observer associated with any given fundamental particle is not the same as the proper time of E.

‡ If this is not the case, the relative motions of the fundamental particles must be regarded as occurring in a space which is not given solely by the fundamental particles themselves. Our assumption is not an *a priori* condition which must be satisfied, but purely a condition characterizing the class of world models to be discussed. Any such model can be regarded as defining the ultimate rigid body 'Alpha' postulated by Carl Neumann in his famous lecture at Leipzig in 1869 (Wilson 1950). It automatically provides the 'compass of inertia', i.e. the aggregate of fixed directions (relative to each fundamental particle) implied by Newton's laws of motion and the properties of rotating bodies. The methods of Robertson and Walker, based on the mathematical theory of continuous groups of transformations, show that our condition is not an independent postulate in addition to homogeneity and isotropy.

common material standards, then conversely the latter can be regarded as shrinking relatively to the size of the universe. As Eddington (1933, p. 90) has remarked, 'The theory of the "expanding universe" might also be called the theory of the "shrinking atom"'.

Next, we assume that an observer A associated with a given fundamental particle assigns distance r and theoretical epoch t to any event E in accordance with the axioms leading to the general homogeneous and isotropic rules obtained on p. 236), namely

$$r = \tfrac{1}{2}\{\xi(t_2) - \xi(t_1)\}$$
$$\xi(t) = \tfrac{1}{2}\{\xi(t_2) + \xi(t_1)\}$$

where t_1 and t_2 are the retarded and advanced times, respectively, of E as recorded by A on a clock which synchronizes with the natural scale of time at A, for example, on an 'atomic clock'. From these formulae it follows that

$$\left.\begin{aligned} \xi(t_2) &= \xi(t) + r \\ \xi(t) &= \xi(t_1) + r \end{aligned}\right\}. \qquad (6.17)$$

Since we have assumed that the universe can be described by A as static, the distance assigned by A to any other given fundamental particle B will be a constant. Therefore, taking event E to be at B it follows that r in equations (6.17) must be independent of time. Consequently, for any t_1 the corresponding values of t and t_2, associated with given B, will satisfy equations (6.17), where r will be a constant depending on the particular A and B concerned and the function $\xi(t)$ is such that its derivative denotes the velocity of light according to A.

If we still retain the smoothed-out universe as a scale of length measurement but consider different auxiliary time scales† associated with A, we shall obtain different expressions for the velocity of light (according to A), corresponding to different functions $\xi(t)$. In particular, if we choose an auxiliary time scale T functionally related to the time scale of equations (6.17) by the formulae

$$T_1 = \xi(t_1), \qquad T_2 = \xi(t_2), \qquad T = \xi(t),$$

we shall obtain the relations

$$T_2 = T + r, \qquad T = T_1 + r \qquad (6.18)$$

in place of (6.17). Hence, in terms of the universe as a hypothetical measuring rod and the auxiliary time variable T, the speed of light

† In general, these will not be natural time scales but will be obtained by mathematical transformations of the natural time scale.

according to A will now be a constant. Since any other fundamental particle B is at a fixed distance from A, it follows from the result established on p. 249 that the value of T theoretically assigned by A to any event E at B must be identical with the value of T' actually recorded by B, i.e. the value obtained by transforming the proper time t' of E at B by the formula $T' = \xi(t')$. Hence, on reverting to natural scales of time at A and B, we find that the time t theoretically assigned by A to any distant event coincides with the proper time t' of the event as recorded on the natural time scale associated with the local fundamental particle. Therefore, we must regard t in formulae (6.17) as being a world-wide time. The velocity of light is thus a universal function $\xi'(t)$ of this cosmic time, and so the spatial distance between any two events is also an invariant for all fundamental observers.

We have found that the relational concept of the universe, according to which there is no independent spatial background against which systematic changes in the geometrical structure of the universe occur, implies the existence of cosmic time.† To proceed further, we now postulate the local validity of the Special Theory of Relativity at all times and places in the smoothed-out universe, since all local gravitational fields have been ironed out into the general world background. The proper time $\mathrm{d}s$ between two neighbouring events in the experience of a Galilean observer when passing a fundamental particle A with relative velocity V is related to the time interval t to $t+\mathrm{d}t$ between these events according to A by the formula $\mathrm{d}s = \mathrm{d}t\sqrt{(1 - V^2/c^2)}$, where $c = \xi'(t)$, the velocity of light at the epoch t. If we write $\mathrm{d}\sigma$ for the distance between the positions of the two events in A's space-like cross-section ($t = \text{constant}$) of space–time, it follows that, since $V = \mathrm{d}\sigma/\mathrm{d}t$,

$$\mathrm{d}s^2 = \mathrm{d}t^2 - \frac{\mathrm{d}\sigma^2}{c^2}.$$

On replacing c by $\xi'(t)$, we obtain the formula

$$\mathrm{d}s^2 = \mathrm{d}t^2 - \frac{\mathrm{d}\sigma^2}{\{\xi'(t)\}^2}. \tag{6.19}$$

Since t, $\mathrm{d}t$, and $\mathrm{d}\sigma$ are invariants for observers associated with all fundamental particles, it follows that $\mathrm{d}s^2$ is also an invariant. Locally, the line element of the spatial cross-section given by the equation $t = \text{constant}$ must be Euclidean, because of the local validity of the Special Theory of Relativity, but when we refer to the frame of reference of any other

† It is interesting to note that when Milne first adopted the kinematic view of the recession of the galaxies he specifically rejected the concept of cosmic time.

fundamental particle this locally Euclidean metric will become the metric of a space of constant curvature, since this is the only type of space which is everywhere homogeneous, isotropic, continuous, and locally Euclidean.

Finally, we retain the cosmic time t but we choose a new scale of length $d\rho$ so that the local velocity of light is always the same constant c. For this to be so we take

$$d\rho = \frac{c}{\xi'(t)}\, d\sigma.$$

If we now write $R(t)$ for $c/\xi'(t)$, so that

$$d\rho = R(t)\, d\sigma,$$

we obtain

$$ds^2 = dt^2 - \frac{1}{c^2} R^2(t)\, d\sigma^2.$$

We have thus derived the standard form of world metric, i.e. equation (6.16), corresponding to recessional motion proportional to $R(t)$. The fundamental particles given by $d\sigma = 0$, i.e. by stationary points in the original spatial framework, are now regarded as embedded in a space of variable scale factor $R(t)$. The light paths are given by the null geodesics $ds = 0$.

Although we have verified that the metric of a homogeneous, isotropic world-model can be expressed by equation (6.16),† we have adduced no evidence so far for assigning a particular form to the scale factor $R(t)$, except for the condition that it must be an increasing function at the present time. Various theoretical arguments have been advanced for adopting a particular form of this function, but none has been generally accepted.

Retaining the general form $R(t)$ for the expansion factor, we can easily calculate the corresponding law of red shifts, i.e. the law correlating Doppler shifts and distances, for the region nearest to the observer. Since proper time ds is zero along a light ray, it follows from (6.16), on replacing $d\sigma$ by dr, that if a photon leaves a fundamental particle P at local time t' and is received by the observer associated with the particle O at the origin at local time t_0, then

$$\int_{t'}^{t_0} \frac{c\, dt}{R(t)} = [r],$$

† Subject to the condition that the model can provide a three-dimensional framework of reference.

where the symbol on the right denotes the *constant* r co-ordinate assigned to P by the observer at O. Consequently, if another photon emitted by P at local time $t'+\delta t'$, where $\delta t'$ is small, is received by O at local time $t_0+\delta t_0$, then

$$\frac{\delta t_0}{R(t_0)}=\frac{\delta t'}{R(t')}.$$

If ν' is the proper frequency of light signals emitted by P and ν_0 is the measured frequency of their reception at O, then, by equating the number received with the number emitted, we have

$$\nu_0\delta t_0=\nu'\delta t'.$$

Hence, we obtain the shift formula

$$\frac{\delta\lambda}{\lambda}=\frac{\nu'-\nu_0}{\nu'}=\frac{\delta t_0-\delta t'}{\delta t_0}=\frac{R(t_0)-R(t')}{R(t_0)}.$$

Consequently, if P is not too far from O and we replace $\delta\lambda/\lambda$ by v/c, where v is the velocity corresponding to this spectral shift, we find that

$$v\sim\left(\frac{\dot{R}}{R}\right)_0 r, \tag{6.20}$$

where $\dot{R}$ denotes the derivative of $R(t)$, the zero subscript indicates the value of the ratio $\dot{R}/R$ at the 'present' epoch, and r is the distance† of P from O.

If the approximate theoretical formula (6.20) is identified with the empirical result (6.15), the Hubble time T_0 must be related to the ratio $(\dot{R}/R)_0$ by the formula

$$\left(\frac{\dot{R}}{R}\right)_0=\frac{1}{T_0}. \tag{6.21}$$

Therefore, if we knew how Hubble's T varied with cosmic time t—instead of knowing only its present value T_0—we could integrate (6.21) to obtain

$$R(t)\propto\exp\int\frac{\mathrm{d}t}{T}. \tag{6.22}$$

In general, T will change with time. However, if it is a true constant of nature and hence independent of epoch, then $R(t)$ must be proportional to $\exp(t/T_0)$, as in the case of the empty de Sitter universe. The same form

† We have substituted the symbol r for $c(t_0-t')$.

of metric also characterizes the steady-state universe (Bondi and Gold 1948, Hoyle 1948) in which a uniform density is everywhere maintained by a postulated process of continual creation of new matter *ex nihilo* as old matter streams away under the influence of cosmic recession.† In both of these world models the total span of cosmic time is infinite, i.e. t can take all values. In the case of many other models depending on other forms of $R(t)$, the total range of time is restricted by the occurrence of singularities. For example, in the case of a uniformly expanding model, namely $R(t)$ proportional to epoch t, there is an initial singularity at $t=0$ when all distances were zero. In this case $T=t$, and so, if we adopt this model, the present value of the Hubble time T is a direct measure of the age of the universe.

Among the various other possibilities, two of particular theoretical interest are due to Einstein and de Sitter, and to Dirac, respectively.

(i) Einstein and de Sitter (1932) introduced an important world model (obeying the laws of General Relativity) in which both the spatial curvature and the cosmical constant are zero. In this model, known as the *Einstein–de Sitter universe*, $R(t)$ is proportional to $t^{\frac{2}{3}}$ and hence $t=\frac{2}{3}T$. Again there is a singularity at $t=0$, so that the present age of the universe would be given by two-thirds of Hubble's T_0.

(ii) Dirac (1938) introduced another model, in which the spatial curvature is zero and $R(t)$ is proportional to $t^{\frac{1}{3}}$. Again there is a singularity at $t=0$, but in this case the present age of the universe would be given by one-third of Hubble's T_0.

Both these models, and also Milne's uniformly expanding model and the steady-state model, are characterized by the fact that in them $G\rho T^2=$ constant, where G, ρ, and T denote the values at any given epoch t of the 'constant of gravitation', the mean local density, and the Hubble time, respectively. This product is dimensionless and will be comparable with unity if we take, in c.g.s. units, $G=6.66\times10^{-8}$, $\rho=10^{-29}$, $T=6\times10^{17}$. This particular order of magnitude of ρ would, however, imply that there must be considerably more matter in the universe than has yet been detected. Since ρ varies inversely as R^3, except in the steady-state theory where it is maintained constant by continual creation, and $T=R/\dot{R}$, it follows, if we assume that $G\rho T^2$ is a universal numerical constant independent of epoch, that G is proportional to $R\dot{R}^2$, i.e. the ratio of G and $R\dot{R}^2$ is independent of epoch. In Milne's model (uniform expansion), R varies as t and hence G varies as t. If, however, we postulate that G is

† In this world model the background space is not defined by the fundamental particles but by the velocity pattern of the model which is fixed in space for all time.

independent of epoch, as is the case in General Relativity, (and in Newtonian theory), then $R\dot{R}^2$ is constant and it follows, by adjusting the zero of t, that $R(t)$ varies as $t^{\frac{2}{3}}$ (Einstein–de Sitter model). In the steady-state model, G, ρ, and T are all independent of epoch, and hence $R(t)$ is proportional to an exponential function of t, as we have already seen.

Dirac's model is obtained by introducing the additional hypothesis that G varies inversely as T. His argument for this was the intriguing fact that the number of chronons (atomic units of time of the order of 10^{-24} seconds) contained in T_0 (the value formerly assigned to the Hubble time, namely 6×10^{16} seconds) is about order 10^{39}.† *This immense number is the order of magnitude of the ratio of the electrostatic force between a proton and electron to their gravitational attraction.* He therefore maintained that this ratio varies as T. Consequently, if we assume that the masses and charges of these elementary particles are constants, it follows that G must vary inversely as T. If we combine this conclusion with the above hypothesis concerning $G\rho T^2$, we find, by adjusting the zero of t, that $R(t)$ must vary as $t^{\frac{1}{3}}$. and hence that $t=\frac{1}{3}T$. (Incidentally, Dirac did not appeal to our postulate concerning $G\rho T^2$ but to an equivalent argument relating to ρ.)

The hypothesis that $G\rho T^2$ is a constant does not, however, apply to all world models. In any homogeneous and isotropic non-static world model *based on the field equations of General Relativity* with cosmical constant Λ, curvature index $k(k=1, 0, -1$ for positive, zero, and negative curvatures, respectively), and negligible pressure‡ it can be shown that

$$\frac{8\pi}{3}\left\{G\rho+c^2\left(\frac{1}{3}\Lambda-\frac{k}{R^2}\right)\right\}T^2=1. \tag{6.23}$$

In general, $G\rho T^2$ will vary with the cosmic time. For example, in the Eddington–Lemaître expanding model (Eddington 1933), in which both Λ and k are positive and which is assumed to have existed at one stage as an Einstein static universe in unstable equilibrium, $G\rho T^2$ is inversely proportional to y^3-3y+2, where y is the ratio of the radius of the model at any subsequent epoch to its radius in the equilibrium phase, when the reciprocal of T was effectively zero. According to this model, whether in

† Taking the present value of the Hubble time (the indicator of the time scale of the universe) to be of the *order* 10^{17} seconds and the unit of neurophysiological time (which presumably controls our thought processes) to be of the *order* 10^{-3} seconds (a millisecond), is it merely a curious coincidence that the ratio of the former to the latter is almost the same as that of the latter to (the maximum value assigned to) the chronon (10^{-24} seconds)?

‡ These world models are often called *Friedmann models* after the Russian scientist A. A. Friedmann (1888–1925) who first investigated them.

the Eddington form which began in the Einstein phase or in the Lemaître form which began as a violently explosive radioactive 'super-atom' and later passed slowly through the Einstein phase, the age of the universe must considerably exceed the current value of T.

Since Einstein's original reason for introducing Λ was to enable him to construct a static world model and most astronomers no longer believe that the universe is static, much more attention is nowadays paid to models with no cosmical constant. If we consider homogeneous isotropic expanding models that obey the field equations of General Relativity with $\Lambda = 0$, we find that a useful term to consider is the deceleration parameter q defined by

$$q = -R\ddot{R}/\dot{R}^2. \tag{6.24}$$

It follows from the field equations that, if the material pressure is negligible, q is related to the local proper density ρ by the simple formula

$$\rho = \frac{3H^2}{4\pi G} q \tag{6.25}$$

where H is the reciprocal of T (so that $\dot{R} = HR$). Also q is related to the curvature index k by the equation

$$2q - 1 = \frac{kc^2}{H^2R^2}. \tag{6.26}$$

The Einstein–de Sitter universe ($\Lambda = 0$, $\mathrm{k} = 0$) corresponds to $q = \frac{1}{2}$. This is easily seen to be a 'critical value'. For $q > \frac{1}{2}$, the curvature is positive and the density is sufficiently high to slow down and eventually reverse the expansion. The model expands from an initial point singularity to a state of finite maximum volume and then collapses to zero volume again. This model is therefore described as 'oscillating', the time for a complete cycle being approximately $2\pi T_0$, i.e. about 10^{11} years for Sandage's value of T_0. On the other hand, for $0 < \mathrm{q} \leqslant \frac{1}{2}$, the curvature is either negative or zero, the model expands from an initial singularity, and its motion is never reversed.

Great efforts have been made to determine empirically the current value of q, since knowledge of this parameter is needed in addition to that of H_0(or T_0) for our understanding of the nature of cosmic expansion and the determination of the age of the universe. According to Sandage and Tammann (1975), there are good reasons for believing that q_0, the current value of q, is less than the critical value 0.5. Indeed, they argue that it is definitely less than 0.28 and in all probability very small.† This conclusion is supported by a completely different investigation due to Peimbert and Torres–Peimbert (1974), according to which $q_0 \leq 0.02$. If this result is correct and indicative of q in the past, it means that the deceleration is negligible. Consequently, the expansion is effectively uniform, as predicted by Milne, and is likely to continue indefinitely. The age of the universe would thus be given by T_0, and so according to the latest estimate of this parameter by Sandage would be nearly 20 thousand million years.

A small, possibly zero, value of q_0 has also been suggested by some recent work of Lynden-Bell. He has sought to explain by means of a 'light-echo' theory the observations of transverse relative velocities in some radio galaxies that *apparently* greatly exceed the velocity of light and so might be thought to conflict with the theory of relativity. He finds that these anomalous *apparent* velocities can be adequately accounted for without infringing the basic assumptions of that theory if Hubble's parameter H_0 is doubled, thus reducing the Hubble time T_0 to about 10 thousand million years. This would rule out the possibility of the universe being of the Einstein–de Sitter type, because its age $2T_0/3$ according to that world model would be too low for stellar evolution theory. Hence, he concludes that $q_0 = 0.5$ is ruled out and that it is probable that the universe is open with a small value of q_0 (Lynden-Bell 1977).

Other more recent investigations, however, point to a value of q_0 that exceeds 0.5. Thus, Bruzuel and Spinrad (1978) have tried to estimate q_0 by examining the possibility of there being a characteristic size of galaxies. They find that the 'best solution' favours a value of $q_0 = 0.25 \pm 0.5$. The large error term makes possible not only the critical value 0.5 but even higher values of q_0. Indeed, a higher value of q_0 has recently been shown to be a definite possibility by Kristian, Sandage, and Westphal (1978) from a sample of galaxies out to red shift $z = 0.4$, *on the assumption that there has been no significant evolution of galaxies in the last* 4–5 *thousand million years.* Photometric studies definitely indicate that there has been no colour change in galaxies over that interval of time, although for larger values of z colours which started off by being monotonically redder become bluer. According to Sandage and his colleagues, if it were known with certainty that there has been no significant change in the luminosities of galaxies during the last 4 or 5 thousand million years or so and if the current value of the Hubble 'constant' H_0 is taken to be 55 kilometres per second per megaparsec, then the present data are 'nearly

† This result would rule out the steady-state universe with continual creation for which $q = -1$ and the oscillating universe based on general relativity (with $\Lambda = 0$) for which $q > \frac{1}{2}$.

good enough to show that the universe is closed', the value of q_0 being about 1.7 ± 0.4. This would mean that the current age of the universe is about 12 thousand million years and that the ultimate collapse of the universe in its contracting phase would occur in about 60 thousand million years time. However, if q_0 could be determined in some other way, so that the Hubble diagram (for z plotted against magnitude) could be used to determine fairly accurately the rate of change of galaxy luminosity with time, then if there were found to be an average dimming of galaxies of only 0.5 magnitudes during the past 5 thousand million years the data would be compatible with an effective zero value of q_0, as previously suggested by Sandage and Tammann (1975). 'It seems possible at present', Kristian, *et al.* write, 'to construct a self-consistent model with $q_0 = 0$ that satisfies the known data, but the case is not yet settled'.

To determine q_0, and hence decide whether distant objects were receding either faster or slower in the past than now and so discover if the expansion of the universe is slowing down, speeding up, or remaining uniform, we need to obtain data from as far away, and hence as far back in time, as possible. Since the intrinsically brightest celestial objects are the *quasars*, they can be observed at very great distances; consequently if they were all of more or less the same absolute brightness, they would be ideal distance indicators for investigating the very distant (and early) universe. Unfortunately, quasars vary in absolute brightness far more than galaxies do. This is not surprising, since their high luminosities make it very unlikely that they could remain active for more than a small fraction of the age of the universe. Consequently, at any given sufficiently remote distance (as indicated by a given red shift), we must expect to see quasars of all ages, newly formed ones coexisting with others nearing the end of their active phase. This is very different from the situation as regards galaxies, since those relatively nearby, and hence with smaller red shifts, are believed to be evolved versions of those farther off which display larger red shifts.

The red shifts of quasars are easily determined, but it is much more difficult to estimate their distances because of the relatively large variations in their intrinsic luminosities. We must therefore look for some different kind of distance indicator for them. Recently, Baldwin appears to have succeeded in this quest. He has found two spectral lines, one due to carbon and the other to magnesium, which seem to be strongly correlated with intrinsic continuum luminosity. This means that the relative absolute luminosities of quasars can be measured in a way that does not critically depend on distance. Consequently by comparing these measurements with the *apparent* luminosities of quasars reasonably reliable estimates of distance can be obtained. In this way, with the aid of observers at Lick Observatory, California, it has been possible to show not only that quasars are definitely very remote objects (this had previously been disputed by some astronomers) but also to obtain a new way of estimating q_0. It appears from this work that q_0 exceeds the critical value 0.5 significantly, and a tentative approximate value of 2 has been derived (Baldwin *et al.* 1978). The implied age of the universe, if H_0 is roughly 50 kilometres per second per megaparsec (T_0 about 20 thousand million years), is approximately 9 thousand million years, which is less than the current lowest estimates for the oldest stars (about 10 thousand million years), but further work along these lines may lead to a somewhat lower value of q_0 and hence to a correspondingly higher estimate for the age of the universe.

If, provisionally, we adopt a round value of 10 thousand million years for the current age of the universe, it would appear from this work that world expansion is slowing down at a rate indicating that after another 20 thousand million years the universe will start contracting and will ultimately collapse completely some 50 thousand million years from now.

6.5. The existence of cosmic time

The concept of the reality of simultaneity on which, in 1905, Einstein based his Special Theory of Relativity at first appeared to eliminate from physics any idea of an objective world-wide lapse of time according to which physical reality could be regarded as a linear succession of temporal states or layers. Instead, each observer was regarded as having his own sequence of temporal states and none of these could claim the prerogative of representing the objective lapse of time. Nevertheless, a quarter of a century later, theoretical cosmologists who made use of the physical ideas and mathematical techniques associated with relativity theory were led, as we have seen, to re-introduce the very concept which Einstein had begun by rejecting.

Commenting on this perplexing state of affairs, Sir James Jeans (1936) pointed out that before 1905 both the scientist and the plain man regarded the occurrence of events with the passage of time as being somewhat similar to the way in which 'the pattern of a tapestry is woven out of a loom'. Einstein's theory had taught us, however, that there was apparently no absolute distinction between past, present, and future. Consequently, the simplest view was that the tapestry was already woven and that we became aware of it bit by bit. In this 'block universe' human consciousness declined to a mere recording instrument. Later, however, time regained a real objective existence on the astronomical scale. This rehabilitation of the traditional concept began with the work of de Sitter in 1917, for, in constructing his world model, he postulated symmetry between space and time and found as a result that the universe must be totally devoid of matter. All later solutions of the cosmological problem in which the density of matter was not zero made a real distinction between space and time. This distinction was therefore evident once again as soon as we abandoned local physics and called the astronomy of the universe to our aid.

From the cosmological point of view adopted by Jeans, it would therefore seem that neither the equivalence of all observers in uniform relative motion (Special Relativity) nor of all observers in any form of relative motion (General Relativity) can be accepted without restriction. The space–time of Special Relativity is an abstract concept strictly applicable only in the absence of gravitional fields, i.e. when the background of the universe can be regarded as if it were effectively empty. Similarly, the equivalence of all frames of reference in General Relativity is compatible with an empty background as is tacitly assumed in problems of local gravitation. However, once the existence of a world-wide distribution of matter, albeit of extremely low mean density, becomes an essential feature of the problem under investigation, then certain frames of reference and observers must be specially distinguished, namely those which move with the mean velocity of matter in their neighbourhood.† In the cosmological models which we have discussed and to which Jeans was referring, the local times of all these 'privileged' observers fit together into one world-wide time called 'cosmic time'. Does it therefore follow that, despite the successes of relativity theory on the local scale, we must revert to the traditional idea of an objective universal time on the cosmic scale?

Surprisingly, Jean's challenge does not appear to have been countered by any follower of Einstein until 1949 when the mathematical philosopher Kurt Gödel (1949b) produced an ingenious new type of world model obeying the laws of General Relativity. Gödel agreed that acceptance of the relativity postulate, namely that all observers are equivalent as regards the formation of the laws of motion and interaction of matter and field, does not preclude the possibility that the particular arrangement of matter, motion, and field in the actual world may offer a more 'natural' or 'simpler' aspect to some observers than to others. However, unlike Jeans, he did not believe that the aggregate of local times associated with such a class of privileged observers must automatically constitute a universal time.

† This does not mean that we cannot postulate the *local* approximate validity of Special Relativity—in the absence of significant local gravitational fields—when constructing a world model. On the contrary, we have in fact already invoked this postulate.

Gödel argued that since the definition of cosmic time depends on the determination of the mean motion of matter in each region of the universe we can obtain only approximations to this concept. 'No doubt,' he wrote, 'it is possible to refine the procedure so as to obtain a precise definition, but perhaps only by introducing more or less arbitrary elements (such as, e.g., the size of the regions or the weight function to be used in the computation of the mean motion of matter). It is doubtful whether there exists a precise definition which has so great merits, that there would be sufficient reason to consider the time thus obtained as the true one' (Gödel 1949a). The answer to this criticism is that the same type of objection could be formulated against the concept of inertial frame. In practice, we have no *absolute* definition of this concept either. Nevertheless, this does not mean that it has no ultimate physical significance. Similarly, the practical difficulty of determining cosmic time precisely is not a compelling argument against its physical reality.

Gödel's main argument for refusing to regard the concept of cosmic time as an essential feature of all theoretical world models was based on his discovery of a homogeneous universe in which the local times of the observers who move with the fundamental particles cannot be fitted together into one universal time. The existence of such a time in the homogeneous world models that had been considered previously depended on the further assumption that in them an observer associated with any fundamental particle would be at a centre of isotropy, or spherical symmetry, so that he would observe no special directions since all directions would be equivalent.† Consequently, in these models there can be no cosmic rotation. From the point of view of an observer associated with an fundamental particle, the directions of cosmic recession are like the spokes of a wheel, except that they form a three-dimensional system. These directions coincide with the directions of inertial motion—i.e. with the directions of free motion not subject to local forces—and the aggregate of these directions from any given point forms a pencil of straight lines without relative rotation. This pencil was called by Gödel the local 'compass of inertia'. The world model that he constructed is observed from each fundamental particle to rotate relative to the local compass of inertia.

The space–time of Gödel's model has a metric of the form

$$ds^2 = a^2(d\tau^2 - dx^2 + \tfrac{1}{2}e^{2x}\,dy^2 - dz^2 + 2e^x\,d\tau\,dy) \qquad (6.27)$$

and it can be shown that Einstein's field equations with non-vanishing

† A world model can be homogeneous without necessarily being isotropic, but a model that is isotropic about every point is also homogeneous (Walker 1944). For homogeneous models in general, see Ryan and Shepley (1975).

cosmical constant Λ are satisfied if

$$1/a^2 = 8\pi G\rho, \qquad \Lambda = -1/2a^2 = -4\pi G\rho, \tag{6.28}$$

where G is the constant of gravitation, ρ is the uniform mean density of matter, and the velocity of light is taken to be unity. This world model is similar to the Einstein universe, in so far as it is both static and spatially homogeneous—since the *space–time* given by formula (6.27) is homogeneous in the sense that, given any two points P and Q in it, there exists a transformation of this space–time into itself which carries P to Q—but it differs from the Einstein universe because it depends on a *negative* cosmical constant and also because it is not isotropic, there being an absolute rotation of matter given by $\sqrt{(4\pi G\rho)}$. Moreover, it is not possible to define an absolute world time in this model.

Although the existence of a universal rotation of matter must be regarded from the extreme relativistic point of view as no less objectionable in principle than the existence of a cosmic time, the most surprising feature of Gödel's model concerns its temporal properties, for, although the world line of each fundamental particle is open, so that no epoch can recur in the experience of an observer anchored to such a particle, it is possible for other time-like lines to exist which are closed. In particular, if P and Q are any two points (instants) on the world line of a fundamental particle and if P precedes Q on this line then there exists a time-like line† connecting P and Q on which Q precedes P. Thus, in this model it is theoretically possible to travel into any region of the past or future and back again, and so make a round trip in time analogous to the round trips in space with which we are all familiar!

This possibility, envisaged by H. G. Wells in his famous story *The Time Machine*, could, however, easily conflict with our conception of causality, for, in principle it would mean that we could travel into our own past and do something to ourselves which, by our own memory, we would know

† By a suitable transformation of co-ordinated, the metric of Gödel's model can be expressed in the form

$$ds^2 = 4a^2\{dt^2 - dr^2 - dy^2 + (\sinh^4 r - \sinh^2 r)\, d\phi^2 + 2\sqrt{2}\sinh^2 r\, d\phi\, dt\}.$$

If $R > \log(1+\sqrt{2})$, so that $\sinh^4 R - \sinh^2 R > 0$, the circle $r = R$, $y = 0 = t$ is everywhere time like ($ds^2 > 0$). Consequently, for α sufficiently small, the space–time locus $r = R$, $y = 0$, $t = -\alpha\phi$ running from the initial point Q (corresponding to $\phi = 0$) to the end point P (corresponding to $\phi = 2\pi$) is also time like. These points are situated on the t line defined by $r = R$, $y = 0 = \phi$, and P *precedes* Q on this line if $\alpha > 0$! Chandrasekhar and Wright (1961) claimed that Gödel's statement about the existence of closed time-like loops in his model was wrong. What their work shows is that Gödel's time-like loops are not geodesics, which in any case Gödel did not claim. For another example of a closed time-like curve in this model, see Ryan and Shepley (1975, p. 130).

had in fact never happened to us.† Gödel calculated that the velocity required in order to make a complete circuit in time in his model must be at least $1/\sqrt{2}$ that of light. Some years later Ozsvath and Schücking constructed a locally rotating world model that did not give rise to the possibility of closed time-like paths and so was free from the conflict with causality that could arise in Gödel's model (Ozsvath and Schücking 1962).

Consequently, if we assume that in a first rough approximation the universe is effectively homogeneous in the large, the crucial factor in deciding whether the idea of universal cosmic time is likely to be well founded is evidence bearing on cosmic rotation. In other words, empirical evidence for world isotropy, and hence for the absence of cosmic rotation, can also be regarded as evidence for the idea of cosmic time.

Impressive support for the assumption of world isotropy has come in recent years from the discovery of what is often called the 'primeval fireball'. In 1965 the radio-astronomers A. A. Penzias and R. W. Wilson, at the Bell Telephone Laboratories in New Jersey, found that some unexpected radiation was leaking into the antenna of their apparatus. They soon discovered that the source of this radiation must be more or less isotropic and that at the wavelength at which they were working (7.4 cm) the radiation appeared to be equivalent to the intensity of a blackbody source at about 3.5 K (Penzias and Wilson 1965). Subsequent work reduced this estimate to approximately 2.7 K. Meanwhile, at Princeton University, R. H. Dicke and his colleagues were arguing on purely theoretical grounds that a universal microwave radiation with similar properties should exist (Dicke *et al.* 1965). Their work was done in ignorance of that carried out by Gamow some fifteen years earlier, who, on the basis of an explosive origin of the expanding universe, had predicted the existence of a primeval fireball with a present temperature of about 5 K (Gamow 1949).

The primeval nature of the observed background radiation is indicated by the blackbody character of its spectrum, which has been well confirmed over the considerable range of wavelengths that has been investigated in recent years.‡ The most remarkable feature of this radiation is its isotropy, the departures from which, as indicated by a very small fluctuation of temperature with direction of not more than about

† It is interesting to note that in Wells's novel the Time Traveller returns from his trip into the future but not from his trip into the past!

‡ According to Alpher and Herman (1975, p. 338), 'one would be hard pressed to argue against the relic blackbody radiation interpretation of the observations in view of the accuracy and consistency of the several different types of data over a wavelength ratio of about 1800'.

one part in a thousand, can be ascribed to the proper motion of the solar system with respect to the local supercluster of galaxies (Conklin 1969). The observed anisotropy is low enough to make the idea that the blackbody spectrum is due to the superposition of radiation from a number of discrete sources seem most improbable (Alpher and Herman 1975, p. 339). Indeed the isotropy is sufficiently precise to exclude the possibility of any local origin of the radiation, since a source restricted to the solar system, our Galaxy, or even the local supercluster would not appear so nearly isotropic to an observer located, as we are, far away from the centres of these systems. Moreover, any large-scale departures from homogeneity or isotropy in the universe would affect the radiation and make it appear anisotropic to us. Consequently, we have strong evidence that the universe as a whole is predominantly homogeneous and isotropic, and this conclusion, as we have seen, is a strong argument for the existence of cosmic time.

6.6. The limits of cosmic time

We must now consider an important, and previously unsuspected, possible limitation of the concept of cosmic time which was foreshadowed by a curious property of de Sitter's world model as first studied by him in 1917.

With his original choice of co-ordinates, this model was static and its metric was given by equation (6.11), namely

$$\mathrm{d}s^2 = \left(1 - \frac{r^2}{R^2}\right)\mathrm{d}t^2 - \frac{1}{c^2}\left\{\frac{\mathrm{d}r^2}{1 - r^2/R^2} + r^2(\mathrm{d}\theta^2 + \sin^2\theta\,\mathrm{d}\phi^2)\right\}$$

where we have now written R for $\sqrt{(3/\Lambda)}$. We note that for a clock, for example an atomic vibrator, at rest at any point (r, θ, ϕ all constants)

$$\mathrm{d}s^2 = (1 - r^2/R^2)\,\mathrm{d}t^2. \tag{6.29}$$

Since $\mathrm{d}s$ represents proper time and $\mathrm{d}t$ the corresponding time interval according to A (the observer at the origin $r = 0$), we see that the apparent duration according to this observer of the time interval between *any* two non-simultaneous events occurring at distance $r = R$ must be infinite. Consequently, in the experience of A there is a horizon at which time appears to stand still, as at the Mad Hatter's tea party where it was always six o'clock. This time horizon is, however, only an apparent phenomenon, like the rainbow, and the time flux experienced by an observer on this horizon will be the same as that experienced by A. However, the time required for light—or indeed for any electromagnetic signal—to travel from this observer to A will be infinite, since the integral

$$\int_0^R \frac{dr}{1-r^2/R^2},$$

deduced from taking $ds=0$, is divergent.

In the original form of metric there was no cosmic time, but we now recognize that the most appropriate metric for the de Sitter universe is given by (6.14) rather than by (6.11), so that this world model is best regarded as a limiting form of the expanding universe. The same metric (6.14) also characterizes the steady-state homogeneous world models which depend on the continual creation of matter. These models involve the concept of cosmic time, but they also involve the idea of an apparent horizon of time in the description of the universe by an observer associated with a fundamental particle.

Although this curious property of time in the de Sitter universe has been known for many years, controversy concerning the properties of the time horizon in steady-state world models revealed a surprising lack of agreement among theoretical cosmologists on the definition of this concept. A systematic discussion was undertaken at my suggestion by my pupil W. Rindler (1956) who did not restrict his investigation to world models associated with de Sitter space–time but considered all models based on the Robertson–Walker metric (6.16).

This metric may be written in full in the form

$$ds^2 = dt^2 - \frac{1}{c^2} R^2(t)\left\{\frac{dr^2 + r^2(d\theta^2 + \sin^2\theta\, d\phi^2)}{(1+kr^2/4)^2}\right\}, \tag{6.30}$$

where t denotes cosmic time, r is a co-moving radial co-ordinate, θ and ϕ are angles measured at any fundamental particle $r=0$, $R(t)$ is the expansion factor, k is the curvature index 0, 1, or -1, and c is the local speed of light. The fundamental particles are given by constant values of (r, θ, ϕ) and r can take all values, except that when $k=-1$ we assume $r<2$, and when $k=1$ (the case of a closed universe of finite spatial cross-section $t=$ constant) each particle on the line of vision corresponds to infinitely many values of r.

It is convenient to introduce the auxiliary variable

$$\sigma(r) = \int_0^r \frac{dr}{1+kr^2/4} \tag{6.31}$$

as an alternative co-moving radial co-ordinate. The proper distance at cosmic time t_1 between the spatial origin A and a fundamental particle with co-ordinate $r = r_1$ is given by

$$l = R(t_1)\sigma(r_1),$$

so that the equation of motion of this particle can be written in the form

$$l = R(t)\sigma(r_1).$$

The equation of motion of a photon emitted in the direction of A at time t_1 by this fundamental particle is given by

$$l = R(t)\left\{\sigma(r_1) - \int_{t_1}^{t} \frac{c\,dt}{R(t)}\right\}. \tag{6.32}$$

Two different types of world horizon must be distinguished:

(i) *event horizon*, for a given fundamental observer A, is a hyper-surface in space–time dividing all events into (a) those that have been, are, or will be observable† by A, and (b) those that are never observable by A;
(ii) *particle horizon*, for a given fundamental observer A and cosmic time t_0, is a surface in the instantaneous space $t = t_0$ which divides all fundamental particles into (a) those already observable by A at time t_0 and (b) those not already observable by that time.

The former type is exemplified by the case of the de Sitter universe and the latter by the case of the Einstein–de Sitter universe. Some world models, for example Lemaître's, possess both types of horizon, whereas some others, for example Milne's, possess neither. As an aid to visualizing the two concepts, we can picture the universe as an expanding balloon.‡ The fundamental particles can be represented by large dots distributed uniformly over the fabric of the balloon. One particular dot may be indicated as representing any given fundamental observer A. Photons can be represented by small dots moving over the balloon along great-circle paths at a constant speed relative to the fabric of the balloon. An event horizon will exist for A, and similarly for all other fundamental observers, in models where the rate of expansion is, and remains, sufficiently great for some of the small dots moving towards A never to reach A. In Eddington's graphic phrase, light is then 'like a runner on an expanding track with the winning post receding faster than he can run' (Eddington 1933, p. 73). On the other hand, a particle horizon will exist for A if, for

† By 'observable' we mean throughout 'observable by an ideal instrument of unlimited sensitivity'.

‡ This analogy suggests a closed universe, but the argument for open universes is similar. Strictly speaking, only the *surface* of the balloon should be imagined.

example, the balloon expands from an initial state approximating to a point and the initial rate of expansion exceeds that of the small dots, so that a finite time must elapse before any given one of them can reach A. *None* will reach A unless the rate of expansion decreases from its initial value, and *some* will never reach A if the rate of expansion, after first decreasing, increases again suitably. This is what happens in a model with both types of horizon, such as Lemaître's universe which explodes violently and later passes slowly through an unstable equilibrium state (Einstein universe) and then expands at an ever-increasing rate.

The necessary and sufficient condition for an event horizon to exist in a given world model is provided by the convergence of

$$\int^{\infty} \frac{\mathrm{d}t}{R(t)},$$

for, at any given cosmic time t_0, the photon emitted towards A by the fundamental particle

$$\sigma = \sigma_0 = \int_{t_0}^{\infty} \frac{c\,\mathrm{d}t}{R(t)}$$

reaches A ($l = 0$) only when t is infinite, as is readily seen from equation (6.32). Photons emitted at t_0 from fundamental particles corresponding to values of $\sigma > \sigma_0$ can never reach A since l will never vanish, whereas photons emitted from fundamental particles with $\sigma < \sigma_0$ will reach A at finite epochs. The event horizon when it exists is a moving spherical light front,† its proper distance from A at time t being given by

$$l = R(t) \int_{t}^{\infty} \frac{c\,\mathrm{d}t}{R(t)}. \tag{6.33}$$

Events beyond this horizon are forever unobservable by A.

From equation (6.32) it follows that photons emitted at time t_1 from a given particle P with r equal to r_1 reach A at time t given by

$$\sigma(r_1) = \int_{t_1}^{t} \frac{c\,\mathrm{d}t}{R(t)}. \tag{6.34}$$

If, for a given r_1 and some $t = t_0$, equation (6.34) has a solution for t_1, then we see that for the same r_1 and for any $t > t_0$ it will always have some solution for t_1. Thus at each epoch later than t_0 there will be a signal from P arriving at A. As t tends to infinity, t_1 will tend to a limiting value which will be the proper time at which P crosses A's horizon. Although, strictly

† Except in the case of the de Sitter metric.

speaking, no particle can pass out of view, its history as observed by A becomes more and more dilated, the event of its crossing the horizon being observable by A only in his infinite future. Consequently, this event at P and all subsequent events at P can never be observed† by A.

It is readily verified that the de Sitter metric satisfies the condition for an event horizon to exist, since in this case

$$R(t) = e^{t/a}, \qquad k = 0.$$

The event horizon is at a constant distance ca from A. The family of world models with expansion factors proportional to t^n includes as special cases Dirac's universe ($n = \frac{1}{3}$), the Einstein–de Sitter universe ($n = \frac{2}{3}$), Milne's universe ($n = 1$), and the uniformly accelerated expanding universe ($n = 2$). The condition for an event horizon to exist for members of this family is given by $n > 1$, and when it exists this horizon expands at a uniform rate.

The necessary and sufficient condition for a particle horizon to exist in a given world model is provided by the convergence‡ of

$$\int_0 \frac{dt}{R(t)},$$

or, in those cases where the definition of $R(t)$ extends to negatively unbounded values of t, by the convergence of the corresponding integral with $-\infty$ as the lower limit, for, from equation (6.32) it follows that, at any given time t_0, all fundamental particles for which

$$\sigma > \int_0^{t_0} \frac{c\, dt}{R(t)}$$

have not yet been observed by A, whereas all others have. Hence, the surface (which is a sphere in the space $t = t_0$) given by

$$\sigma = \int_0^{t_0} \frac{c\, dt}{R(t)} = \psi(t_0), \tag{6.35}$$

say, divides all fundamental particles into those which can be observed by A at or before $t = t_0$ and those which cannot be so observed. The proper

† All fundamental particles other than A that are some time within A's event horizon (if this horizon exists) must eventually pass beyond it *in their own histories*. Nevertheless, from the point of view of A, if a fundamental particle is once visible then it always remains visible, *provided that A has an instrument of unlimited sensitivity*.

‡ The lower limit of the integral refers to those models with a singularity (creation) at a finite epoch which is taken to be zero epoch.

distance of this horizon from A is given by

$$l = R(t) \int_0^t \frac{c\,dt}{R(t)}, \tag{6.36}$$

where the lower limit must be replaced by $-\infty$ if the model is defined for t negatively infinite.

Since $R(t)$ is positive and finite, it follows that

$$\psi(t) = \int_0^t \frac{c\,dt}{R(t)}$$

is an increasing function of t. Hence, from (6.35) we deduce that more and more particles become visible to A as time goes on. If $\psi(t)$ converges as t tends to infinity, all fundamental particles for which

$$\sigma > \int_0^\infty \frac{c\,dt}{R(t)}$$

are entirely outside A's possibility of observation, and the model possesses an event horizon as well as a particle horizon. As before, a fundamental particle once seen always continues to remain visible.

Of the models with expansion factors proportional to t^n only those with $n<1$ possess a particle horizon, in particular, Dirac's universe and the Einstein–de Sitter universe. *Of this class of expanding universes we may therefore say that those with an increasing rate of expansion* ($n>1$) *possess event-horizons and those with a decreasing rate of expansion* ($n<1$) *possess particle horizons. The uniformly expanding model* ($n=1$) *possesses neither type.*

In formulating the definitions of horizon it has been assumed that the observer A remains anchored to a particular fundamental particle, so that he himself keeps cosmic time. If this restriction is relaxed and the observer can cruise through the universe with local speed less than c, then the class of events observable by him is increased. Nevertheless, if the model possesses both an event horizon and a particle horizon, as previously defined, there are still events absolutely unobservable by A, however he may move. Moreover, if the model possesses an event horizon for A before he moves, then he can never move so as to be able to observe every event in the universe. His time horizon will change but can never be wholly abolished.

Thus, Locke's assertion, quoted at the beginning of this chapter, that every part of space is in every part of duration and every part of duration is in every part of expansion, can no longer be accepted unreservedly; for, although the relativity of time is only a local phenomenon, we must reckon with the possibility that events can occur in the universe, knowledge of which can never be brought, even in principle, to a given observer, however long he lives, and so can never enter his temporal experience.

6.7. The existence of singularities in space–time

The homogeneous and isotropic world models that are based on General Relativity with no cosmical constant and negligible pressure are all characterized by the presence of an initial state of infinite density and infinite curvature. This state is called a 'singularity' and at it the ordinary concept of space–time and the usual laws of physics break down. It is natural to ask whether the existence of an initial singularity is a purely theoretical consequence of the mathematical simplifications that have been introduced to make the cosmological problem tractable. In particular, assumptions have been made concerning the geometrical symmetry of world models, and the simultaneous compression of the whole universe into a single point at a finite epoch in the past might well be considered a property of an idealized world model that does not correspond to any actual physical situation in the past.

Similarly, if we consider an oscillating world model (Friedmann universe with $\Lambda = 0$ and $q > 0.5$) in its contracting phase, all the matter in it eventually condenses to a single point and becomes infinitely dense. If, however, the inward motion is slightly perturbed so that it is not all directed to a single point, we would still expect the density to become very high, but it might not become infinite. If this were the case, then the universe might pass through a phase of maximum density and eventually start to expand again. Consequently, we could envisage the possibility that the present expanding state of the universe was preceded by a contracting phase. Such a possibility has, however, been rendered unlikely by some remarkable theorems due to Penrose and Hawking.

The first theorem about singularities of space–time that did not involve any assumption about symmetry was due to Penrose (1965). It arose in connection with the gravitational collapse of a star. If we consider the ultimate state in the evolution of a star after it has exhausted all possibilities of generating radiation by nuclear transmutations, so that its mechanical equilibrium under gravity can no longer be sustained by thermal pressure and radiation pressure, the first possibility is that it becomes a white dwarf star of density about a million times that of water and of radius comparable with that of a planet such as the earth. However, as was first shown by Chandrasekhar in 1931, this can only be a stable end-state if the star is not too massive, for the entire structure of a white dwarf is determined by its mass—the larger the mass the smaller the radius—but if the mass is about 1.4 times that of the sun the corresponding radius is zero. This upper limit for the possible mass of a white dwarf star is known as the *Chandrasekhar limit.* In these stars gravity is balanced by electron degeneracy pressure. For the fate of a more massive star there are believed to be two possibilities. One is that during its contraction it will become so heated that it will explode and become a supernova. If it thereby sheds enough mass, it can develop into a neutron star of density about 10^{14} (the same as that of the atomic nucleus) and radius about 10 kilometres. In such a star gravity is balanced by neutron degeneracy pressure. The greatest mass of a stable neutron star is about 0.7 times that of the sun. This is known as the *Oppenheimer–Volkoff limit.* If, however, the original mass of the star is greater than about five solar masses, it cannot reach equilibrium as either a white dwarf star or a neutron star, since it cannot generate sufficient internal pressure to withstand continual contraction under its own gravitation. In due course it will be of radius less than $2GM/c^2$, where M is its mass. In the case of a spherically symmetrical star, the external metric is the Schwarzschild metric (given by equation (6.6) on p. 278) and at distance $2GM/c^2$ from the centre the velocity of escape is equal to the velocity of light. Points at this distance constitute what is known as the *Schwarzschild surface.* From the point of view of a remote observer this surface is analogous to the cosmological event horizons discussed in §6.6, for, if a spherical body is of radius less than or equal to its Schwarzschild radius, no material particles or photons can be emitted from it to reach any external observer and it becomes what is known as a 'black hole'. If there is exact spherical symmetry, then once the body has shrunk within its Schwarzschild surface all the matter within it will fall radially inwards to encounter a space–time singularity at the centre, where $r = 0$. To a potential observer situated on the surface of such a body this would happen in a finite time. Depending on the mass of the body, this time could be very short. For a black hole of solar mass the time that would elapse after shrinking to its Schwarzschild surface and before reaching the singularity would be only about 10 microseconds! The local curvature of space–time would then become infinite, and time itself would come to a full stop.† On the other hand, to an observer situated well away from such a body its collapse would appear to be slowed down by time dilatation, so that it

† The theory of 'gravitational collapse' was first investigated in a classic paper by J. R. Oppenheimer and H. Snyder (1939). Although they neglected the effect of pressure, they deduced what we now recognize to be the basic features characterizing this bizarre phenomenon. In particular, they found that the total proper time of collapse of a star initially of mass and density similar to that of the sun would be of the order of a day.

would only approach its Schwarzschild surface asymptotically and consequently would only attain it after an infinite time.†

So far, it has been assumed that the collapsing body is spherically symmetrical, and it is by no means obvious that the same phenomena would occur if there were irregularities or if the body were rotating. Penrose (1965) has shown, however, that similar results do arise even in the absence of exact spherical symmetry. This is because he was able to define an analogue of the Schwarzschild surface for an asymmetrical body which he called a *trapped surface*. This is any closed space-like two-dimensional surface for which the paths of all future-directed light rays orthogonal to it, both outgoing as well as incoming, converge towards each other.‡ In Schwarzschild space–time this convergence can be attributed to the intense pull of gravity which draws the photons into a central singularity. In general, once a trapped surface has been formed collapse would seem likely to follow. Nevertheless, it is not yet known whether a very massive star will automatically develop such a surface or instead explode into fragments of small enough mass to produce stable white dwarfs or neutron stars, but it is thought that some black holes do in fact exist, a notable instance being the X-ray star Cygnus X-1 discovered in 1972.

In a series of ingenious theorems that deal only with the topological structure of space–time Hawking and Penrose (1970) have proved that, so long as the energy and pressure of contracting matter do not contravene some very plausible conditions (validity of General Relativity, positive energy, absence of closed causal time-like trajectories, etc.), the existence of a singularity becomes inevitable once a trapped surface is formed.§ We have seen that in the spherically symmetrical case (Schwarzschild metric) an event horizon occurs at a distance of $2GM/c^2$ from the centre. This shields the rest of the universe from any influence (other than gravitational attraction) due to the singularity at the centre. Penrose (1969) has extended this idea to other cases of collapse by his 'cosmic censorship'

† In fact, the star would become too faint to be seen after a finite time, for, although the red shift of its light would at first increase only very gradually, eventually it would increase exponentially. For example, a collapsing star of mass 10^8 times that of the sun and initial radius 100 light years will appear to have a red shift of order 10^{-3} for about 10^5 years and then an e-folding time of about a minute, so that as Weinberg (1972) remarks it is as if the collapsing sphere were suddenly cut off from the rest of the universe, although it would still continue to exert its gravitational force.

‡ The outermost trapped surface is an event horizon similar to the Schwarzschild surface.

§ Detailed discussion of the relevant theorems will be found in the book by Hawking and Ellis (1973). A *singularity* is said to occur when at least some time-like geodesics (paths of particles) cannot be extended to infinite proper time of the particles concerned. The singularity theorems can be regarded as indicating the limits of *classical* (non-quantal) General Relativity.

hypothesis. This is a hypothetical law of nature forbidding the formation of singularities unless they are contained within an event horizon.

The Schwarzschild metric applies only to non-rotating bodies that are spherically symmetrical. Most stars possess angular momentum, and their collapse will produce rotating black holes. The rigorous mathematical treatment of these has been greatly facilitated through the discovery by Kerr (1963) of an axially symmetric external solution of the field equations of General Relativity in the presence of a rotating body. The resulting metric is now known as the Kerr metric. It was generalized by Newman *et al.* (1965) to include the effect of electric charge as well as of angular momentum and mass. It was later shown by Wald (1971) that the generic final state of gravitational collapse is a Kerr–Newman black hole. In other words, once a black hole has been formed, the only properties of the original body that can survive are its mass, angular momentum, and electric charge.

Remarkable though the singularity theorems of Hawking and Penrose are, they tell us very little about the precise nature of the singularity concerned, for they are only existence theorems. Nevertheless, it seems that the space–time trajectory of any particle which lies within a trapped surface cannot re-emerge through it. Moreover, it seems that space–time itself comes to a full stop at a singularity. Consequently the proper time kept by anything that reaches it must end there too. However, in some of the idealized special cases which have been studied matter succeeds in avoiding the singularity and passes on to other regions of space–time connected with our own only through the black hole. These regions are hidden from us behind the event horizon formed by the outermost trapped surface, but it should be noted that these theoretical investigations do not take quantum theory into account, although on the scale of the Planck length (10^{-35} metres) quantum effects cannot be ignored.†

The expansion of the universe can be regarded as the time reversal of collapse to a singularity. We have seen that, if the universe is of Friedmann type with zero cosmical constant, there was a singularity in the past associated with the universe as a whole, besides possible singularities in the future associated with collapsing massive stars. In a Friedmann universe the initial singularity occurred when the density was

† Hawking (1975) has shown that quantum jumps (or 'tunnelling') can take particles and photons out of the black-hole region of a collapsing body, and hence in due course the black hole can evaporate completely. The process is, however, very slow (from the point of view of an external observer) and only microscopic black holes can completely evaporate in a time comparable with that of the Hubble time for the whole universe. (These small black holes are not formed by stars but may have been generated by the collapse of highly compressed regions in the hot dense early universe.) On the other hand, a black hole of the mass of the sun will last about 10^{66} years, and more massive ones longer still.

infinite, and in the case of a closed model when all the matter was at the same point. Moreover, even if Λ does not vanish and the effect of pressure is not neglected, the initial singularity cannot be avoided. If the universe were more irregular than a Friedmann model, it might be thought possible that it could avoid having an initial singularity, but in this case too it has been found by Hawking and Ellis (1965) that, subject to some reasonable assumptions about high density, there must exist at least one singularity.

Novikov and Zel'dovich (1973) have pointed out that, although 'bounce' (change-over from contraction of the universe to expansion) can occur in a Friedmann model with $\Lambda = 0$, if it is 'open' ($k \leq 0$) it can only occur once. In the case of a (finite) 'closed' model ($k > 0$), although an unending number of cycles persisting indefinitely into the future is possible, this was not so in the past, for, as was shown by Tolman (1934), each cycle involves irreversible changes of entropy.† If the baryon number remains constant, the total mass and pressure must increase from cycle to cycle and hence the maximum radius must also increase from cycle to cycle. 'The multicycle model therefore has an infinite future but not an infinite past' (Novikov and Zel'dovich 1973).

Consequently, as previously indicated by our analysis of Kant's first antinomy in Chapter 1 (pp. 27–32) and particularly by the observational evidence concerning cosmic background radiation discussed in §6.5 (pp. 306–7), it would seem that there is a strong case for believing in the existence of a cosmological origin of time corresponding to an origin of the universe in the finite past, beyond which the laws of physics are silent and time itself stood still.

References

ALEXANDER, S. (1920). *Space, time and deity*. Macmillan, London.

ALPHER, R. A. and HERMAN, R. (1975). *Proc. am. phil. Soc.* **119,** 325–48.

ARP, H. (1973). Evidence for discordant redshifts. In *The redshift controversy* (ed. G. B. Field, H. Arp and J. N. Bachall), pp. 15–58. W. A. Benjamin, Reading, Mass.

BALDWIN, J. A., BURKE, W. L., GASKELL, C. M. and WAMPLER, E. J. (1978) *Nature, (Lond.)* **273,** 431–5.

BONDI, H. and GOLD, T. (1948). *Mon. Not. r. astron. Soc.* **108,** 252–70.

† This result has been confirmed by the more recent investigation of entropy in an oscillating universe by Landsberg and Park (1975). They have pointed out that the entropy increases from phase to phase, even when the model is contracting. They claim that this one-way trend of entropy thus provides a consistent arrow of time for such a universe. Nevertheless, as we have previously remarked, there is some doubt about the concept of entropy in this context.

BROAD, C. D. (1923). *Scientific thought*, p. 458. Routledge and Kegan Paul, London.

BRUZUEL, G. A. and SPINRAD, H. (1978) *Astrophys. J.* **220,** 1–7.

CAJORI, F. (1934). *Newton's Principia* (transl. A. Motte), p. 420. Cambridge University Press, Cambridge.

CHANDRASEKHAR, S. and WRIGHT, J. P. (1961). *Proc. nat. Acad. Sci. U.S.A.* **97,** 341–7.

CLIFFORD, W. K. (1879). *Lectures and essays*, p. 245. Bohn, London.

—— (1885). *Commonsense of the exact sciences*, p. 225. Bohn, London.

CONKLIN, E. (1969). *Nature (Lond.)* **222,** 971.

DICKE, R. H., PEEBLES, P. J. E., ROLL, P. G. and WILKINSON, D. (1965). *Astrophys. J.* **142,** 414–9.

DIRAC, P. A. M. (1938). *Proc. R. Soc. Lond. A* **165,** 199–208.

EDDINGTON, A. S. (1920). *Space, time and gravitation.* Cambridge University Press, Cambridge.

—— (1933). *The expanding universe.* Cambridge University Press, Cambridge.

—— (1939). *The philosophy of physical science*, p. 76, Cambridge University Press, Cambridge.

EINSTEIN, A. (1917). Kosmologische Betrachtungen zur allgemeinen Relativitaetstheorie. *Sitzungsber. K. preuss. Akad. Wiss.*, 142–52.

EINSTEIN, A. and DE SITTER, W. (1932). *Proc. nat. Acad. Sci. U.S.A.* **18,** 213.

FORBES, E. G. (1961). A history of the solar red shift problem. *Ann. Sci.* **17,** 129–64.

GAMOW, G. (1949). *Rev. mod. Phys.* **21,** 367.

GIBSON, J. (1931). *Locke's theory of knowledge and its historical relations*, pp. 245–54. Cambridge University Press, Cambridge.

GÖDEL, K. (1949a). A remark about the relationship between relativity theory and idealistic philosophy. In *Albert Einstein: philosopher–scientist* (ed. P. A. Schillp), p. 560. Library of Living Philosophers, Evanston, Illinois.

—— (1949b). *Rev. mod. Phys.* **21,** 447.

GUNN, J. A. (1929). *The problem of time*, p. 68. Methuen, London.

HAWKING, S. W. (1975) Particle creation by black holes. In *Quantum Gravity* (ed. C. J. Isham, R. Penrose and D. W. Sciama), pp. 219–67. Clarendon Press, Oxford.

HAWKING, S. W. and ELLIS, G. F. R. (1965). *Phys. Lett.* **17,** 246–7.

—— (1973). *The large-scale structure of space–time.* Cambridge University Press, Cambridge.

HAWKING, S. W. and PENROSE, R. (1970). *Proc. R. Soc. Lond.* A300, 187–201.

HOYLE, F. (1948). *Mon. Not. r. astron. Soc.* **108,** 372–82.

INFELD, L. and SCHILD, A. (1945). *Phys. Rev.* **68,** 250.

JEANS, J. H. (1936). Man and the universe. In *Scientific progress* (Sir Halley Stewart Lecture, 1935), pp. 11–38. Allen and Unwin, London.

KERR, R. P. (1963), *Phys. Rev. Lett.* **11,** 237–8.

KEYNES, J. M. (1951). *Essays in biography.* Rupert Hart-Davis, London.

KRISTIAN, J., SANDAGE, A., and WESTPHAL, J. A. (1978). *Astrophys. J.* **221,** 383–94.

LANDSBERG, P. T. and PARK, D. (1975). *Proc. R. Soc. Lond. A* **346,** 485–95.

LEMAÎTRE, G. (1925). *J. math. Phys.* **4,** 188.

LOCKE, J. (1690). *An essay concerning human understanding* (ed. A. S. Pringle-Pattison, 1924), p. 98. Clarendon Press, Oxford.

LYNDEN-BELL, D. (1977). *Nature (Lond.)* **270,** 396–9.

MCCREA, W. H. (1956). *Proc. r. Ir. Acad. A* **57,** 173.

MCVITTIE, G. C. (1970). *Astron. J.* **75,** 287–96.

MILNE, E. A. (1933). *Z. Astrophys.* **6,** 1–95.

—— (1937), *Proc. R. Soc. Lond. A* **158,** 324.

MINKOWSKI, H. (1923). In A. Einstein and others, *The principle of relativity.* (transl. W. Perrett and G. B. Jeffery), p. 76. Methuen, London.

MÖSSBAUER, R. L. (1958). *Z. Phys.* **151,** 124.

NEWMAN, E. T. *et al.* (1965). *J. math. Phys.* **6,** 918–9.

NOVIKOV, I. D. and ZEL'DOVICH, YA. B. (1973). Physical processes near cosmological singularities. *A. Rev. Astron. Astrophys.* **11,** 387–412.

OPPENHEIMER, J. R. and SNYDER, H. (1939). *Phys. Rev.* **16,** 455–9.

OZSVATH, L. and SCHÜCKING, E. (1962). *Nature (Lond.)* **193,** 1168.

PEIMBERT, M. and TORRES-PEIMBERT, S. (1974). *Astrophys. J.* **193,** 327–33.

PENROSE, R. (1965). *Phys. Rev. Lett.* **14,** 57–9.

—— (1969). *Riv. del Nuovo Cimento,* 1, 252–76.

PENZIAS, A. A. and WILSON, R. W. (1965). *Astrophys. J.* **142,** 419.

POUND, R. A. and REBKA, G. A. (1960). *Phys. Rev. Lett.* **4,** 337.

POUND, R. A. and SNIDER, J. L. (1964). *Phys. Rev. Lett.* **13,** 539.

RINDLER, W. (1956). *Mon. Not. r. astron. Soc.* **116,** 662–77.

ROBERTSON, H. P. (1928). *Phil. Mag.* **5,** 835.

—— (1929). *Proc. nat. Acad. Sci. U.S.A.* **15,** 822.

RUSSELL, B. (1948). *Human knowledge,* p. 477. Allen and Unwin, London.

RYAN, M. P. and SHEPLEY, L. C. (1975). *Homogeneous relativistic cosmologies.* Princeton University Press, Princeton, N.J.

SALECKER, H. and WIGNER, E. P. (1958). Quantum limitations of the measurements of space–time distances. *Phys. Rev.* **109,** 571–7.

SANDAGE, A. and TAMMANN, G. A. (1975). *Astrophys. J.* **197,** 265–80.

—— (1976). *Astrophys. J.* **210,** 7–24.

SHAPIRO, I. I. (1964). *Phys. Rev. Lett.* **13,** 789.

SHAPIRO, I. I., PETTENGILL, G. H., ASH, M. E., STONE, M. L., SMITH, W. B., INGALLS, R. P., and BROCKELMAN, R. A. (1968). *Phys. Rev. Lett.* **20,** 1265.

DE SITTER, W. (1917). *Proc. K. Ned. Akad. Wet.* **19,** 1217 (see also *Mon. Not. r. astron. Soc.* **78,** 3).

TOLMAN, R. C. (1934). *Relativity, thermodynamics and cosmology,* pp. 440 *et seq.* Clarendon Press, Oxford.

WALD, R. M. (1971). *Phys. Rev. Lett.* **26,** 1653–5.

WALKER, A. G. (1937). *Proc. Lond. math. Soc.* **42,** 90.

—— (1944). *J. Lond. math. Soc.* **19,** 219–26.

WEINBERG, S. (1972). *Gravitation and cosmology*, p. 348. Wiley, New York.

WEYL, H. (1922). *Space–time matter* (transl. H. L. Brose), p. 217. Methuen, London.

—— (1923). *Phys. Z.* **24,** 230.

WHITROW, G. J. (1959). *The structure and evolution of the universe*, p. 103. Hutchinson, London.

WILSON, W. (1950). *Sci. Prog.* **38,** 622–36.

7

THE NATURE OF TIME

7.1. Time reversal and the asymmetry of time

We have seen that in the case of world models with a particle horizon fundamental particles suddenly come into the view of an observer associated with any given fundamental particle. On the other hand we have not encountered the reverse phenomenon of particles suddenly disappearing from view, even in the case of models with an event horizon. It might be thought that this asymmetry is associated in some way with the fact that the universe is expanding. That this is not the case can readily be seen by considering the effect of time reversal on these models.

To each of these models we can assign a dual by replacing t by $-t$. Thus, to a model with the scale function $R(t)$ defined over the whole range $-\infty<t<\infty$ and possessing an event horizon given by equation (6.33), namely

$$l=R(t)\int_{t}^{\infty}\frac{c\,\mathrm{d}t}{R(t)}, \tag{7.1}$$

there corresponds a dual model with scale function $R(-t)$ which, by comparison with equation (6.36) has a *particle horizon* given by

$$l=R(-t)\int_{-\infty}^{t}\frac{c\,\mathrm{d}t}{R(-t)}. \tag{7.2}$$

Since equation (7.2) is also obtained simply by replacing t by $-t$ in equation (7.1), we deduce that *on time reversal a model with an event horizon is transformed into one with a particle horizon and vice versa.*

Strictly speaking, we have considered only the case in which the model has no finite origin of time. If the model has such an origin, on time reversal this initial time singularity (world creation) is transformed into a final time singularity in the future (world annihilation). The particle horizon is transformed into an event horizon in the sense that events occurring beyond it will not be observed in the finite stretch of time allowed to the observer before annihilation.

The surprising consequence of this analysis is that time reversal *cannot* lead to the disappearance of particles from view in the actual finite experience of the observer. The temporal asymmetry revealed by the absence of this possibility cannot therefore depend on the way in which the scale of the universe changes with time, for when time is reversed expansion becomes contraction. Instead, this asymmetry depends on the fact that the time reversal contemplated does not affect the primary condition that the observer's attention is confined in all cases to *incoming* light.

As we have previously pointed out (see p. 8), this is not the only temporal asymmetry associated with the transmission of light, for, although spherical light waves spreading outwards equally in all directions can and do exist, we never encounter the converse phenomenon of spherical waves closing in isotropically to a precise point of extinction. Moreover, as already mentioned, this feature of our experience is not confined to light but extends to other types of phenomena, for example water waves generated by a disturbance at a particular place. From this consideration Popper (1958) argues, I believe convincingly, that only such conditions can be causally realized as can be organized from one centre. A contracting wave would have the character of a physical miracle unless controlled by an expanding signal from the centre. Otherwise, it would be like a conspiracy undertaken by many people, each carefully acting so as to support all the others but without anything like a prepared plan.

Attempts have, of course, often been made to describe a world in which time 'runs backwards', i.e. one which would be the time reflection of our actual universe, just as in a mirror a left hand appears as the space reflection of a right hand. The philosopher F. H. Bradley argued that in such a world, i.e. for beings whose lives would run in the opposite sense to our own, 'Death would come before birth, the blow would follow the wound, and all must seem irrational' (Bradley 1902, p. 215). As in Lewis Carroll's Looking-Glass world, punishment would precede the trail and the crime would come last of all. Smart, however, disputed this view of the matter. He claimed that if *all* processes in the world were reversed then instead of memory we should have precognition, like Lewis Carroll's White Queen who 'remembered' best the things that happen the week after next! However, as all things would now seem to occur in reverse order, there would be nothing peculiar in this. Indeed, everything could be reversed in time and the world both be and seem exactly as it is. Hence, he concluded that 'the temporal asymmetry is not due to the properties of time itself (which is purely formal) but is due to an asymmetry in the content of the world' (Smart 1954).

However much one may sympathize with Smart's view that time should not be reified and that, strictly speaking, it is misleading to speak of it as 'flowing' in a definite direction, his discussion of time reversal must be regarded as either fallacious or, at best, trivial, for he contemplated a double reversal, namely of all events *and* of our sense of before and after. Such a double reversal would clearly leave everything the same, and it is hardly necessary to justify this conclusion by detailed discussion! His mistake was to assume that in a time-reversed world memory would be replaced by precognition. On the contrary, to a being for whom events on earth occurred in the reverse order to ours, those events of whose occurrence he was already aware would still be assigned by him to the *past* so that for him 'the blow *would* follow the wound', just as during the war the sound of the approaching supersonic rocket was heard after it had exploded on impact. For, whatever the laws of nature, the direction of time in our personal experience is the direction of increasing knowledge of events. Events for which we have information of their actual occurrence are in the past and not in the future. A world in which events occurred in the reverse order to those in our world can be imagined, but a reversal of our sense of before-and-after would imply a state of mind in which we began with maximum information of the occurrence of events and ended with minimum, and this is self-contradictory; for it is a fact of experience that we are not aware of everything at once and the order of our individual time is the order of our awareness, i.e. of the growth in our information of what occurs. By definition, any event which leaves a 'trace' of its occurrence is in the past. Therefore, it is *not* a contingent fact but merely a consequence of this definition that there is no such thing as a future analogue of a trace.

7.2. The causal theory of time

One consequence of a reversal of our sense of before-and-after would be that causes and effects would be interchanged. Indeed, efficient causes would appear to be transformed into final causes and scientific inquiry would be drastically affected. Thus, instead of the stone falling to the ground because my hand has released it, the stone would fly upwards in order that my hand should grasp it. Whether or not we agree with the widely held modern view—in contrast to the Aristotelian doctrine of teleological explanation—that only by the study of 'efficient causes' is the physical world accessible to inquiry, we all recognize the close connection between time order and causal processes.

This connection was exploited by Hume who sought to reduce causal order to temporal order. In his view, the only possible test of cause and effect is their 'constant union', the invariable succession of the one after the other. Unfortunately, this condition is neither necessary nor sufficient. We may keep two chiming clocks in the same room and always hear one begin to strike before the other, but this would not signify that there was any causal connection. Hume's condition is therefore not sufficient. To show that it is also not necessary, consider the death of Aeschylus who is supposed to have been killed by an eagle in flight dropping a tortoise on to his bald head. Although an invariable law of nature was involved—the law of gravity—this strange event essentially depended on a peculiar 'initial condition', namely that the eagle let go of the tortoise at a particular time at a particular place, with the result that Aeschylus suffered a unique fate. Or, again, consider the destruction of Pompeii and Herculaneum. It would be fantastic to doubt that this was caused by the eruption of Vesuvius in August 79 simply because this particular succession of events is not repeatable. Consequently, we cannot accept Hume's criterion as a necessary condition for causality.

These examples refer to unique historical events. On the other hand, it may be argued that, in the natural sciences, we are concerned with repeatable sequences of events. It is assumed that, if experiments are repeated at different times and different places *under the same conditions*, then the same results will follow. This is what is meant by the 'principle of causality'. Philipp Frank (1949), however, has shown that, strictly speaking, this methodological principle functions only as a conventional definition. For how can we be sure that the experiment in question has been repeated under precisely the 'same conditions'? Frank argued that there is no exact method except to find out whether the experiment yields the same result. Consequently, the law of causality is merely a rule for defining what we mean by 'under the same conditions'.

Instead of trying to reduce causal order to temporal succession, Platt (1956) has suggested that our intuitive conception of cause and effect should be analysed in terms of the biological process of stimulus and response. In discussing this process he drew attention to the vital role of *amplification*. For example, in the retina the energy of an incident light signal may be multiplied thousands of times to produce a single neural reaction which operates a kind of biological relay and may then be multiplied thousands of times more in the energy of a motor response. Because of this amplification the direction of stimulus response is irreversible. Cause and effect are as asymmetric in time as our sense amplifiers which cannot emit their input and our motor amplifiers which cannot respond to their output. The irreversibility of our consciousness is the same as that of our amplifiers, and Platt concluded that 'perhaps that is what we mean by the direction of time'. Thus, so far from reducing causality to time, he inverted the problem.

The idea of reducing time order to causal order, often called *the causal theory of time*, was orginally suggested by Leibniz, but was first developed in detail by Kant. He pointed out that we discover time order by examining causal order as distinct from perceptual order. Thus, in the case of the noise of firing of a distant gun and the sound of the nearby impact of the shell, we infer the temporal order from the causal order and not from the order in which we perceive these events. However, in justifying his contention that a judgment about an objective time sequence of events is possible only through a causal judgment, so that it is therefore the irreversibility of causal sequence which constitutes the irreversibility of time, Kant was faced by the difficulty that most efficient natural causes appeared to be simultaneous with their effects. He therefore tried to argue that it is the 'order' of time and not the *lapse* of time with which we have to reckon. 'The time between the causality of the cause and its immediate effect may be a *vanishing* quantity, and they may thus be simultaneous, but the relation of the one to the other will always remain determinable in time. If I view, as a cause, a ball which impresses a hollow as it lies on a stuffed cushion, the cause is simultaneous with the effect. But I still distinguish the two through the time-relation of their dynamical connection. For if I lay the ball on the cushion, a hollow follows upon the previous flat smooth shape; but if (for any reason) there previously exists a hollow in the cushion, a leaden ball does not follow upon it' (Kemp Smith 1934).

To this we may object that, if the causal connection *is* instantaneous, then we cannot also regard it as directed in *time.*

The cogency of the causal theory of time was, however, greatly enhanced by Einstein's Special Theory of Relativity. According to this theory, the time order of events is invariant (under a Lorentz transformation) for different observers if, and only if, the events in question can† be connected by signals, i.e. by causal chains, which 'travel' at speeds not exceeding that of light *in vacuo.* Only when no causal connections can possibly exist (because the speed of the connection would have to exceed that of light *in vacuo*) can the temporal sequence of two events be reversed by choosing an observer with the appropriate motion.‡ In this case the two events are essentially indeterminate in their time-order. As Reichenbach (1957) has emphasized, this result can be used to enlarge Einstein's concept of simultaneity, since we may define *any* two events which are indeterminate in their time order as simultaneous. From this point of view, as distinct from Kant's, *simultaneity means the exclusion of causal connection.* Hence, the classical idea that there could be at every place a unique event simultaneous with a given event ('here' and 'now')

† On the causal theory of time it is not *necessary* for there to be an actual causal connection between two events in a temporal sequence. Only causal connectibility, rather than actual causal connectedness, is required.

‡ For a formal proof of this result see pp. 352–5.

would be permissible only if there were no *finite* upper limit to the speed of causal connection, contrary to Einstein's theory.

Important as Einstein's theory is for our understanding of time, it presupposes that the observer is aware of temporal succession in his experience. If we wish to define time order in terms of causal order, instead of making it solely dependent on human awareness, we cannot define causal direction, i.e. distinguish cause from effect, in terms which presuppose knowledge, or awareness, of temporal direction. Instead, we must find some criterion other than temporal order for making this distinction unambigously and without circularity. Reichenbach suggested what is known as the 'mark-method': *if E_1 is a cause of E_2, then a small variation (a mark) in E_1 is associated with a small variation in E_2, whereas small variations in E_2 are not necessarily associated with small variations in E_1*. Thus, if we denote by a starred symbol an event which has been slightly varied (marked), we shall find according to this criterion that only the combinations $E_1 E_2$, $E_1^* E_2^*$, $E_1 E_2^*$ can occur, but not the combination $E_1^* E_2$. Because of this asymmetry, Reichenbach argued that we can automatically derive an ordering relation for E_1 and E_2 which we can define as the time relation between them.

At first sight this appears to be an extremely ingenious definition. Unfortunately, it involves a hidden *petitio principii*, for we must be able to recognize whether or not a particular pair of events (with different suffixes) can be combined. However, if we are given only the *individual* events E_1, E_2, E_1^*, and E_2^*, each of which it is assumed can occur, how can we decide which *combinations* are allowed and which not, other than by tacit appeal to temporal considerations?

Although attempts have been made to rescue the causal theory of time, the assumption that the concept of causality is more primitive than the concept of time has been strongly attacked by Zwart (1976, pp. 91–2). He maintains that, if the accustomed order of events were suddenly to change so that those events which we normally regard as causes of other events were to succeed the latter instead of preceding them, we would exchange the terms cause and effect and not the temporal order. For, whereas all our knowledge of causal relations is essentially empirical in nature, being based on our experience, it is logically impossible for the temporal order to change, since this would mean that what came first would be later and what came second would be earlier and it is inconceivable that we should perceive the earlier event later and the later event earlier. Since the meaning of the concepts of before and after cannot change, they must be regarded as completely elementary and not reducible to still simpler concepts. Zwart therefore concludes that the temporal order is more fundamental than the causal order.†

† Criticism of the causal theory of time has also been made by Earman (1973).

Although I agree with Zwart's argument and accept his conclusion that the concepts of before and after must be regarded as elementary, I believe that the basic difficulty confronting the causal theory is that the very essence of time lies in temporal succession, and therefore any theory which endeavours to account for time ought at least to throw some light on why everything does not happen at once, but unless the existence of *successive* states of phenomena is tacitly assumed it is impossible for the theory to yield temporal succession. Thus, in the case of a spherical wave produced by a point source we associate a definite direction of time with the sequence of positions of the wave as it travels outwards. However, unless we already have some sense of temporal succession we cannot distinguish between the case in which these positions of the wave form a sequence in time (finite wave velocity) and the case in which they are regarded as simultaneous (infinite wave velocity). In either case the role of the point source could be interpreted as 'causal', in so far as its existence is regarded as a necessary condition for the wave to exist, but this causal interpretation would not enable us to draw any conclusions concerning time if we were not independently aware of its existence.

7.3. The statistical theory of time

The Austrian physicist Ludwig Boltzmann suggested that the concept of time, in particular its direction ('time's arrow'), depends on the concept of entropy, which he interpreted statistically.

Originally, as stated in Chapter 1, entropy was defined differentially, like potential energy, change of entropy being equal to heat received divided by the temperature at which the heat is received. The Second Law of Thermodynamics, as formulated about 1850 by Kelvin and Clausius, states that the entropy of an isolated (or closed) system never diminishes; it is a generalization of the statement that heat does not flow from lower to higher temperatures.† From the point of view of thermodynamics regarded as a subject *sui generis*, this is a universal law

† If any physical system is divided into two parts, energy can flow from one part to the other but the total energy of the system cannot be increased or decreased thereby (*First Law of Thermodynamics*). Accompanying this internal flow of energy there will also be a flow of entropy from one part to the other, but only for reversible processes will there be no resulting change in the entropy of the system. For irreversible processes there must be an increase in this *total* entropy. This is the essential content of the *Second Law of Thermodynamics*.

It must emphasized that these laws apply only to closed systems. The thermodynamics of open systems, involving a flow of matter and energy from outside, is a modern development due to I. Prigogine and others. Living organisms, growing crystals, etc. are open systems and can therefore lose entropy by interaction with their environments, but if an open system and its environment together can be regarded as a closed system there is no conflict with the Second Law.

admitting no exception. Unfortunately, it is by no means a clear law. For example, as expounded by Kirchhoff, entropy could be measured only by means of a reversible process and therefore, strictly speaking, could not be applied to the irreversible processes which are the subject of the law. A determined attempt to resolve this difficulty was made by Max Planck in his doctoral thesis. He realized that to clarify the statement that the process of heat conduction cannot be completely reversible it is essential to have a proper definition of reversibility and irreversibility. In his view it was insufficient to define an irreversible process as one which cannot run backward, for it is possible that although a process cannot run backward the original situation may in some way be restored. He therefore defined an irreversible process as one which cannot be cancelled, i.e. one for which no counter-process restoring the original situation is possible. Consequently, Planck believed that the question of whether a process is reversible or not depends only on the nature of the initial and final states. In an irreversible process nature has a 'preference' for the final state and the entropy of Clausius is a measure of this principle.

This interpretation of the Second Law of Thermodynamics was essentially teleological and in accord with Planck's interpretation of other fundamental laws of physics, notably the principle of least action in dynamics and Fermat's principle of least time in optics, as evidence for the existence of 'purpose' in the universe as the correlative of causality. This point of view has been rejected by most physicists. The variational (stationary integral) principles in physics which Planck regarded as the formal expression of certain goal-seeking tendencies in nature are, in fact, deduced from causal laws (described in terms of differential principles). Laws of either type are mathematically convertible into laws of the other. We therefore regard the variational principles of physics as showing no preference for either causality or finality.†

Boltzmann, however, realized that this symmetry of causality and finality disappears *automatically* when we consider phenomena of mixing or separation involving *large numbers* of particles or other primary parts. A familiar example from everday life is the effect of stirring when cream is poured into a cup of coffee. In a short while we obtain a liquid of uniform colour, and however long we continue to stir after that, however much we interchange the order of the parts, we never find the contents of the cup

† Two essential points should be borne in mind concerning Planck's teleological claim:

(i) we cannot *deduce* the existence of a purposive economy in nature merely from the fact that certain physical laws can be formulated as integral principles;

(ii) correlatively, we cannot *disprove* the existence of purpose in the world merely as a consequence of (i).

reverting to their original state in which coffee and cream were clearly separated. This state may be described as *orderly*, whereas the state in which coffee and cream are thoroughly mixed is *disorderly*. A similar situation arises in the shuffling of a pack of playing cards. Now the number of possible arrangements of the cards of a pack exceeds 8×10^{67}, so that if we laid them out at the rate of three different arrangements a second it would take about a billion billion billion billion billion years to exhaust all the possibilities. On the other hand, there are only 48 possibilities of finding the cards in the same strict numerical order, from lowest to highest or highest to lowest, in each suit. Consequently, if we begin with an orderly pack, shuffling will make it disorderly and, in general, continued shuffling will not make it orderly again. Although artificial, the shuffling process may be regarded as the analogue of the natural motion of the molecules in a vessel containing a mixture of liquids and gases, and like this motion may be considered reversible. The unidirectional effect of its operation is due essentially to *statistical* considerations based on the inexorable law of large numbers. Boltzmann sought, therefore, to reformulate the concept of entropy (for any given closed physical system) in terms of probability.

Statistical mechanics as developed by Boltzmann (and also by Willard Gibbs) provided a mechanical interpretation of thermodynamics. The concepts required to supplement those of classical dynamics (the particles and rigid bodies of which have no intrinsic heat, temperature, and entropy) were obtained by considering large aggregates of moving particles (representing molecules etc.) and inventing statistical analogues of thermodynamic concepts. Thus, Boltzmann obtained, in 1872, his famous formula

$$S = k \log P,$$

where S denotes entropy, k is a constant now known as Boltzmann's constant, and P is the number of different 'microscopic complexions' of a given macroscopic system, i.e. states in which the velocities, positions, and—according to modern views—quantum states of all the constituent atoms and molecules are specified. Thus, if we consider two bodies (1 and 2) in contact, so that they can exchange only heat, both being immersed in a common insulating container, then at any instant each will have a definite energy, say E_1 and E_2. By the law of conservation of energy, $E_1 + E_2$ is the same at all instants. However, associated with E_1 there will be P_1 complexions of body 1, and with E_2 there will be P_2 complexions of body 2, P_1 being a function of E_1 and P_2 a function of E_2. The total number of complexions of the whole system will be $P = P_1 P_2$, i.e.

the product† of P_1 and P_2, since each complexion of body 1 can be associated with each complexion of body 2. The most probable energy distribution is that for which P is a maximum, and this is found to correspond to the equalization of the temperatures in 1 and 2. The Second Law of Thermodynamics was therefore interpreted by Boltzmann as signifying that any closed, or isolated, system (i.e. any system insulated against accretion of energy from, or loss of energy to, anything outside), automatically tends towards an equilibrium state of maximum probability, associated with equalization of temperature, pressure, etc., if not already in such a state. Since the probabilities of occurrence of ordered arrangements of molecules (e.g. where the molecules in one part of the system are at one temperature and those in the remaining part are at another) are far less than those of random or disordered arrangements (where no sorting occurs) the law signifies that ordered arrangements tend to degenerate into disordered ones.‡

Boltzmann's interpretation meant, however, that the principle of the increase of entropy, now regarded as a measure of the disorder of a thermodynamic system, ceased to be an invariable law of nature but became a statistical one. From this point of view, the reverse process of a thermodynamical trend to a state of lower entropy was no longer impossible but only extremely improbable. Consequently, part of the water in a kettle on the fire could freeze while the rest boils, although it is so unlikely that this will happen as to be virtually impossible in our experience.

Boltzmann believed that his statistical interpretation of the law of entropy as the tendency to maximize P automatically accounted for the directional character of time. Despite the power and cogency of his theory, Boltzmann's statistical explanation of time's arrow was soon criticized as being logically unsound. Thus, Loschmidt (1876) argued that the symmetry of the laws of dynamics with respect to past and future should imply a corresponding reversibility of molecular processes, in contradiction with the law of increasing entropy. For, since the probability of a molecule having a given velocity is independent of the sign of the

† The fact that P is multiplicative whereas S is additive accounts for the appearance of the logarithmic function in Boltzmann's formula. The number of microscopic complexions, or microdistributions, corresponding to a given microscopic thermodynamic state is a measure of the probability of occurrence of that state.

‡ This explains why that part of the energy of a material system which is available for doing useful work should tend to diminish, since this energy is orderly energy, whereas heat, because it is associated with the random motion of large numbers of molecules, is disorderly energy. The energy of a system thus tends to become less and less available for mechanical work as more and more of it is converted into heat and the disorderliness of the molecules increases.

velocity, because it depends on the square of the velocity, the principle of dynamic reversibility implies that to each state of motion of a given (isolated) system there will correspond another state of motion when it passes through the same configuration with all velocities reversed in direction. Since these two states of motion are equally probable, in the course of time separation processes should occur as frequently as mixing processes. According to Loschmidt, the observed occurrence of only entropy-increasing processes must therefore be a consequence of the particular initial conditions in the world and not of the laws governing molecular motions.

Besides Loschmidt's *reversibility objection*, another objection, known as the *periodicity objection*, was formulated later by Zermelo (1896). He appealed to a theorem of dynamics, proved by Poincaré in 1890, which asserts that, under certain conditions concerning the finite nature of the motion of a system, the initial state of the system will recur infinitely often.† Zermelo claimed that molecular processes must, therefore, be cyclical, again in apparent contradiction with the Second Law of Thermodynamics and with the hypothesis that entropy increase is the ultimate explanation of time's arrow.

These difficulties were considered by P. and T. Ehrenfest in a famous article contributed to the German mathematical encyclopaedia (Ehrenfest and Ehrenfest 1911). They pointed out that Boltzmann's statistical proof of the Second Law of Thermodynamics (his celebrated H-theorem) concerned only the *average* variation of the entropy of an isolated system and therefore did not preclude the possibility of decreases in its value. This averaging was the expression of our ignorance concerning the actual microscopic situation. Fluctuations do occur, but Loschmidt's and Zermelo's objections effectively apply only to very small‡ fluctuations around a state of thermodynamic equilibrium since macroscopically significant fluctuations will normally only be detectable after immensely long intervals of time. For example, for air at 300 K and a density of 3×10^{19} molecules per cubic centimetre the mean recurrence time of a 1 per cent fluctuation in the number of oxygen molecules in a sphere of radius 5×10^{-5} centimetres is of the order of 10^{68} seconds, or 3×10^{60} years, but if we reduce the radius of this sphere by no more than a factor 5 to 10^{-5} centimetres the mean recurrence time is reduced to 10^{-11} seconds

† Poincaré's theorem will be found in his *Oeuvres*, vol. 7, p. 262. The Poincaré time of recurrence depends on the initial conditions. Boltzmann (1898, § 88) estimated that, for two gases in a can of volume 0.1 litre and initially unmixed, the time before they are unmixed again will be 'enormously long compared to $10^{10^{10}}$ years'.

‡ Fluctuations are continually occurring on a very small scale in the phenomena associated with the Brownian movement.

(Smoluchowski 1916). Consequently, *if* the direction of time is to be defined by means of internal entropy increase we must conclude that many microsopic phenomena can have no intrinsic time direction.

The conclusion that unidirectional time does not exist on the microscopic scale is also a consequence of the theory of time reversal in particle physics formulated by Stückelberg (1941) and developed by Feynman (1949). They claimed that in studying the behaviour of elementary particles there were advantages in regarding antiparticles as ordinary particles travelling backwards in time. For example, a positron (the antiparticle of the same mass as the electron but of opposite charge) could be regarded as an ordinary electron with reversed time direction, the physical effects of this reversal being counteracted by the change in the sign of the electric charge. This idea was invoked to 'explain' the curious phenomena of pair production and pair annihilation observed on photographs taken in a Wilson cloud chamber: a γ-ray suddenly turns into an electron and a positron; the latter usually soon meets another electron and these two then vanish and leave behind a new γ-ray starting from the point of collision. According to Feynman, the situation illustrated in Fig. 7.1(a) in which the two world lines of an electron and a positron meet and annihilate each other can be re-interpreted as the world line of a single electron† travelling forwards and backwards in time, as illustrated in Fig. 7.1.(b).

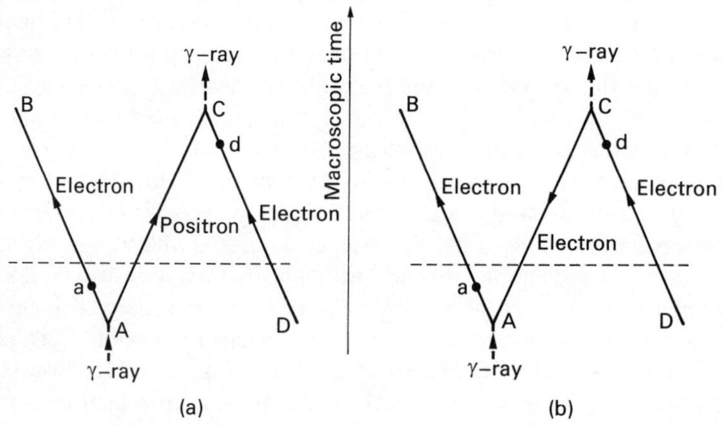

FIG. 7.1

† Following the charge rather than the particle corresponds to considering the world line as a continuous whole instead of breaking it up into parts. Feynman drew the analogy: 'It is as though a bombardier flying low over a road suddenly sees three roads and it is only when two of them come together and disappear again that he realizes that he has suddenly passed over a long switchback in a single road'.

The observational basis of this theory is a photograph depicting a row of small drops of water. We believe these to have been produced by a rapidly moving particle which collides with more massive particles, thereby generating local condensations of water vapour along its path. The travelling particle, however, is not directly observed but is inferred and so is an example of what Reichenbach has aptly called an 'interphenomenon'. In the usual description, as illustrated in Fig. 7.1(a), the interphenomenon is simply a positively charged particle travelling 'forward' in time. In Feynman's description, illustrated in Fig. 7.1(b), it is a negatively charged particle travelling 'backwards' in time. Thus, according to Feynman, the collisions experienced by the particle travelling between A and C occur, from the point of view of the particle, in the opposite sequence to that in which we regard them as occurring in macroscopic time. From the 'worm's-eye view' of the particle, there are no anomalies of pair production and pair annihilation because there is only one particle, but the elimination of these anomalies is achieved only by introducing the further anomaly of time reversal.

This local (microscopic) time reversal is essentially different from the cosmical time reversal which would be simulated by a film of the world shown in reverse. For, when we endeavour to identify a positron with an electron travelling 'backwards' in time, only one causal chain is reversed and others are unaffected. Consequently, the between-relations of temporal order are altered. Thus, in Fig. 7.1(b) event A is causally between C and B, and event C is causally between D and A, whereas neither of these statements applies to the events in Fig. 7.1(a). As Reichenbach (1956, p. 266) pointed out, Feynman's interpretation leads to the possibility of closed causal chains. For example, consider the case in which, according to the normal interpretation, electron 1 at event a experiences a collision accompanied by the emission of a γ-ray which travels faster than the positron and collides with electron 2 at event d. When we introduce Feynman's interpretation, the light ray ad will *not* be reversed. Consequently, the sequence of events dCAad will now form a closed causal chain, i.e. a closed cycle in time. Reichenbach commented that although such processes have not yet been observed and appear rather improbable', nevertheless 'their possibility cannot be denied'. In his view, a closed causal line *at this sub-atomic level* does not conflict with our customary concept of causality, because he regarded the latter as essentially a macroscopic concept.

Nevertheless, the identification of a positron with an electron in 'negative' time does conflict with our common-sense notion of genidentity.†

† This term is used by Reichenbach (1956, p. 38) to denote physical, as distinct from logical identity, i.e. the identity of a thing throughout its history.

For, although we believe that the same thing can be at the same place at two different times we find it difficult to imagine that it can be at two different places at the same time. Quantum statistics has made us familiar with the idea that, for example, photons of the same frequency are indistinguishable so that the concept of material genidentity must be abandoned. Nevertheless, the rule was still retained that no two *simultaneous* states could be associated with the same particle. However, if we adopt Feynman's interpretation, electrons and positrons need not obey even this rule, because to interpret a positron as an electron moving 'backwards' in time is equivalent to allowing it to be in more than one place at the same time: in Fig. 7.1(a) the simultaneity cross-section $t = t_0$ intersects the world line of the electron at least three times. Thus, although from the point of view of the electron, i.e. in its proper time, the events represented by these intersections occur in a definite sequence, for the macroscopic apparatus recording the events (the Wilson cloud chamber) they will appear to be simultaneous. Moreover, as we have already seen, the apparent order of some events will even be reversed. Consequently, Feynman's interpretation of pair production and pair annihilation leads to the conlusion that not all sequences in time can be subsumed under a universal time order.

Reichenbach (1956, p. 268) regarded this as 'the most serious blow that the concept of time has ever received in physics'. He argued that time as we normally understand it, i.e. macroscopic time, must therefore be essentially statistical in character. It cannot be traced in the elementary phenomena which give rise to it but is 'born anew at every moment from the atomic chaos as a statistical relationship'. Ordered and directed time arises in this way only because positrons (and other antiparticles) are short lived in the presence of those particles, such as electrons, which conform to the rules of ordered and directed time. The statistical preponderance† of the latter particles is decisive. Reichenbach therefore concluded that the existence of an asymmetrically directed macroscopic time order is due to the asymmetry of negative and positive electric charge in the world.

A trenchant criticism of Feynman's interpretation has, however, been made by Zwart (1976, pp. 155–9). He advances two important objections. First, he points out that only the time of the electron is reversed while the time of the environment retains its original direction, and this is

† At present we have no theoretical explanation of this asymmetry, and it may be an 'accident of nature' that negative electrons and positive protons are so enormously more numerous than their counterparts of opposite charge which we know can, and do, exist.

objectionable because in this case the electron interacts with its environment (the γ-ray and the water vapour in the cloud chamber). The electron going from C to A (Fig. 7.1(b)) is going to meet an approaching γ-ray, but since it is moving backwards in time it follows that from the point of view of the γ-ray and of positive time generally it is moving away! Moreover, how does the electron manage to reverse its direction in time in the particular circumstances concerned when otherwise it never does so? The other objection is that, if the time of the electron were truly reversed at C, it would move back along the path it traversed previously and so would move back to D and not to A. In view of these arguments and those concerning genidentity and closed causal chains mentioned above, it appears that Feynman's interpretation raises more difficulties than it resolves and that the original interpretation in terms of pair production and pair annihilation is far less puzzling than any interpretation based on the idea of time reversal.

The arguments against adopting Feynman's interpretation are typical of the difficulties encountered by the hypothesis that different time directions coexist. Conflict with our customary ideas of causality, genidentity and the impossibility of closed circuits in time will arise whatever the level (sub-atomic or otherwise) at which the hypothesis is made. It is therefore preferable to retain the concept of unidirectional time at all levels, although this means that we are obliged to reject Boltzmann's statistical theory of time based on the concept of entropy increase because of the inevitability of small-scale fluctuations. However, before we go on to conclude that the arrow of time is a feature of the world that cannot be explained in terms of statistics, or of anything else more fundamental, we must examine more generally the relations between it and the basic laws of physics.

First, we note that none of the laws governing the fundamental interactions of physical particles can be used to establish the existence of time's arrow. On the contrary, not only the long-range forces of gravitation (whether expressed in terms of Newton's theory or of Einstein's) and of electromagnetism, but also the 'strong' interactions or short-range forces that occur between nucleons, are unaffected if we imagine a reversal of time. Only in the case of the short-range 'weak' interactions, involving neutrinos, is there any evidence against time-reversal invariance. Even here the crucial experiment (Christenson *et al.* 1964) did not test this invariance directly but only through a theorem based on special relativity known as the CPT theorem. According to this theorem all physical laws governing systems of elementary particles must remain valid under the combination of the following three transformations: (i) reversal of the signs of all electric charges (C reversal); (ii) reversal of parity (P reversal), i.e. replacement of a system by its mirror image;† (iii) reversal of time (T reversal). Hence, all such laws of physics are said to be CPT invariant. The experiment revealed a violation of CP invariance in the decay of the neutral K meson, which implies that if the CPT theorem is to remain valid there must be a compensating violation of T invariance. This violation of time-reversal symmetry is, however, only observed on about one occasion in a hundred of the decay of the K meson. Moreover, K mesons are found only in experiments in high-energy physics. They are not constituents of ordinary matter and they play no part in any of the macroscopic processes that are associated with time's arrow (see, for example, Davies 1974, p. 176).

Apart from this one minor loophole, it does not seem that the laws governing the fundamental forces and reactions between elementary particles can provide any indication of the unidirectional nature of time.‡ Instead, if we are to account for this on the basis of some physical law, we must consider systems involving many particles. For, even as regards quantum mechanics, the most convincing arguments for a preferred time direction depend on the interaction of quantum-mechanical systems with macroscopic systems, notably in the actual process of observation in the laboratory. In this case, we find that Schrödinger's equation behaves asymmetrically with respect to past and future. The role of time in relation to quantum mechanics is, however, revealed most clearly when we analyse the nature of physical observation in general. According to the modern theory of information, as expounded for example by Brillouin (1962), an observation is essentially an irreversible process. Whether we take the thermodynamic point of view or whether we agree to include the conventionalized measure of 'information' in the entropy, an increase of entropy is inevitable whenever an observation is made. Moreover, as has been emphasized in quantum theory by von Neumann and others, strictly one should never speak of a system as being in a certain state unless a measurement of some quantity involved in the description of the state is either made or assumed to be made. However, since the process of measurement itself automatically influences the future behaviour of the

† Although it has been known for many years (notably since the pioneer researches of Pasteur) that the distinction between a system and its mirror image is often significant at the molecular level, for example the occurrence of dextro- but not of laevo-tartaric acid in fermenting grapes, it was generally believed that at the more fundamental level of elementary particles no difference of this type could occur. However, in 1957 it was found that parity is not conserved in weak interactions.

‡ In particular, as Weyl (1949, p. 264) has remarked, elementary laws cannot account for the temporal asymmetry revealed by the fact that there is spontaneous emission but not spontaneous absorption of photons by atoms.

system, its effect is irreversible. Like Heisenberg's uncertainty principle, the 'negentropy' principle—which asserts that any information resulting from a physical observation must be paid for by an increase of entropy in the laboratory—is a fundamental limitation of physical measurement; however, unlike Heisenberg's principle it is not effectively restricted in practice to the microscopic level. Nevertheless, because it essentially involves the participation of the observer, it cannot be employed to derive an objective time sequence of phenomena.

Finally, we must consider whether we can circumvent the difficulty previously encountered in connection with small-scale fluctuations so as to make it possible after all to base the unidirectional nature of time on the statistical concept of entropy. The most thorough attempt to do this is due to Reichenbach (1956). He accepted the reversibility objection as a decisive reason for rejecting any definition of time direction in terms of the entropy of an *isolated* system. Instead of the history of a single system, he considered statistically a large number of what he called 'branch systems'. These are subsystems relatively isolated from the main system of the universe in so far as the energy exchanges which occur within them are large compared with their reactions with the rest of the universe. A cube of ice initially placed in a glass of hot water is a typical example. A branch system can be put initially into a state of low entropy (while the entropy of a more inclusive system is increased), but in general we find thereafter that the relative entropy of the branch system tends to increase. Reichenbach showed that the probability that a low-entropy state will be followed by a high-entropy state is greater than the probability that the same low-entropy state was itself preceded by a state of high entropy. He thus arrived at the following definition: *the direction of positive time is given by the direction in which most thermodynamic processes occur in isolated systems.* Reichenbach believed that this definition was free from the reversibility objection because the statistical criterion now refers to a large number of branch systems (a 'space ensemble') and not to the sequence of states (the 'time ensemble') of a single system.

Unfortunately, as Reichenbach himself realized, this definition does not immediately allow us to identify increasing time with increasing entropy, for it is an essential presupposition of Reichenbach's analysis that the subsystems branch off in their low-entropy states and this possibility depends on the main system being itself in a relatively ordered (or low-entropy) configuration and consequently on the upgrade of its entropy curve. If, in accordance with the intention of the Boltzmann–Reichenbach theory of time, we take a purely statistical view of the problem, we must assume that the main system is passing through an enormous and 'highly improbable' fluctuation from its 'most probable' equilibrium state.† This assumption was, in fact, adopted by Boltzmann (1898) who believed that the universe as a whole is so vast (in both space and time) that 'our part of it' is experiencing such a fluctuation and is at present in a state of sufficiently low entropy, and hence not too much disorganized, for our own existence to be possible. However, at the beginning of such a fluctuation entropy must decrease, and therefore Boltzmann argued that there must also be regions of the universe in which the direction of time is opposite to that in our own, although these regions could be separated from us by immense distances (of empty space) and long periods of time. Nevertheless, even if we accepted his argument—and Boltzmann was, of course, quite ignorant of modern knowledge concerning the structure and evolution of the universe—we should still be faced by the difficulty that, if our own region is now experiencing a fluctuation, its entropy cannot continually yield a consistent direction of time within it.

Consequently, since it depends on the entropy curve of a 'main system' which is in a state of fluctuation, Reichenbach's theory of branch systems fails to achieve its object, and if we attempt to salvage it by referring to larger and larger systems we can end only by referring to the whole universe. Indeed, if at some stage there are no branch systems because no systems can be sufficiently isolated, then the only way left to define the arrow of time at such an epoch would be by direct appeal to world-entropy increase. However, as we have already seen in Chapter 1, there are grave difficulties in formulating this concept. These difficulties are accentuated by the fact that there is at present no general agreement concerning the extent of the universe, i.e. whether it is finite or infinite, and also by the hypothesis that it is expanding.

The failure not only of Boltzmann's original attempt to invert the Second Law of Thermodynamics so as to provide a statistical explanation of time's arrow but also of the later refinement of his theory by Reichenbach is therefore further evidence for believing that time cannot be derived from more fundamental concepts that involve no implicit appeal to it. Time, as Leibniz said, is the order of succession of phenomena, and if there were no succession there would be no time. The basic objection to attempts to deduce the undirectional nature of time from concepts such as entropy is that they are attempts to reduce a more fundamental concept to a less fundamental one. As Bridgman (1955) has said, 'If it

† Classical thermodynamics is concerned only with the equilibrium states and states very near to equilibrium of closed systems. For any arbitrarily chosen non-equilibrium state of such a system, statistical mechanics indicates an overwhelming probability that the system is undergoing a fluctuation from equilibrium.

were found that the entropy of the universe were decreasing, would one say that it was a law of nature that entropy decreases with time? It seems to me that in any operational view of the meaning of natural concepts the notion of time must be used as a primitive concept, which cannot be analyzed but must be accepted, so that it is meaningless to speak of a reversal of the direction of time'. He concluded that the underlying operations cannot be formulated without assuming as understood the notion of earlier or later in time.

7.4. Cosmology and the order of succession of phenomena

It is natural to ask whether the different manifestations of temporal order in the world as revealed to us by our fundamental sense of earlier and later can be correlated with some characteristic temporal property of the universe as a whole, or whether, because the general laws of physics do not reveal unambiguously the one-way trend of time, we must conclude that time's arrow is a purely subjective phenomenon peculiar to the human mind. I do not believe that we are obliged to adopt the second alternative, since the *laws* of physics do not tell us everything about the world. For, in order to apply the laws to a given physical system, we must impose certain constraints. Laws indicate what is possible but not what actually occurs. To determine this we need to know certain additional items of information, such as the positions and velocities at some chosen instant of the constituent parts of the system studied. These are often referred to as 'initial conditions' or 'boundary conditions'. They are not given by the laws themselves, which are of a general character and so are applicable to many different systems. If there is a deep connection between time and the universe, it may be because time's arrow is associated in some way with the particular initial conditions that determined the universe that actually is, as distinct from any other universe that might have existed in accordance with the same physical laws.

In order to gain a deeper understanding of the temporal order of phenomena various attempts have been made to correlate the different manifestations of time's one-way trend, particularly the electromagnetic arrow associated with the existence of retarded radiation processes, the thermodynamic arrow defined for entropy-generating processes in closed systems, the historical arrow associated with information-generating processes in certain open systems and the cosmological arrow indicated by the expansion of the universe.

The electromagnetic arrow was the subject of a lively discussion by Einstein and Ritz in 1909. Einstein (1909) argued that the retarded and advanced descriptions of radiation processes occurring in any finite region are equivalent, since the equations of wave propagation are symmetric with regard to time, but that the auxiliary conditions giving the precise circumstances of emission and absorption are very different. In the retarded description it is sufficient if all the macroscopic sources are known whereas in the advanced description all the absorption processes must be known, but unlike the former they must be fully specified in microscopic detail. In practice, we do not have this information concerning the absorption processes and so we are obliged to use the retarded description. On the other hand, Ritz (1909) asserted that only the retarded waves have any physical significance, since advanced waves are not experimentally observed. The initial conditions characterizing the source (or sources) of the radiation are the causes of its transmission and consequently are responsible for the special role played by retarded waves.†

Einstein's point about the difference between macroscopic sources and microscopic absorptions partly anticipated the important argument, due to G. N. Lewis and extended by K. R. Popper (see pp. 8–9), that the temporal inverse of the expansion of a spherical wave from a point source would, in general, require the co-operation of a large number of sources scattered throughout space so as to produce a coherent wavefront that converges to annihilation at a particular point at a particular time. This coherence is so highly improbable that it could not as a rule occur spontaneously.‡

The idea of the conditional independence of influences emanating from different parts and directions of space was postulated by Penrose and Percival (1962) in an attempt to account for the common direction in time of various irreversible physical processes. This assumption of conditional independence was introduced as a fundamental asymmetric condition supplementing the temporally symmetric laws of dynamics so as to account for the irreversiblity of most physical systems. According to this hypothesis, influences coming 'from infinity' in different directions are statistically independent. In adopting this hypothesis they were guided by the principle of retarded action formulated by Costa de Beauregard (1958) which states that, if an otherwise isolated system reacts with its surroundings or with another system at time t_0, the effect of the interaction is felt after time t_0 but not before. From their principle Penrose and Percival derived the time direction of some irreversible physical processes. For example, they based the irreversibility of recording processes on the assumption that before a certain epoch the system producing a record and the recording instrument are uncorrelated. Correlation then occurs

† According to Brillouin (1964), Ritz was the first to make this point.

‡ This does not, of course, mean that the implosions are impossible.

and its effect persists in the form of the record concerned. In particular, the direction of subjective time can be understood on this basis, since human memory is a registering influence capable of recording the past but not the future. Among other examples, they derived the irreversibility (associated with retarded potentials) of outgoing electromagnetic waves radiated from a dipole.

In making their hypothesis precise, Penrose and Percival restated the principle of causality statistically, defining the term 'effect' as a correlation between a system and its surroundings. Their 'statistical principle of causality' asserted that a system which has been isolated throughout the past is uncorrelated with the rest of the universe. This principle is asymmetric. To allow for the fact that such isolated systems do not occur naturally, they generalized the principle by replacing 'the system' and 'the rest of the universe' by two disjoint regions A and B in space–time separated by a 'statistical' barrier C. An essential feature is that C extends infinitely into the past, or if the universe began with a singularity at a particular instant they assumed that all spatially separated regions of space become uncorrelated in the limit as we extrapolate back to that epoch. The direction of time was thus made to depend on what is essentially a cosmological postulate concerning the existence of real independent events. For a world model with a particle horizon, for example the Einstein–de Sitter universe (see p.312), it is the case that initially distant regions of space are not causally connected, but we do not know whether the universe is of this type.

The dependence of the electromagnetic arrow of time on the thermodynamic arrow was a feature of the absorber theory of radiation formulated by Wheeler and Feynman (1945). In an attempt to produce a theory of charged elementary particles which avoided the difficulties that had beset previous theories of their interaction with electromagnetic fields, they introduced the hypothesis that every photon has an absorber as well as an emitter. In their theory an accelerated charged particle emits radiation equally into the past and future. In other words, retarded and advanced waves are generated symmetrically. If the radiation is confined to an opaque enclosure, so that all of it is absorbed, the waves striking the walls will cause the charged particles therein to radiate likewise into both the past and the future. Wheeler and Feynman showed that, if the enclosure is *fully opaque*,† the advanced waves emitted by the walls will just cancel those from the source particle and only the retarded waves will be left. The origin of this time asymmetry lies in the mechanism of absorption; the waves absorbed by the walls are converted into heat

† If the enclosure is not fully opaque, bizarre effects involving time reversal of causality will occur, in contradiction with experience.

which is dissipated through the walls in accordance with the law of entropy.

Although in this theory the electromagnetic arrow of time appears to depend on the thermodynamic arrow, cosmology plays a significant role too. The absorber theory is applicable only to a universe that is like a perfectly opaque enclosure, the possibility that all advanced radiation will be absorbed depending on the state of the universe in the remote future. Wheeler and Feynman confined their discussion to the hypothetical case of an infinite uniform static universe. Later Hogarth (1962), Sciama (1963), and Hoyle and Narlikar (1964) considered the viability of the absorber theory in expanding world models. It was found that, because of the ever-decreasing density of matter, Friedmann universes that continue to expand forever ultimately cease to be opaque. In other words, in these universes not every emitted photon can eventually be absorbed. Consequently, in these world models the Wheeler–Feynman theory cannot reduce to conventional electrodynamics. On the other hand, in the steady-state models of the expanding universe, just as in a static universe, every photon can ultimately be absorbed. The only Friedmann universe to which the Wheeler–Feynman theory can be successfully applied is the cyclic model in which expansion is eventually reversed so that ultimately the model collapses to a point singularity. Consequently, as far as Friedmann models are concerned, the absorber theory carries definite implications as to how the universe will end, so that as remarked by Davies (1977), 'local behaviour of electromagnetic radiation enables us to look into the future and foretell what will happen to the cosmos'. However, not only is the observable universe far from opaque, since light can travel to us from remote galaxies without encountering much absorbing matter on the way, but the current evidence in favour of an ever-expanding universe, if correct, must be regarded as evidence against the absorber theory. Moreover, since all attempts to produce a quantum-mechanical version of the absorber theory lead to the same difficulties as previous theories of the interaction of charged particles with the electromagnetic field, there is no strong argument in its favour and in fact its original proponents have abandoned it.

It has been suggested by Gold (1958, 1962) that the thermodynamic arrow of time is a direct consequence of the expansion of the universe. Because of the contrast between the hot stars and cold empty space, the universe is far from being in a state of thermodynamic equilibrium. Gold argued that this situation is due to the universe being an almost perfect cold sink for radiation, and this is so because it is expanding. In his view the expansion of the universe in some way automatically maintains the thermodynamic disequilibrium of the world, so that if expansion should eventually cease and be succeeded by a contracting phase the time direction of thermodynamic and electromagnetic processes would be reversed. However, as has been pointed out by Davies (1974), this conclusion is difficult to accept. For example, although contraction could restore the temperature of the cosmic background radiation (since this varies inversely as the expansion factor R), it could not return all the starlight to the stars against their temperature gradients. Indeed, the flow of radiation from a hot star into cold surrounding space involves the production of photons and is irreversible. Gold's argument also conflicts with the result obtained by Tolman (1934) that repeated cycles of expansion and contraction of the universe would increase the entropy of the cosmic material.†

The problem of relating the different arrows of time with the cosmological arrow has been tackled in a different way by Layzer (1970, 1976). Since he believes that the electromagnetic arrow is essentially determined by the thermodynamic arrow (Layzer 1970, p. 457), he has directed his attention to the latter, as defined by entropy processes in closed systems, and to the historical arrow, as defined by information-generating processes in certain open systems. To explain how irreversible processes in different closed systems can define the same arrow, he argues that thermodynamics needs to be supplemented by cosmological considerations. In his view, the most conspicuous class of unidirectional processes are those that give rise to evolutionary records, since they all point in the direction of increasing information, thereby automatically indicating the direction of the historical arrow. Moreover, these records are produced not only by biological systems and human memory, but even by inanimate bodies. For example, the Moon's cratered surface provides a record of its past, the changing internal structure and chemical composition of a star records the process of its ageing, and the forms of spiral galaxies reflect the evolutionary processes that shaped them. Layzer defines the historical arrow through the statement that 'the present state of the universe (or any sufficiently large subsystem of it) contains a partial record of the past and none of the future'. He rejects the widely held belief that the universe was initially more highly ordered than it is now and that the primeval order has gradually been dissipated by irreversible processes. Instead, he considers a universe that initially is in a very simple undifferentiated state without any structure, being statistically homogeneous and isotropic without any microscopic 'information' and no statistical means of specifying a direction or position in space. He argues that thermodynamic equilibrium of the whole universe may be expected to

† See also Landsberg (1975).

prevail only close to a singular state of infinite density, for since cosmic expansion is not adiabatic it leads to departures from thermodynamic equilibrium. As this expansion continues galaxies and stars are generated, and this production of more complex systems from simpler ones means that cosmic expansion generates information. Consequently, the historical arrow coincides with the cosmological arrow. At the same time irreversible processes generate entropy.

Although at present there is no generally accepted theory concerning the correlation of the different arrows of time, the various attempts that have been made to elucidate the nature of time and its unidirectional character indicate that ultimately time must be regarded cosmologically. It is a fundamental aspect of the relationship between the universe and the observer. Our notion of earlier and later is a primitive concept that cannot be reduced to anything more basic, but we can relate it to the whole cosmological process. For, although world expansion may seem at first sight a singularly inappropriate signpost for local time, our current knowledge of physics and astronomy reveals definite links between them. Many cosmologists now believe that the universe began with primeval matter and radiation in an extremely dense high temperature state following an initial 'explosion'. As the universe expanded, the temperature decreased inversely with the expansion factor. In the first three minutes or so it fell sufficiently for the principal elementary particles and the simplest atomic nuclei to be generated (Weinberg 1977). These were followed much later by the creation of stable atoms and in due course, after the decoupling† of matter and radiation, by the formation of galaxies and stars. The expansion of the universe thus provided the time-direction signpost of inorganic evolution (including that of radioactive decay), and this in turn ultimately gave rise to that of biological evolution and of all organic processes, culminating in our conscious awareness of unidirectional time.

7.5. The transitional nature of time

The idea that time relations are ultimate and irreducible is one that many philosophers, mathematicians and physicists have been unwilling to accept. Even though it has seldom been denied that time is 'real' in the sense that it is a phenomenon of our experience—and indeed, in Leibniz's phrase, a 'phenomenon *bene fundatum*'—thinkers as diverse in their general outlook as, for example, Plato and Kant, and Bradley and Weyl

† 'Decoupling' means that matter and radiation were no longer in thermal equilibrium. This occurred when the temperature had sunk to about 3000 K.

have repeatedly argued that the temporal mode of our perception has no *ultimate* significance. Although this point of view is primarily associated with the long line of idealist philosophers going back to Parmenides, it has also been accepted by so empirically minded a thinker as Bertrand Russell. In his essay on *Mysticism and logic*, after dismissing the idealist arguments for the contention that time is unreal, he admitted that 'Nevertheless, there is some sense—easier to feel than to state—in which time is an unimportant and superficial characteristic of reality. Past and future must be acknowledged to be as real as the present, and a certain emancipation from slavery to time is essential to philosophic thought' (Russell 1917). As a historian of philosophy has remarked when commenting on this passage, any philosopher who approaches philosophy through logic is likely to argue in this way, since implication is not a temporal relation (Passmore 1957).

The aspect of time which is most often objected to by scientists as well as philosophers is its transitional nature, i.e. the concept of 'becoming' and its relations to past, present and future. Scientists often argue that this aspect plays no role in physics and that the passage of time is merely a feature of our conscious awareness that has no objective significance. Some philosophers, however, have gone further than this and have contended that the transitional nature of time is a self-contradictory concept. This was the standpoint adopted early this century by the Cambridge philosopher J. M. E. McTaggart who maintained that the statements that an event E is now present, has been future and will be past are mutually incompatible.

McTaggart distinguished the changing A series, as he called it, of past, present and future from the static B series in which events are related in the order 'earlier than' or 'later than' (McTaggart 1927, p. 10). The B series is a permanent series in the sense that if the statement that 'P occurs before Q' is true, then it is always true. For example, the statement that the Battle of Hastings occurred before the Battle of Waterloo is a permanent truth. The B series is the way in which we normally *contemplate* a sequence of events in time. It is a method of ordering analogous to numerical ordering and is compatible with the 'block universe' idea. It is the aspect of time which is required in much of theoretical science. On the other hand, the A series characterizes the way in which we actually *experience* events. Unlike the B series, it is an ever-changing series and gives meaning to the concept of 'becoming' or 'occurrence'. The fact that it is a changing series—that what happens now was once future and will be past—leads us to make statements that are not permanent truths. This has been the source of much perplexity. McTaggart argued that, although the A characteristics of events are an essential feature of the ideas of time and change, they involve a contradiction which can be circumvented only by an infinite regress. He therefore calculated that in the final analysis the contradiction cannot be resolved and that time is an illusion.

The foundation of McTaggart's detailed and intricate argument was his contention that an event can never cease to be an event. 'Take any event—the death of Queen Anne, for example—and consider what changes can take place in its characteristics. That it is a death, that it is the death of Anne Stuart, that it has such causes, that it has such effects—every characteristic of this sort never changes'. McTaggart pointed out that from the dawn of time, the event in question was the death of a Queen and that it always will be this. In every respect but one it is equally devoid of change. 'But in one respect it does change. It was once an event in the far future. It became every moment an event in the nearer future. At last it was present. Then it became past, and will always remain past, though every moment it becomes further and further past'. (McTaggart 1927, p. 13). McTaggart argued that, although past, present, and future are incompatible determinations, every event must have them all. To the obvious retort that events do not have these characteristics simultaneously but successively, McTaggart countered with the argument that our statement that an event E is present, will be past and has been future means that E is present at a moment of present time, past at a moment of future time and future at a moment of past time. However, each of these moments is itself an event in time and so is both past, present, and future; in other words, the difficulty breaks out all over again and we are launched on a vicious infinite regress.

The answer to McTaggart's argument has been clearly formulated by Broad (1938), who has pointed out that we do not say that the Battle of Hastings precedes the Battle of Waterloo but that it preceded the latter, and that generally the copula in propositions which assert temporal relations between events is not the timeless copula of logic but the temporal copula 'is now', 'was', or 'will be'. For example, the sentence 'It has rained' does *not* mean that, in some mysterious non-temporal sense of 'is', there *is* a rainy event that momentarily possessed the quality of presentness which it has since lost and acquired instead the quality of pastness. What the sentence means is that raininess has been, and no longer is being, manifested in my neighbourhood. Similarly, the sentence 'It will rain' does not mean that, in some mysterious non-termporal sense of 'is', there *is* a rainy event that now possesses the quality of futurity which it will eventually lose and acquire momentarily the quality of presentness. Instead, it means that raininess will be, but is not now being, manifested in my neighbourhood.

The essence of McTaggart's argument is that the happening, or occurrence, of an event is regarded as if it were a form of qualitative change. Our rejection of his argument depends on our refusing to accept this assumption. *Time is not itself a process in time.*† The occurrence of an event is not itself a further event, and there is no infinite regress of the type contemplated by McTaggart.‡ Events happen and do not exist in any other sense.

McTaggart's great merit, as compared with other idealist philosophers such as F. H. Bradley, was that, not content merely with denying the reality of time, he attempted to explain how we come by the illusion that makes us attribute temporal characteristics to existents. His explanation was based on the ingenious hypothesis that there is a third series, the C series, which we misperceive as a temporal series. The two basic relations of this series, like those of the B series, are transitive and asymmetrical, and one is the converse of the other, just as 'earlier' in the B series is the converse of 'later'. McTaggart decided that the relations 'included in' and 'inclusive of' fulfilled the intricate set of conditions which he believed the C series should satisfy. Apart from the criticism that, without some implicit appeal to the idea of time, there does not seem to be a conclusive case for his correlation of 'inclusive of', rather than its converse, with 'later on', the fact remains that the C series is insufficient for a complete account of our concept of time since it does not explain how we come by the A series which, as McTaggart recognized, is an essential aspect of that concept. The C series is analogous only to the B series, and in the later parts of McTaggart's analysis the B series does duty almost exclusively for time and the A series is strangely neglected. As Cleugh (1938) has commented, 'the passage from the B series to the C series is successful insofar as the B series is *not* temporal', for although the terms which this series relates are events the series as a whole is not temporal. Consequently, she continues, 'As long as the B series is taken as a series all is well; but as soon as reference is made to the specifically *temporal* connotation of the series, trouble begins. The ghost of time cannot permanently be laid'.

McTaggart's theory of time and the criticism to which it has been subjected are not matters of concern for professional philosophers alone, for, the distinction that he drew between the A series and B series is reflected in the views of many physicists and others who accept the latter

† We are reminded of Zeno's paradox concerning space, for, if everthing that exists has a place, it follows that place too will have a place and so on *ad infinitum* (Lee 1936).

‡ A similar criticism can be made of the theory of serial time due to J. W. Dunne, but whereas McTaggart regarded the infinite regress as 'vicious', Dunne believed that 'an infinite regress is, after all, the proper and valid description of mind's relation to its objective universe' (Dunne 1934).

but reject the former as purely subjective.† As already remarked, this distinction has a direct bearing on the hypothesis of the 'block universe', and in the opinion of many this hypothesis has been powerfully reinforced by the space–time interpretation of relativity. From the point of view taken by Einstein and Weyl, 'The objective world simply *is*, it does not *happen*. Only to the gaze of my consciousness, crawling upward along the lifeline of my body, does a section of the world come to life as a fleeting image in space which continuously changes in time'. (Weyl 1949, p. 116). In other words, the claim is made that the relativistic picture of the world recognizes only a difference between earlier and later and not between past, present, and future. Indeed, we can even draw an analogy between the terms of McTaggart's C series and the successive backwards-directed light cones with vertices along the world line of an observer in the Minkowski diagram.

Nevertheless, as has been stressed by Eddington (1935) and Reichenbach (1956 *passim*), the theory of relativity does not provide a complete account of time. Despite what Weyl has said, the theory is not incompatible with the happening of events but is neutral in this respect. At a given instant E on the world line of an observer A (who need not be regarded as anything more than a recording instrument), all the events from which A can have received signals lie within the backwards-directed light cone with its vertex at E. As will be shown in § 7.6, there is an objective time order for all these events, and the anomalies of time ordering that Weyl had in mind when he made the statement quoted above concern only events that lie outside this light cone. Signals from these events can only reach A after the event E, and when they do reach A they will then lie within A's backwards-directed light cone at that instant. The passage of time corresponds to the continual advance of this light cone. As far as the theory of relativity is concerned, we can consider either the set of all these light cones or the continual transition from one to another. The theory is compatible with either point of view and does *not* invalidate the concept of temporal transition.

Some physicists and philosophers, for example Bondi (1952) and Reichenbach (1953), have claimed that support for the transitional nature of time is provided by quantum theory, for in quantum mechanics we find that the past history of an individual system does not determine its future in any absolute way but merely the probability distribution of possible

† An influential paper by a philosopher who argues against the objectivity of the A series is 'The myth of passage' by Williams (1951). The objectivity of the A series has been defended by Čapek (1965) in a paper bearing the title 'The myth of frozen passage: the status of becoming in the physical world'. Some of the many papers on this controversy will be found in Gale (1968).

futures. In general there is no conceivable set of observations that can provide enough information about the past of a system to give us complete information as to its future. The future is a mathematical construction that can be changed by an observation (Watanabe 1953). Consequently, the past is the determined, the present is the moment of 'becoming' when events become determined, and the future is the as-yet undetermined.

This elucidation of the distinctions between the past, present and future of physical events is not, however, regarded by all philosophers of science as establishing the objectivity of 'becoming' and the transitional nature of time.† According to Smart (1964), indeterminism can be assigned to the B series and means no more than that successive temporal states are not related in a perfectly deterministic way. Similarly, Grünbaum (1963) has argued that the transition from an undetermined to a determinated state of affairs has always occurred and therefore this is an inadequate method of defining the present since it does not distinguish one present from another. He supports his argument by a reference to Bergmann (1929), who rejected Reichenbach's contention that the 'now' has objective significance. Bergmann argued as follows. Consider the descriptive phrase, 'the present state of the planetary system'. Which 'now' is intended? That of the year 1800, say, or 2000, or which other one? According to Bergmann, Reichenbach's definition of 'now' as the instant of transition from the undetermined to the determined is circular, since it explains the present 'now' only by reference to itself, and he claimed that all we can say is that 'now' is the temporal mode of our experience. Consequently, the distinctions we make between past, present, and future are mind dependent and do not apply to the physical world.

This conclusion, however, does not necessarily follow, for we must be careful to distinguish between the terms 'the present' and 'present'. Whereas 'the present' or 'now' always refers to the current phase of our experience,‡ *any* event when it occurs is 'present', irrespective of our being aware of it, or even of our existence at the time concerned. Consequently, although we can accept the argument that 'the present', or 'now', cannot be defined except by reference to itself, since any phase of our experience can be so designated only when it occurs, this does not commit us to the view that past, present, and future are mind dependent. One can agree with Davies (1974, p. 21) that the concept 'now' is irrelevant in *basic* physics, but it does not follow that we must therefore

† A particularly lucid discussion of the transitional aspect of time has been given by Denbigh (1975).

‡ Nevertheless, 'now' is objective to the extent that all human beings at a given place normally agree on the same 'now'.

reject the concept of the transitional nature of time. In some branches of science it is essential. For a meteorologist engaged in forecasting the weather the distinctions between past, present, and future are vital. Similarly, for the palaeontologist studying the fossil record in terrestrial rocks the distinction between past and present is a characteristic feature of the world. At a given instant an enduring object, whether animate or inanimate, has a past that comprises all that has happened to it so far. As Zwart (1976, p. 48) has remarked, 'How could one speak of the fatigue of metals, of the weathering of crystals, of the ageing of glass if these systems had no past'? In short, a concept of physical time that is restricted to the static relations before, after, and simultaneous with it is an impoverished concept that does not comprise all that is involved in the occurrence of events.

If past, present, and future did not apply to events in the physical world but only to mental events, a peculiar difficulty would arise when we consider the interaction of these kinds of events. For, whereas physical events would neither come into existence nor cease to exist but would just be, mental events certainly come to be and cease to be in our personal experience. This difference would have the most peculiar consequences for cause and effect. In purely physical causation an effect would not actually be produced by its cause; it would merely be further on in time. However, mental causation of a physical event—such as deciding to drop a stone into a pond—would mean that a cause (in this case, the decision to drop the stone) suddenly comes into being, but the effect (the splash when the stone strikes the water) would not; it would just be. This strange difference between cause and effect would be very difficult to understand.

Two concepts that are closely associated with the transitional nature of time are indeterminism and the emergence of novelty. For, despite Smart's claim that indeterminism can be assigned to the B series, the indeterminism of the future as revealed by quantum theory is more naturally associated with the transitional idea of time. Indeed, there may well be some deep connection between this idea of time and the existence of an incalculable factor in the universe. On the other hand, the concept of the block universe in which events do not happen but just are is more naturally associated with determinism, since determinism implies that all events are unalterably fixed and that which we call the future is just as unalterable as that which we call the past. As Denbigh (1975) has emphasized, if determinism is correct it would seem that nothing essentially new could ever come into existence. If, however, the world is evolving in time and the future is never wholly predictable, since the present state of the universe does not contain enough information to define any future state, the B series fails to provide a fully adequate concept of time.

The hypothesis that the B series, in which events are related by the concepts of 'earlier than', 'later than', and 'simultaneous with', is more fundamental than the A series of past, present and future, has been submitted to critical analysis by Sellars (1962). He has argued that, so far from the concepts used in the B series being logically independent of the distinctions between past, present and future, they are in fact bound up with them. He has remarked that 'The idea of a tenseless existence of events related by *earlier than*, has a flavour of absurdity, if not of self-contradiction' (Sellars 1962, p. 560), for as he points out in a later passage (p. 574) 'The earlier-later relation has its primary mode of being as earlier-later in the context of a specific past-present-future'. In other words, detensed pictures of the world have their roots in tensed pictures; they result from our constructing a more abstract pattern of events than that which we actually experience but one based upon it.

Although Sellars was careful to talk only about 'tenseless existence', the term that is often applied to events in the B series by those who regard that series as more fundamental than the A series is 'tenseless occurrence'. This term, however, is a blatant self-contradiction, for the words 'occur' or 'happen' when applied to an event signify that its existence is limited to the time at which (or during which) it is present. Before then it was future and afterwards it was past. None of these distinctions apply to the terms in the B series. *That series consists of a network of permanent relationships between the epochs we assign to events, whereas the A series concerns the actual events or occurrences themselves.*

Sellars (1962, p. 557) has raised the pertinent question 'can we describe in nontensed terms what a world would be like for tensed talk to be appropriately used in it'? It is difficult to see how we could construct the A series given only the B series, whereas given the former we can readily construct the latter. Moreover, if physical events are not subject to the distinctions between past, present and future, why do we have the illusion that they are? Although it cannot be strictly proved, it is more plausible to assume that these distinctions are not illusory. Surely, we have the faculty of temporal awareness of successive phases of sensory experience leading us to regard time as transitional because our minds are adapted to the world we live in and this is a constantly changing world. If we adopt this point of view, instead of temporal transition being a purely subjective phenomenon the ultimate significance of time is to be found in its transitional nature.†

† What Broad (1959) calls 'Absolute Becoming', which manifests itself as the continual supersession of what was the latest phase by a new phase, 'seems to me', he writes, 'to be the rock-bottom peculiarity of time, distinguishing *temporal sequence* from all other instances of one-dimensional order'.

7.6. The Minkowski diagram and the nature of time

We have seen that the 'universal' time of physics is a much more complex concept than was at one time imagined. For, although according to most current cosmological theories the bulk distribution of matter throughout the observable universe is compatible with the idea of a 'world-wide' cosmic time, this time does not pertain to frames of reference moving rapidly relative to the local mean distribution of matter. Moreover, if the expansion of the universe is non-uniform, i.e. if the relative radial motion of nebular clusters is accelerated, it may happen that there are events in distant regions which can never be detected, even in principle, by observers in our region. These conclusions depend on the hypothesis that the local speed of light in free space sets a theoretical upper limit to the rate at which signals can be transmitted. This hypothesis leads us to abandon the picture of physical time advancing as a vast knife-edge, and if we wish to retain some such mental image we must visualize instead a complex of advancing light cones in space–time, the track of each vertex being the world line of a potential observer.

Although in the Minkowski diagram† associated with a given inertial frame of reference A and an event E (chosen as space–time origin of the frame) any point (t, x, y, z) represents a potential event, only those events P which lie *inside* or *on* the forward light cone LEM ($c^2t^2 \geqslant x^2+y^2+z^2$, $t>0$) can be said unequivocally to lie 'in the future' relative to E, and similarly those events P′ which lie *inside*, or *on*, the backward light cone L′EM′ ($c^2t^2 \geqslant x^2+y^2+z^2$, $t<0$) can be said unequivocally to lie 'in the past' relative to E. For, *they are the only events that can stand in the corresponding causal relations to* E.

To establish this important theorem‡ we begin by recalling that, if an event lies inside either light cone (LEM or L′EM′ in Fig. 7.2), it can be connected with E in the appropriate causal order by a signal or particle travelling (relative to A) with speed less than c. On the other hand, if Q is an event which lies outside both light cones ($c^2t^2 < x^2+y^2+z^2$), then anything which travels from Q to E, if $t<0$, or from E to Q, if $t>0$, must have speed greater than c.

However, before we can regard the proof as complete, we must consider the relation between E and Q from the point of view of any other frame B having the same space–time origin E but moving with any uniform speed V ($<c$) in any direction relative to A. We can always

† When gravitational fields are taken into account we must confine attention to the immediate neighbourhood of E, i.e. we must replace t, x, y, z by the corresponding differentials.

‡ In Robb's theory it functions as a definition of conical order.

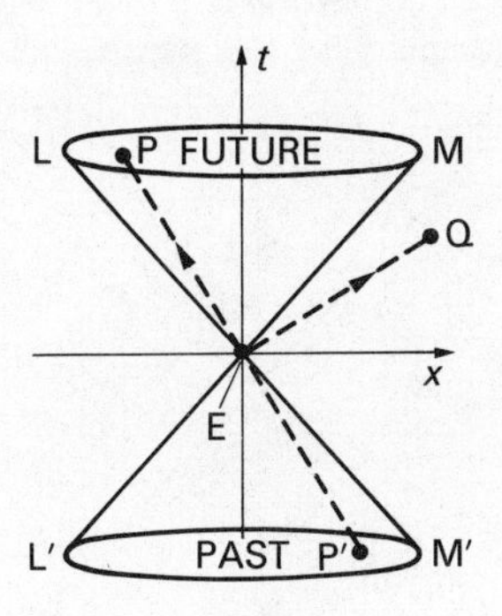

FIG. 7.2

choose the spatial axes of A so that B moves along the x axis, and we shall assume this to be the case. We shall also assume that the x', y', and z' axes of B coincide with the respective x, y, and z axes of A when the origins of the two frames coincide at E. If we choose units of measurement so that c becomes unity, the Lorentz formulae correlating the space–time coordinates (t', x', y', z') assigned by B to any given event with the coordinates (t, x, y, z) assigned by A will be of the form

$$t' = \beta(t - Vx), \qquad x' = \beta(x - Vt), \qquad y' = y, \qquad z' = z$$

where $\beta = 1/\sqrt{(1 - V^2)}$. Consequently, in the Minkowski diagram of A, although the y', z' axes of B will lie along the y, z axes of A, the t' and x' axes of B will lie in the (t, x) plane of A along lines equally inclined to the t and x axes, respectively. Moreover, the line in this diagram representing the t' axis will lie *inside* the light cones LEM and L′EM′, whereas that representing the x' axis will lie *outside* these light cones† (see Fig. 7.3). Similarly, the (x', y', z') hyperplane will also lie outside these cones, the x' axis being the line in which this hyperplane intersects the (t, x) plane. By noting whether an event lies above or below this hyperplane we can immediately decide whether the t' co-ordinate assigned to it by B will be positive or negative. We see that if, according to A, P lies in the future relative to E (i.e. $t > 0$) inside the light cone LEM, then it also lies in the future, relative to E, according to B (i.e. $t' > 0$). Similarly, if, according to A, P′ lies in the past relative to E ($t < 0$) and inside the light cone L′EM′, then it also lies in the past relative to E, according to B ($t' < 0$). Consequently, if any event lies inside the light cones it will lie either in the future or in the past relative to E, *irrespective of the frame of reference*

† When c is unity the lines LM′ and L′M in which the light cones intersect the (t, x) plane are equally inclined to the t axis and also to the x axis.

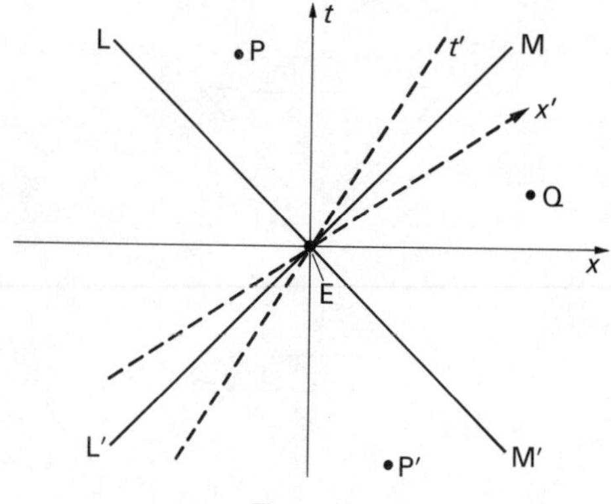

FIG. 7.3

considered.† However, if it lies outside both light cones, its temporal relation to E will depend on the frame of reference adopted. Thus, in Fig. 7.3, Q is in the future with respect to E, according to A, but it is in the past with respect to E, according to B. If, however, the speed V, of B relative to A, were sufficiently slow, then Q would lie above the (x', y', z') hyperplane and so would be in the future with respect to E, according to both A and B. Similarly, if Q is in the past relative to E, according to A, and lies outside both light cones, then, depending on V, it can be in either the past or the future relative to E, according to B. Furthermore, if V were such that the (x', y', z') hyperplane passed through the event Q, then, from the point of view of B, both events E and Q would be simultaneous‡ ($t' = 0$). Thus, if an event lies outside the light cones at E, the temporal relation between it and E will depend on the frame of reference. This ambiguity is irreconcilable with any objective criterion of causality connecting the two events, and the theorem is established.

The space–time region lying inside (and on§) the forward light cone

† Since $c^2t^2 - (x^2 + y^2 + z^2)$ is a Lorentz invariant, it follows that, when this form is positive, zero, or negative according to A, the form $c^2t'^2 - (x'^2 + y'^2 + z'^2)$ is correspondingly positive, zero, or negative according to B, provided that the relative speed of B is less than c. Hence, so long as the frames considered have relative velocities less than that of light, an event lies either inside, or on, or outside the light cones at E, irrespective of the particular frame of reference adopted.

‡ It is easily proved that in this case E and Q are closer together in space for B than for any other inertial frame of reference. Schrödinger has suggested that this minimum distance be called the *simultaneous distance* between E and Q. According to B, anything moving from E to Q would be in two different places at the same time and so its velocity would be infinite.

§ In the case of events *on* the light-cones, although the proper time between such events is zero, we have to distinguish between E and all other events lying on the light cones of E. These events occur both elsewhere and in either the absolute future or the absolute past with respect to E.

LEM may be called the *absolute future* with respect to E, and the region lying inside (and on) the backward light cone L′EM′ may be called the *absolute past* with respect to E. The region lying outside both light cones may be called the region of *potential simultaneity with E.* It is the relativistic analogue of the world-wide simultaneity of Newtonian physics.

Events such as P and P′ which lie inside the light cones at E are said to be in *absolute temporal sequence.* The relation of being in absolute temporal sequence can be shown to be transitive; in other words, if E_3 is absolutely later than E_2, and if E_2 is absolutely later than E_1, then E_3 is absolutely later than E_1. This theorem can be established easily with the aid of Fig. 7.4, in which LE_2M is the forward light cone at E_2, and $L'E_2M'$ the backward light cone. It is clear that, if E_1 is *any* event inside $L'E_2M'$ and E_3 is *any* event inside LE_2M, then the line joining E_1 and E_3 must be parallel to a line through E_2 which lies inside the light cones shown. Consequently, this line lies inside the respective light cones at E_1 and E_3. Hence, E_3 is absolutely later than E_1, and the transitive property is established.

On the other hand, the relation of being 'potentially simultaneous' is not transitive,† for, it is possible for events E_1^* and E_2 to be potentially simultaneous and also for E_2 and E_3^* to be potentially simultaneous, and yet for E_1^* and E_3^* to be in absolute temporal sequence. This situation is

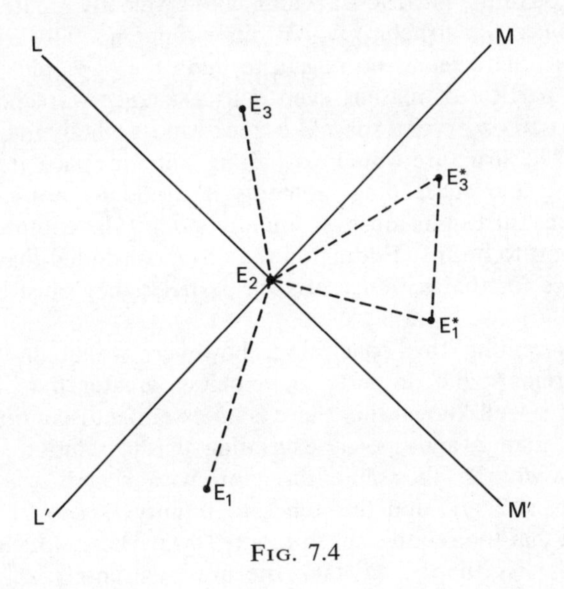

FIG. 7.4

† In this respect, *potential simultaneity* is similar to *overlap* in the case of durations in a single time experience (see p. 207).

illustrated in Fig. 7.4, where the line joining E_1^* and E_3^* is parallel to a line through E_2 which lies inside the light cones at E_2. Hence, $E_1^* E_3^*$ lies inside the light cones at E_1^* and E_3^*.

A particle of matter is represented in the Minkowski diagram associated with any event E in its history by a line which lies (strictly) inside the light cones at E. Any direction pointing from E to the interior of these light cones is called *time like*, because it can represent a sequence of instants in the history of a material particle. We can therefore regard a particle of matter as a structure that is represented in the Minkowski diagram by a world line that is everywhere time like. Similarly, a photon (in free space) is represented by a world line, or segment of a world line, lying along a generator of a light cone.

A world line lying in that part of the Minkowski diagram which is outside the light cones (at E) is called *space like*, because it can represent a set of simultaneous events according to a suitably chosen observer who is himself represented by a time-like world line. *Is there any kind of physical structure corresponding to such a world line*? This question was raised many years ago by Eddington (1923 p. 23). In a fascinating passage in his famous treatise on relativity he wrote, 'A particle of matter is a structure whose linear extension is timelike. We might perhaps imagine an analogous structure ranged along a spacelike track. That would be an attempt to picture a particle travelling with a velocity greater than that of light; but since the structure would differ fundamentally from matter as known to us, there seems no reason to think that it would be recognized by us as a particle of matter, even if its existence were possible. For a suitably chosen observer a spacelike track can lie wholly in an instantaneous space. The structure would exist along a line in space at one moment; at preceding and succeeding moments it would be non-existent. Such instantaneous intrusions must profoundly modify the continuity of evolution from past to future'. Eddington therefore concluded that in default of any evidence for the existence of such particles they must be impossible structures.

Before accepting this conclusion, however, we ought to take into account a remarkable property of velocities greater than that of light. Although it is well known that there is an essential discontinuity between speeds less than c and speeds exceeding c (the relative speed of two particles moving in the same direction with speeds $c+\epsilon$ and $c-\epsilon$, respectively, is $2c^2/\epsilon$, and this tends to infinity as $\epsilon \to 0$), it is seldom pointed out that the relative speed of any two particles which travel faster than light is *less* than c. To take the most extreme case, suppose the particles move in exactly opposite directions with speeds u_1 and u_2, respectively. According to Einstein's velocity-addition law, their relative speed is

$$u_3 = \frac{u_1 + u_2}{1 + u_1 u_2 / c^2}.$$

If we take units of measurement so that $c = 1$, it follows that u_3^2 will be less than unity if

$$(1 - u_1^2)(1 - u_2^2) > 0.$$

However, this will be the case not only when $u_1 < 1$ and $u_2 < 1$, but also when both u_1 and u_2 exceed unity, i.e. are greater than the speed of light. For example, if u_1 is infinite† (as for a particle travelling from E to Q, in Fig. 7.3, when considered with respect to an observer B with x' axis along EQ), if follows that, re-introducing the symbol c, $u_3 = c^2/u_2$, and hence $u_3 < c$, since $u_2 > c$.

Consequently, if we consider all conceivable rectilinear world lines through E in the Minkowski diagram, corresponding to particles meeting at E and moving relative to each other in all directions with speeds ranging from zero to infinity, we find that there is a reciprocal relation between the family of world lines lying strictly inside the light cones at E and the family of world lines which lie strictly outside these light cones. According to an observer associated with any member of the first family, all velocities of pseudo-particles whose world lines belong to the second family exceed the speed of light, whereas all velocities of particles whose world lines belong to the first family are less than that of light. Similarly, to a hypothetical observer associated with a world line of the second family, the velocities corresponding to world lines of the first family are all greater than the speed of light, whereas those associated with world lines of his own family are all less than this critical speed. According to all observers associated with members of either family, the light cones will be the same but the regions which will be regarded as respectively 'inside' and 'outside' will depend on the particular family to which the observer's world line belongs, for each observer will regard his own world line as lying *inside* the light cones, and the world lines of all observers to whom he assigns speeds exceeding c will appear to him to be *outside* these cones.

† If both u_1 and u_2 are infinite, u_3 will be zero, i.e. *relatively to each other* two pseudo-particles moving (with respect to an ordinary particle) in opposite directions with infinite speeds will be at rest. These 'particles' are two superimposed 'lines', or 'rays', and so the result is perhaps not surprising. A more unexpected consequence is that, if we imagine two pseudo-particles moving in opposite directions with very high speeds (far in excess of c) relative to an ordinary particle, they will have only a very low speed (negligible compared with c) relative to each other!

According to an observer A whose world line belongs to either one of these two families, the proper times of all particles, etc. whose world lines lie on the same side of the light cones as his does will be 'real', although they will, in general, he subject to the time-dilatation factor. However, the proper time of anything whose world line lies on the other side of the light cones will be 'imaginary', i.e. its square will be negative. On the other hand, the time assigned by A to the passage of such an object between two events, for example between E and Q in Fig. 7.3, will of course be real. (Similarly, the proper length of such an object will seem to A to be imaginary, but he will assign a real relative length to it.)

In the intermediate case of a particle whose world line is a generator of the light cones, i.e. a photon, the proper time is zero. To a hypothetical observer moving with a photon the whole span of our time would pass in a flash, so that for him there would not even be

> 'One Moment in Annihilation's Waste,
> One Moment of the Well of Life to taste....'

The usual interpretation of this curious result is that we cannot associate a 'clock', i.e. a time-keeping system similar to that used by A, with anything moving with the critical speed *c*. Similarly, we cannot associate any such clock with any object or observer whose relative speed exceeds *c*. Just as photons must be sharply distinguished from material particles, so the objects, if any, which move faster than photons cannot be composed of ordinary matter. Nevertheless, the fact that the two regions into which the light cones (at any event) divide space–time are mirror images of each other, i.e. are perfectly reciprocal in the sense discussed, leads one to ask whether the universe is in fact asymmetrical in so far as one region is populated and the other completely empty.

When Eddington raised the question of whether a space-like track could be the world line of anything, there was no evidence that any physical objects other than particles of ordinary matter and photons could exist. However, physicists now believe that to every charged elementary particle there corresponds an antiparticle of the same mass but opposite charge. The reason why we do not normally come across these antiparticles is that when they collide with their counterparts, for example when a positron meets an electron (as discussed on p. 332), they destroy one another and generate a γ-ray. Nevertheless, although antiparticles cannot survive close encounters like ordinary particles, there is complete symmetry between the two, so that antimatter (constructed of antiparticles in exactly the same way as ordinary matter is composed of ordinary particles) could presumably exist in bulk so long as it does not come into contact with ordinary matter. The symmetry of the Minkowski diagram with respect to the light cones suggests an analogy with the symmetry of matter and antimatter, but the velocities of antiparticles that have been determined experimentally are less than the velocity of light and so the analogy fails.

Instead, therefore, of associating antimatter with 'imaginary' proper time, some physicists in recent years have considered the possibility of yet another kind of matter consisting of particles that always travel faster than light. Feinberg (1967) has introduced the name *tachyons* for such particles, from the Greek word meaning 'swift'. He has pointed out that, although no body can be accelerated from a speed less than that of light to a speed which exceeds that of light, we find that in subatomic physics particles can be easily created or destroyed and that in their mutual interactions their energies and other properties can change discontinuously. He has therefore argued that we can imagine the creation of particles moving faster than light without having to be accelerated from velocities less than that of light. Such particles once created would always move relative to ordinary particles with speeds exceeding that of light. For them, as for ordinary particles, the speed of light would constitute an impenetrable velocity barrier.

Provided that tachyons cannot be used to convey information at a speed exceeding that of light their existence would not contradict Special Relativity. However, if there is to be no violation of the causality principle that cause precedes effect it is essential that no interactions take place between tachyons and ordinary matter. For, as was pointed out by Tolman (1917), if such interactions could occur then it would be possible to signal into one's own past. This is easily shown. Let A and B be two observers moving with uniform velocity V from coincidence at time zero on the clock of each. At time t_1 on his clock let A emit a tachyon moving with infinite velocity in the direction of B. Call this event E_1 and the arrival of the tachyon at B the event E_2. The space and time co-ordinates of E_2 according to A are (Vt_1, t_1). By the Lorentz transformation the space and time co-ordinates of E_2 according to B are $(0, t_2')$, where

$$t_2' = \surd(1 - V^2/c^2)t_1. \tag{7.3}$$

Now let B immediately send a tachyon of infinite velocity to A, and let its arrival at A be the event E_3. The space and time co-ordinates of E_3 according to B are $(-Vt_2', t_2')$. By the Lorentz transformation the co-ordinates of E_3 according to A are $(0, t_3)$ where

$$t_3 = \surd(1 - V^2/c^2)t_2' = (1 - V^2/c^2)t_1. \tag{7.4}$$

This means that event E_3 occurs before the event E_1, since $t_3 < t_1$.

Consequently, in this way A could signal into his own past, which is impossible. Despite this conclusion, however, various experiments have been devised to detect tachyons, but no sign of any interaction of tachyons with matter has been observed.

Does this mean that there are no tachyons in the universe? The kinematic symmetry of the two regions into which the light cones divide the Minkowski diagram may not correspond to anything in the actual universe. In the case of matter and antimatter the precise symmetry of physical properties strongly suggests that each should exist in equal amounts. Nevertheless, all the empirical evidence available points to the overwhelming preponderance of ordinary matter. So far there has been no sign of an appreciable amount of antimatter anywhere in the universe. The cosmic rays that enter the earth's atmosphere contain particles that we believe have come from remote regions of our Galaxy and possibly some have even come from outside, but hardly any antimatter is found in them. Nor do we observe the photons that would result from the annihilation of matter and antimatter on a cosmic scale. It has been suggested (Goldhaber 1956) that the primeval universe may have split into two independent regions which flew apart with high relative speed, one region containing matter and the other antimatter. Similarly, we might argue on the basis of symmetry about the light cones in the Minkowski diagram that there should be just as many tachyons in the universe as there are particles with velocities less than that of light. However, there is no empirical evidence for either of these hypotheses, and in the case of tachyons there is no indication so far that any exist.

This negative result has an important bearing on the question of whether there is a universal time order. The concept of cosmic time is associated, as we have seen, with the bulk distribution of matter in the universe. This is the time kept by the observers who move with the fundamental particles of the world model that represents the basic large-scale structure of the universe. It is modified by the time-dilatation factor for observers in uniform motion of velocity less than c with respect to the fundamental particles, but this is only a scaling factor. Apart from this, there is a universal time order for all events lying inside the observers' light cones. The proper time of tachyons, however, would be another mode or dimension of time related to ordinary proper time by the square root of -1, the two kinds of proper time being like the axes in the Argand diagram. Consequently, the non-existence of tachyons may be considered as corroborative evidence for the hypothesis that there is only one dimension of time and that, irrespective of time dilatation and time horizons, there is a universal time order for all pairs of events that can be causally related.

7.7. Time and the fundamental constants of physics

As explained in Chapter 6, we have reason to believe that the universe is characterized by a cosmic time scale. Although the existence of such a scale gives meaning to the concept of the age of the universe, it leaves open the question of whether there is a *unique* uniform time in nature. For, as mentioned on p. 43, if we apply to a given scale of time t a transformation of the form

$$\tau = f(t),$$

where $f(t)$ is a monotonically increasing function of t, we obtain another scale of time τ that keeps pace with t if, and only if, $f(t)$ is a linear function of t. In other words, if τ is not a linear function of t, then processes that are periodic according to the t scale will not be periodic according to the τ scale. Similarly, if there are any natural processes that are periodic according to a τ scale that is not a linear function of t, they will not be periodic on the t scale. Consequently, two scales of time can only be regarded as effectively the same (except for possible differences of time zero and time unit) if they are linearly related. Otherwise, they are essentially distinct. We are therefore faced with the question raised at the end of Chapter 1: is the uniform time defined, for example, by the Rutherford–Soddy law of radioactive decay effectively the same as that implied by the laws of dynamics or gravitation, or is each the basis of a different scale of cosmic time?

This type of question was first raised about 1935 by Milne in exploring the properties of his world model (see p. 292). He found that, if t is the time scale of uniform, i.e. unaccelerated, expansion and τ is a logarithmic function of t, then in terms of the τ scale the model is static. On the τ scale the universal constant of gravitation G is a secular invariant, but on the t scale it increases linearly with t. Nowadays, the usual way in which questions of this kind are discussed is in terms of the various fundamental constants of physics and the possible secular variation of some of them in terms of the postulated invariance of others.

As previously mentioned (see p. 298), Dirac suggested in 1937 that the ratio of the electric to the gravitational force between an electron and a proton may be changing (Dirac 1937, 1938). This would mean that in terms of the charges and masses of these particles G is varying. The basis of his argument was what he has since called the 'Large Numbers' hypothesis (Dirac 1973, 1974). This asserts that all very large dimensionless numbers occurring in nature are connected with the present epoch expressed in terms of the atomic unit of time, or chronon, of order 10^{-24} seconds. The ratio of the electric to gravitational force between a proton and electron is of the same order of magnitude as the current value of the Hubble time T expressed in these units. In Dirac's theory G therefore varies inversely with the time, assuming that the charges and masses of the proton and electron do not change. The rate of change of G predicted by Dirac is of the order of one part in 10^{10} *per annum.*

To test the possibility of a variation in G, Pochoda and Schwarzschild (1964) calculated the effects on the evolution of the sun. They computed evolutionary models under various assumptions of the type G varies as t^{-n}. In accordance with our current ideas of the age of the earth and solar system, they assumed that the sun was in its initial hydrogen transmutation, or 'Main Sequence', stage about 4.5×10^9 years ago. For G varying inversely as t, i.e. $n = 1$, they found that the present state of the sun could only be accounted for satisfactorily if the age of the universe were 15×10^9 years or more. This limitation arose because they found that the present state of the sun would have occurred relatively earlier in the history of the universe when, under the assumption that G varies inversely as t, G would still have been rather high. Since the luminosity of the sun depends on the eighth power of G, its initial luminosity would have been so high that the rate of transmutation of hydrogen in the sun's thermonuclear core would have been so great that this hydrogen would have been exhausted and the sun would have left the Main Sequence before now, i.e. before 4.5×10^9 years had elapsed, and presumably would be a red giant. In other words, the sun could not now be in the state in which it is.

In 1964 it seemed that this conclusion conflicted with Dirac's hypothesis, because the age of the universe according to the then accepted value of Hubble's constant was thought to be less than 15×10^9 years, but the more recent value obtained by Sandage and Tammann (see p. 287) implies a possible age of 15×10^9 years or more, so that on the basis of solar evolution calculations a variation of G of the type envisaged by Dirac cannot be ruled out.

Dirac's hypothesis concerning the rate of variation of G has been supported by Van Flandern (1975). He claimed that a careful comparsion of atomic time and ephemeris time in the timing of occultations of stars by the moon between 1955 and 1973 indicated that G is decreasing secularly by about one part in 10^{10} *per annum.* It remains to be seen whether this result will be confirmed. If it is, the theoretical consequences will be far reaching, especially for our ideas of gravitation, since a varying gravitational constant is incompatible with general relativity.

In recent years considerable attention has been devoted to the geophysical and astronomical consequences of a secular variation in the value of G. Attempts have been made to establish limits on this variation by studying the effects it would produce on the interior of the earth, the main influence being an expansion of the earth due to a decreasing gravitational constant and hence a diminishing gravitational pull on the surface layers (Wesson 1973). However, van Diggelen (1976), by studying counts of the growth bands of fossil coral, has concluded that no expansion of the earth has taken place in the past 500 million years.† An astronomical consequence for the earth if G were diminishing would be a gradual increase in the size of the earth's orbit around the sun because of a decrease in the sun's force of attraction. It is this effect that leads to a slow change in the relation between ephemeris time and atomic time, since the former is based on the earth's motion whereas the latter is based on the oscillations of the caesium atom, which are assumed to be constant.

Any firm conclusion concerning the possibility of a secular variation in the value of G will necessarily depend on precise measurements. The development of direct radar ranging to the planets seems the most hopeful way of attaining the required precision. Shapiro *et al.* (1971) analysed the results of a long series of radar-echo time delays, primarily between the earth and Mercury, and obtained an upper limit on the fractional variation of G of 4 parts in 10^{10} *per annum.* They predicted that in due course this uncertainty could be reduced to 3 parts in 10^{11}. So far, however, despite Van Flandern's claims, there is no conclusive argument in favour of any secular variation in the value of G.

Strictly speaking, neither lunar occultation timings nor radar-echo time delays provide a test of the constancy or variation of G alone. They test the ratio of G to the rate of an atomic clock. The latter depends on Planck's constant h, the velocity of light c, the electron charge e, and the electron mass m. All standards of time and distance depend on these fundamental quantities because clocks and measuring rods are scaled by the size of atoms and the frequencies they emit. Recently, Baum and Nielsen (1976) reported the results of an experiment that directly tested the secular behaviour of the product hc. As we have seen, the fundamental assumption made by most cosmologists is that the extragalactic red shifts are due to the Doppler effect associated with recessional motion. In cosmologies based on this interpretation of the spectral shifts it is assumed that old photons have the same relation between energy E and wavelength λ as young photons, namely

$$E\lambda = hc$$

† Blake (1977) has deduced from the fossil data studied by van Diggelen that the rate of decrease of G is less than 5 parts in 10^{11} *per annum.* A still smaller upper limit to this rate has been obtained by McElhinny, Taylor, and Stevenson (1978) from a study of the planet Mercury.

where Planck's constant h and the velocity of light c are taken to be true constants that do not vary with epoch. To test this hypothesis, Baum and Nielsen compared light from remote galaxies with light from nearby galaxies by means of a special photomultiplier tube with which incident photons of the same wavelength but different energies can be distinguished. For example, if old photons of selected wavelength have more energy than young photons of the same wavelength, they will eject more energetic photoelectrons from the cathode. The observational results showed, however, to an accuracy of a few parts in 10^{12} *per annum*, no sign of any red-shift dependence of the product $E\lambda$ for photons. In other words, old photons have the same energy as young photons of the same selected wavelength, and hence the product hc, to within the errors of measurement, can be regarded as a secular invariant. This means that in units of measurement based on a standard spectral wavelength, the mass of the electron, and the velocity of light c, we can regard any secular variation in the value of Planck's constant h as negligible over ranges of past time that are significant on the cosmological scale.† A similar experiment to that of Baum and Nielsen has been made by Solheim, Barnes, and Smith (1976) who found an upper limit for the variation of h of less than one part in 10^{12} *per annum.*

Despite the claim made by Van Flandern concerning the gravitational constant, there does not seem to be any compelling reason for discarding the hypothesis of a unique uniform time in terms of which the fundamental constants and laws of physics are secular invariants. This conclusion was reached after a very detailed survey of the problem by Dyson (1972), who pointed out that, although there are other hypotheses consistent with the available evidence, the data do not conflict with the orthodox view that all laboratory-measured constants are truly constant. Investigations that have been made since 1972 tend, on the whole, to reinforce this conclusion.

7.8. Precognition and the nature of time

Although we reject the view of Bradley (and other idealist philosophers) that time has no ultimate significance, our analysis of space-like tracks in the Minkowski diagram helps us to appreciate his point that we cannot automatically assume that phenomena exist only if they are in temporal

† Incidentally, Baum and Nielsen have drawn attention to the fact that the aberration constant is the same for distant galaxies as for stars in our own galaxy, because photographs of distant clusters of galaxies on different days of the year show no relative displacements between galaxies and star fields. Consequently, photons of all ages are arriving with the same velocity c.

relation with our world: 'For there is no valid objection to the existence of any number of independent time-series. In these the internal events would be interrelated temporally, but each series as a series, and as a whole, would have no temporal connection with anything outside. I mean that in the universe we might have a set of diverse phenomenal successions. The events in each of these would, of course, be related in time, but the series themselves need not have temporal relations to one another' (Bradley 1902, p. 211).

Bradley did not discuss world lines in the Minkowski diagram but drew attention to the time sequences of dreams: each has its own internal temporal connections, but when considered one with another they do not appear to have any common unity in time. Nevertheless, although this seems to be true in general, claims have been made, notably in recent times by J. W. Dunne, that occasionally in dreams future events in our waking life are experienced as pre-presentations. To account for these, and other alleged phenomena of precognition, which may be defined as *non-inferential knowledge of future events*, he formulated a theory of 'serial', or multi-dimensional, time (Dunne 1934). This was an ingenious development of the hypothesis originally put forward by C. H. Hinton (1887) that the world is a four-dimensional spatial manifold and particles are 'threads' in it. Human beings are only perceptually aware at any instant of a three-dimensional cross-section of this manifold, but as time goes on they become aware of different cross-sections, so that in effect they seem to be 'travelling' along the fourth dimension. This 'travelling' is, however, merely a progressive transference of awareness to one cross-section after another, producing the illusion that there is a three-dimensional world enduring in time and that parts of it are in motion. According to this hypothesis the world is static and the illusion of time arises from the continual change of the observer's attention.

Dunne, however, realized that this continual transfer of attention is itself a temporal process and so could not produce the time necessary for its own occurrence. To account for this time, he postulated that the manifold has a fifth spatial dimension and that a second consciousness 'travels' along it. However, as the same difficulty now breaks out all over again, he was obliged to postulate an infinite number of extra dimensions and a corresponding number of observers. Precognition is possible in such a world because time is unreal. Everything is already laid out before us and the problem is purely one of cognition.

Dunne's theory was criticized by Broad (1935) who showed conclusively that the infinite regress contained in it was wholly unnecessary. Instead of fallaciously arguing as if time were itself a process in time and therefore could only be eliminated by invoking an infinite number of spatial dimensions and a hypothetical observer at infinity 'who would plainly have to be the last term of a series which, by hypothesis, could have no last term', Broad (1937) endeavoured to account for precognition and 'temporal displacement' (as disclosed in numerous experiments in extra-sensory perception) by postulating a two-dimensional time. The point of his proposal was to suggest that, although an event α precedes an event β in the familiar time dimension, β might precede α in another time dimension. Consequently, if α were the precognitive impression of the event β, it would be intelligible to say that β determines α.

Broad's hypothesis was sympathetically, but acutely, criticized by Price (1937) who maintained that it would commit us to the puzzling notion of a 'double now', for what is 'now' in one respect could be 'past' or 'not yet' in another. Worse still, it would entail the even more curious notion of 'partial becoming'. 'Suppose that I precognize an event which is to occur next Saturday. In one respect this event has not yet come into being: it is still future and does not yet exist. But in another respect it is past and so *has* come into being. It is so to speak *half-real*; it has *partially become* but not wholly. When next Saturday arrives, but not before, it will receive its second instalment of being, and will then be completely real.—Yet will it? For these two halves of its being are so to speak out of step'. For when it begins to be in the one dimension to time it will already be along past in the other!

Tyrrell (1946, p. 94) has drawn attention to an interesting alternative hypothesis due to Saltmarsh. Recognizing that the fundamental process of extra-sensory perception occurs not at the conscious level but at the *subliminal* (i.e. beneath the threshold of consciousness), Saltmarsh suggested that the specious present of the subliminal mind may cover a much longer period than that of the conscious mind.† Consequently, in the co-present of the subliminal mind there might exist knowledge of two events one of which would be in the future for consciousness. Therefore, if knowledge of any event in the subliminal specious present could be passed to consciousness, it would be possible for the conscious self to become aware of an event which, to it, would be in the future.

As Tyrrell pointed out, this theory does not explain how a future event can be directly apprehended before it happens. To circumvent this difficulty, he maintained that not everything happens in the world order with which we are familiar. Nature does not come to an end where our

† 'There is some ground for supposing that the length of the specious present may vary in certain circumstances, such as concentration of attention, fatigue, hypnosis, and the influence of drugs, for example *Cannabis indica*; there is therefore no *a priori* objection to holding that the length of the subliminal specious present may be greater than that of the normal consciousness' (Saltmarsh 1938).

senses cease to register it and our minds become incapable of dealing with it. He suggested that, just as in telepathy A's subliminal self and B's subliminal self are in some kind of cognitive relation that transcends space, so in precognition the cognitive relation transcends time.† 'It is very hard', he wrote, 'to resist the view that the subliminal self exists outside temporal conditions as we know them, or at any rate exists in a different kind of time. Time, as we know it, may be a special condition applying only to the physical world or to conscious appreciation of it' (Tyrrell 1946, p. 96).

Be that as it may, genuine precognition of events, as distinct from shrewd forecasting, would violate our customary ideas concerning cause and effect. On this Flew (1967) has made the following forthright comment, with which I completely agree. The statement that 'A cause must precede or be simultaneous with its effect is a necessary truth'. For, 'It is no more possible to discover an effect preceding its cause than to light upon a bachelor husband'.‡ If we postulate a precognitive faculty that, contrary to retrocognitive memory as we usually experience it, is in fact always right, then all our efforts to prevent anything precognized from actually happening, should that be our wish, would necessarily be completely ineffective. However, as Flew rightly asserts, if predictions about future human situations were *known* to the people involved, it is sheer common sense to believe that, in some cases at least, they would be able to, and in fact would, take avoiding action to prevent themselves from being involved in these situations—for example, by refraining from travelling in an aircraft that is going to crash if they had precognized their own deaths in this catastrophe, for 'to be forewarned is to be forearmed'. Flew takes a less dramatic example to illustrate the same point, citing the predictions of British psephologists before the General Election in 1955. These predictions themselves appear to have been at least one factor making for the falsification of the psephologists' 'arrogantly overconfident forecasts'. Flew says that these effects are due to the dual nature of man, and he quotes a famous saying of Niels Bohr: 'It must never be forgotten that we are both actors and spectators in the drama of existence' (Flew 1959).

We may therefore conclude that unless we live in a 'block universe' in which everything in our future is pre-ordained so that we are mere

† Tyrrell (1946) believes that not only the evidence of precognition but also the phenomena of mysticism indicate that the subliminal self possesses something more like pan-awareness than does the conscious self. Consequently, he regards the conscious mind as struggling 'to string into a temporal sequence thoughts which seem to be present in the subliminal self in a kind of altogetherness'.

‡ *See* also the articles by Dummett (1954) and Flew (1954) on the question: 'Can an effect precede its cause'?

automata completely unable to influence any of our future actions—a situation that seems to conflict with our actual experience—a future event is nothing but an unrealized possibility until it happens and therefore cannot give rise to genuine precognition.

7.9. The dimensionality of time

We have seen that in attempting to account for alleged cases of precognition Broad was led to consider a two-dimensional concept of time. His former pupil Dobbs also used the same concept to account for certain phenomena in quantum physics (Dobbs 1951) and in the theory of visual perception, especially in connection with the 'specious present' (Dobbs 1972). In particular, he used the concept in analysing the well-known reversal of perspective (in the third spatial dimension) that occurs when we fix our attention for some seconds on the (two-dimensional) drawing of the 'Necker Cube'. Alternately, the face ABCD seems to be in front of the face EFGH and *vice versa*, the change-over occurring spontaneously after a few seconds. Any continuous transformation between three-dimensional congruent but non-superposable counterparts (for example, a pair of gloves) can only take place in a manifold (space) that is at least four dimensional. Dobbs (1972, p. 285) argued that, if it is necessary for the field of perception to be at least four dimensional at each moment of a perceptual situation in which reversal of perspective occurs, then presumably the momentary structure of *all* human visual perceptual situations is four dimensional. Since we do not see more than three dimensions (height, length, and breadth), how can we identify this extra dimension? He suggested that we should regard it as an 'imaginary' time dimension (in the mathematicians' technical sense of the adjective), since in the Minkowski diagram time is treated as an 'imaginary' quasi-spatial dimension. Together with the usual temporal dimension we thus obtain a

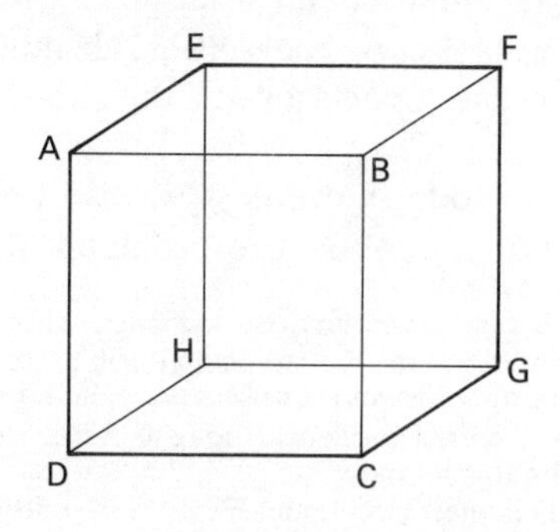

FIG. 7.5

two-dimensional concept of time, or a complex time variable. Dobbs called the usual time dimension 'transition time' and the other dimension 'phase time', and he regarded the 'specious present' to have extension in the latter.

We have already seen in our study of the Minkowski diagram (§7.6) that there may be another mode of time related to ordinary cosmic time by the square root of −1, the two being associated in much the same way as the 'real' and 'imaginary' axes in the Argand diagram. We did not find it possible to correlate the 'imaginary' time dimension with antimatter but only with tachyons, provided that such hypothetical superluminal particles actually exist.† The hypothesis of multi-dimensional time has occasionally been considered by others besides Broad and Dobbs, in particular by Eddington. For example, in his well-known treatise on relativity (Eddington 1923, p. 25) he raised the question of whether the metric of space–time must necessarily be everywhere *locally* Minkowskian, i.e. expressible in the form

$$ds^2 = c^2\,dt^2 - dx^2 - dy^2 - dz^2.$$

He considered the possibility that in some region it might be of the form

$$ds^2 = c^2\,dt^2 + dx^2 - dy^2 - dz^2,$$

and he argued that the change-over would have to take place in a transitional region where

$$ds^2 = c^2\,dt^2 - dy^2 - dz^2.$$

In this region space would be only two-dimensional, but there would be no barrier to passage through it. Nevertheless, consitions on the far side where time becomes two dimensional 'defy imagination'. Again in his final, posthumously published, book (Eddington 1946) he maintained that, according to his theory a 'uranoid' (smoothed-out universe) composed entirely of charged elementary particles of the same sign would occupy a three-dimensional space and a two-dimensional time, and he made the comment that this 'scarcely imaginable result' was not surprising, because the hypothetical system considered was entirely beyond experience. Eddington's discussion was, to use his own expression, merely 'a

† Incidentally, there is a point of contact between the condition that the proper time of tachyons is 'imaginary' (according to observers moving with velocities less than c) and Feynman's hypothesis that a positron can be regarded as an electron travelling backwards in time, or, in other words, with negative *speed*. For, because to an observer B whose (t', x') axes are depicted as in Fig. 7.3 the event Q is earlier than the event E(t' is negative for Q), although to A it is later than E, it follows that to some observers a particle whose world line is regarded as running from E to O (and therefore outside the light cones at E) will appear to travel backwards in time and its speed v will be negative (in the range $-\infty < v < -c$).

theoretical exercise'. Bunge (1955), however, introduced into the theory of the electron a complex time $t + i\tau$, where τ denoted a *constant* intrinsic time (of the order of 10^{-21} seconds) given by $h/4\pi mc^2$, h being Planck's constant and m the mass of the electron. He interpreted τ as the period of electron spin.

Despite these speculations there is no convincing reason for regarding time as having more than one dimension. Dobb's argument, that only by invoking a two-dimensional time variable can we explain the reversal of perspective which occurs when we fix our attention on drawings like that of the Necker Cube, depends on the assumption that there is a continuous transformation of appearance when the reversal occurs. This reversal, however, occurs suddenly and there is no evidence of a continuous change of appearance. Consequently, a two-dimensional time variable is not needed to account for it. Nor does it seem necessary to regard the specious present as existing in any other temporal dimension than the usual one. The linear nature of our stream of consciousness† and the one-dimensional nature of our perceptual time are, no doubt, dependent on the one-dimensional nature of physical time. We have already seen that the failure so far to detect tachyons is corroborative evidence that time has only one dimension. Moreover, it is difficult to believe that our system of physics, based on the concept of a unidimensional time variable, could be as successful as it is if in fact we inhabit a world in which time has two or more dimensions. Consequently, although it has not been possible to prove that physical time must necessarily be restricted to one dimension, there appears to be no reason so far for doubting that it is.

10. Conclusions

At the beginning of this book I said that the history of natural philosophy is characterized by the interplay of two rival philosophies of time—one aiming at its 'elimination' and the other based on the belief that it is fundamental and irreducible. The central point of dispute concerns the role of time in the relation of man to the external world.

According to Kant, time (like space) pertains only to the perceiving mind and not to things in themselves. According to McTaggart, series which in themselves are non-temporal appear to us as temporal: in principle, the same set of objects are eternally (i.e. 'timelessly') there, the only change being in our consciousness from less (and more confused) to greater (and clearer) awareness. I maintain, however, that our conscious awareness of time is neither a necessary condition of our experience, in

† See the footnote on p. 65.

the sense intended by Kant, nor a simple sensation, as Mach believed, but an intellectual construction that depends not only on our physical surroundings but also on the particular type of culture in which we happen to live. Unlike McTaggart, most scientists believe that our perception of time is based on an objective factor that provides an external control for the timing of our physiological processes. This external factor is what we call *physical time*, but what is the nature of this 'universal' time?

Our sense of time involves not only some awareness of duration and of temporal ordering but also of the distinctions that we habitually make between past, present and future. There is some evidence that our awareness of these distinctions, and hence of the transient, or transitional, nature of time, is one of the most important faculties distinguishing man from all other living creatures. Indeed, it would seem to be closely connected with our faculty of self-consciousness. But are these distinctions between past, present and future merely a pecularity of the way in which our minds happen to work, or is there some external factor corresponding to them? On this question, as we have seen, expert opinion is divided. Many philosophers and scientists believe that, although there exists an external time, it consists solely in the before-and-after sequence of events and is not concerned with the distinctions that we make between past, present and future. In other words, to use McTaggart's terminology, external events can be associated with the B series but not with the A series. It is frequently alleged that support for this point of view is provided by the theory of relativity, in particular by the fact that in the Minkowski diagram there is a large class of potential events for which there are no objective time ordering relations, the same for all potential observers irrespective of their relative motions. However, the theory does yield an objective time order for all events that an observer† can be aware of, or causally influenced by, at any given time, assuming that no signal, or transmitter of information, can travel faster than the velocity of light *in vacuo*. Moreover, in a universe that is characterized by the existence of a cosmic time, relativity is reduced to a local phenomenon, since this time is world-wide and independent of the observer. In fact, both relativity and cosmology are *neutral* as regards the objectivity or otherwise of past, present and future. Furthermore, the possible existence of events that cannot be temporally related in the same way for all potential observers is just as relevant, or irrelevant, an objection to the objectivity of the B series as it is to that of the A series.

Against those philosophers who deny that time is 'real' and those theoretical physicists who seek to prove that time's arrow is a derivative

† An 'observer' could, in principle, be simply an inanimate recording machine.

concept, to be explained in terms of some allegedly more fundamental concept such as entropy, we can assert that *the very essence of time is its transience,* and that this is a fundamental concept that cannot be explained in terms of something still more fundamental. *Time is the mode of activity, and without activity there can be no time.* Consequently, time does not exist independently of events, but is an aspect of the nature of the universe and all that comprises it. Activity involves transition, for it implies that there are different states of that which is active that are not co-present. A distinction can be drawn between those states that have been present (and are now past) and those that have not so far been present (and are now future). The former are those of which, in principle, enduring traces of their occurrence could have been recorded when they occurred and can also be co-present with the present state. That there are states that are not co-present is an empirical fact; their division into these states of which genuine traces can be co-present with our present state, and those states for which this is not possible, even in principle, must also be regarded as an empirical fact, i.e. a fact about the nature of the external world and not only about the nature of the mind of man. To those who deny this, and argue that the transience of time is solely mind-dependent, we can reply in the words of Lotze, that 'we must either admit Becoming or else explain the becoming of an unreal appearance of Becoming' (Lotze 1884). And to those who believe in what William James called the 'block universe' and what Milič Čapek (1965) calls 'the myth of frozen passage', we can put the question: if events are eternally (or tenselessly) there and we merely come across them in the course of our experience, how do we get the illusion of time's transience without presupposing transient time as its origin?

On the contrary, I believe that we have the faculty for temporal apprehension of successive phases of sensory experience because, in the course of evolution, our minds have been adapted to the world we live in and this is a constantly changing world.

Our actual perception of time is a complex process. Beneath the level of consciousness beat the innumerable clocks of cellular and physiological activity. Although in the course of evolution man has become less dependent than other forms of life on biological rhythms, he is not entirely emancipated from them, as is evident, for example, in the modern complaint of jet-lag fatigue. Our cognitive time-sense, however much it may be controlled by other factors, is superimposed on the rhythms of the biological clocks that beat silently within us, and these have been selected in the course of our evolution because of their close chronometric relation to external influences of an astronomical nature associated with 'universal time'.

Despite overwhelming evidence for the existence of biological clocks in both animals and plants, no generally accepted identification of a single such clock has yet been made although a great deal of fascinating and valuable knowledge has been obtained. Notwithstanding the elusiveness of the temporal mechanisms controlling them, it is clear that biological rhythms are a well-nigh universal and fundamental characteristic of living organisms, but their importance is only now coming to be recognized.

Our conscious awareness of temporal phenomena involves psychological and sociological factors that overshadow the physiological. It depends on processes of mental organization uniting thought and action. It is dominated by the tempo of our attention and is acquired by the process of 'learning'. Whereas, so far as we know, all animals live, like very young children, entirely in the here and now, man has gradually learned to transcend the limitations of the 'eternal present'. First, he may have become conscious of the future state through becoming aware of his own mortality. Man's sense of the past and its relevance for the present was probably a later development, since coherent memory is not just a simple re-excitation of mental 'traces' but depends on the imaginative reconstruction of past events. The evolution of our sense of time is revealed by the increasing importance of tense in the development of language. Nevertheless, it is only in recent centuries that time has come to be a concept of major importance in human thought and our way of life, particularly through the introduction of ever more precise time-keeping instruments.

Although it has long been recognized that the practical measurement of time depends on counting, the modern scientific concept of time is based on an analogy with the geometrical concept of the continuous straight line. Nevertheless, although the linear continuum of point-like instants implied by the time-variable t is an invaluable mathematical tool, there is no evidence that it corresponds to anything in nature. Our discussion of Zeno's paradoxes supports this conclusion and implies that there are physically, as distinct from mathematically, indivisible 'instants', or chronons, although their magnitude is still uncertain. The psychological present is also not a strictly durationless instant. The continuous time-variable t can be constructed from overlapping durations of perceptible time only if certain hypotheses of continuity are involved, and must therefore be regarded as a logical abstraction like the straight line in geometry.

In the present century the classical idea of universal time as a simple moving knife-edge covering all places and observers at the same instant has been replaced by Einstein's strict localization of the concept of simultaneity and his postulate of the world-wide invariance, at least in the absence of an appreciable gravitational field, of the velocity of light *in vacuo*, c. In the treatment of relativity in Chapter 5, instead of postulating the invariance of the velocity of light, a number of elementary assumptions have been made concerning simple electromagnetic signalling in order to establish criteria for the timing of 'distant' events. This method has been used in the subsequent chapter to obtain the metric of the homogeneous and isotropic 'smoothed-out' universe. In such a universe each observer associated with a fundamental frame of reference with respect to which the expansion appears isotropic keeps the same cosmic time, and any observer moving relative to the local fundamental frame experiences the appropriate time-dilatation, which is significant only if his velocity is an appreciable fraction of the velocity of light.

Theorems concerning the existence and range of cosmic time have been supplemented in recent years by remarkable and powerful theorems due to Penrose, Hawking and Geroch concerning the occurrence of physical singularities in space-time. Although these theorems do not specify the nature of the singularities, they tend to strengthen belief in the possible existence not only of 'black holes' but also of an initial singular state of the universe, as is suggested by the cosmic 'fire-ball' explanation of the origin of the isotropic 3K background radiation discovered by Penzias and Wilson in 1965.

The existence of a cosmic time-scale with a finite origin in the past gives meaning to the concept of 'age of the universe'. Although the observational data so far obtained do not yield a unique form for the expansion factor $R(t)$, nor even the current value of the acceleration term, the view adopted in this book is that cosmic time is a fundamental characteristic of the universe. This universal time is basic, but although it cannot be derived from other concepts we can still seek to explain the correlation of the electromagnetic, thermodynamic, historical, and cosmic 'arrows' of time. The fact that so far no transluminal particles, or tachyons, have been detected is evidence in favour of a universal *time-order.* Although the question, originally raised in 1936 by Milne, of there being two (or more) physical *time-scales* not linearly related to each other has not yet been finally answered, so far there seems to be no need to abandon the classical assumption of a unique universal scale of time, modified where necessary by the demands of relativity. In other words, there is no reason yet to doubt that the fundamental physical constants are truly invariant in time as well as in space and that there is a unique basic rhythm of the universe.

The possibility of there being a hierarchy of distinct concepts of time, increasing in richness of content as we proceed from the level of elementary particles, through complex physical systems composed of many such particles, to the mind of man, has been advocated by J. T. Fraser (1975):

time at the most primitive level being reversible, at the next level irreversible, and at the highest level transient, and therefore *ipso facto* irreversible as well. The view adopted in this book, however, is that at all levels time is essentially the same, although certain aspects of it become increasingly significant the more complex the nature of the particular object or system studied.

The idea that time is also ultimate and irreducible does not, however, commit us to the unnecessary hypothesis that time is absolute in the Newtonian sense, for moments do not exist in their own right, but are merely sets of co-existent events. Nor is time a mysterious illusion of the intellect. It is an essential feature of the universe.

References

BAUM, W. A. and NIELSEN, R. F. (1976). *Astrophys. J.* **209,** 319–29.

BERGMANN, H. (1929). *Der Kampf um das Kausalgesetz in der jüngsten Physik*, pp. 27–8. Vieweg, Braunschweig.

BLAKE, G. M. (1977). *Mon. Not. r. astron. Soc.* **178,** 41P–43P.

BOLTZMANN, L. (1898). *Vorlesungen über Gastheorie*, II, PP. 257–8. Leipzig. (English translation by S. G. Brush (1964). *Lectures on gas theory*, pp. 446–8. University of California Press, Berkeley and Los Angeles, California.).

BONDI, H. (1952). *Nature* (*Lond.*) **169,** 660.

BRADLEY, F. H. (1902). *Appearance and reality*. Swan Sonnenschein, London.

BRIDGMAN, P. W. (1955). *Reflections of a physicist*, p. 251. Philosophical Library, New York.

BRILLOUIN, L. (1962). *Science and information theory*, pp. 231–2. Academic Press, New York.

—— (1964). *Scientific uncertainty and information*, p. 63. Academic Press, New York, London.

BROAD, C. D. (1935). *Philosophy* **10,** 168.

—— (1937). *Proc. Aristotelian Soc.* **16** (suppl.), 177 *et seq.*

—— (1938). *An examination of McTaggart's philosophy*, vol. II, part I, p. 316. Cambridge University Press, Cambridge.

—— (1959). A reply to my critics. In *The philosophy of C. D. Broad* (ed. P. A. Schilpp), p. 766. Tudor, New York.

BUNGE, M. (1955). *Nuovo Cimento*, **1,** 977.

ČAPEK, M. (1965). *Boston stud. phil. sci.* **2,** 441–61.

CHRISTENSON, J. H., CRONIN, J. W., FITCH, V. L., and TURLAY, R. (1964). *Phys. Rev. Lett.* **13,** 138.

CLEUGH, M. A. (1938). *Time*, pp. 164–5. Methuen, London.

COSTA DE BEAUREGARD, O. (1958). *Cah. Phys.* **12,** 317.

DAVIES, P. C. W. (1974). *The physics of time asymmetry*. Surrey University Press, Leighton Buzzard, London.

—— (1977). *Space and time in the modern universe*, p. 185. Cambridge University Press, Cambridge.

DENBIGH, K. G. (1975). *An inventive universe*, pp. 44–53. Hutchinson, London.
van DIGGELEN, J. (1976). *Nature* (*Lond.*) **262,** 675–6.
DIRAC, P. A. M. (1937). *Nature* (*Lond.*) **139,** 323.
—— (1938). *Proc. R. Soc. A* **165,** 199–208.
—— (1973). *Proc. R. Soc. A* **333,** 403.
—— (1974). *Proc. R. Soc. A* **338,** 439–46.
DOBBS, H. A. C. (1951). *Br. J. phil. Sci.* **2,** 122.
—— (1972). The dimensions of the sensible present. In *The Study of Time, I* (ed. J. T. Fraser *et al.*), pp. 274–92. Springer, Berlin.
DUMMETT, M. A. E. (1954). *Proc. Aristotelian Soc.*, **28** (suppl.), 27–44.
DUNNE, J. W. (1934). *An experiment with time*, Faber, London.
DYSON, F. J. (1972). Fundamental constants and their time variation. In *Aspects of quantum theory* (ed. A. Salam and E. P. Wigner), pp. 213–36. Cambridge University Press, Cambridge.
EARMAN, J. (1973). Notes on the causal theory of time. In *Space, time and geometry* (ed. P. Suppes), pp. 72–84. Reidel, Dordrecht.
EDDINGTON, A. S. (1923). *The mathematical theory of relativity*, p. 23. Cambridge University Press, Cambridge.
—— (1935). *The nature of the physical world*, p. 75. Dent, London.
—— (1946). *Fundamental theory*, p. 126. Cambridge University Press, Cambridge.
EHRENFEST, P. and EHRENFEST, T. (1911). *Encykl. Math. Wiss.*, IV, part 2, II, 6. Teubner, Leipzig. (English translation by M. J. Moravcsik (1959). *The conceptual foundations of the statistical approach in mechanics.* Cornell University Press, Ithaca, N.Y.)
EINSTEIN, A. (1909). *Phys. Z.* **10,** 185.
—— and RITZ, W. (1909). *Phys. Z.* **9,** 323.
FEINBERG, G. (1967). *Phys. Rev.* **159,** 1089–1105.
FEYNMAN, R. P. (1949). *Phys. Rev.* **76,** 749.
FLEW, A. (1954). *Proc. Aristotelian Soc.* **28** (suppl.), 45–62.
—— (1959). Broad and supernormal cognition. In *The philosophy of C. D. Broad* (ed. P. A. Schilpp), p. 435. Tudor, New York.
—— (1967). Precognition. In *The encyclopedia of philosophy* (ed. P. Edwards), vol. 6, pp. 436–41. Macmillan, New York.
FRANK, P. (1949). *Modern science and its philosophy*, pp. 53–60. Harvard University Press, Cambridge, Mass.
FRASER, J. T. (1975). *Of time, passion and knowledge.* Braziller, New York.
GALE, R. M. (1968). *The philosophy of time: a collection of essays.* Macmillan, London.
GOLD, *T.* (1958). *In La structure et l'evolution de l'univers* (ed. R. Stoops), p. 95. Institut International de Physique Solvay, Brussels.
—— (1962). *Am. J. Phys.* **30,** 403.
GOLDHABER, A. (1956). *Science* **124,** 218.
GRÜNBAUM, A. (1963). *Philosophical problems of space and time*, p. 322. Knopf, New York.
HINTON, C. H. (1887). *What is the fourth dimension?* Allen and Unwin, London.

Hogarth, J. E. (1962). *Proc. R. Soc. A* **267,** 365.
Hoyle, F. and Narlikar, J. V. (1964). *Proc. R. Soc. A* **277,** 1–23.
Kemp Smith, N. (1934). *Immanuel Kant's Critique of pure reason*, p. 130. Macmillan, London.
Landsberg, P. J. (1975). *Proc. R. Soc. A* **346,** 485–95.
Layzer, D. (1970). Cosmic evolution and thermodynamic irreversibility. In *Proc. Int. Conf. on Thermodynamics, Cardiff, 1–4 April 1970*, pp. 457–68. Butterworths, London.
—— (1976). *Astrophys. J.* **206,** 559.
Lee, H. D. P. (1936). *Zeno of Elea*, p. 37, Cambridge University Press, Cambridge.
Loschmidt, J. (1876). *Wien. Ber.* **73,** 135, 366.
Lotze, H. (1884). *Metaphysics* (transl. B. Bosanquet), p. 105. Clarendon Press, Oxford.
McElhinny, M. W., Taylor, S. R. and Stevenson, D. J. (1978). *Nature (Lond.)* **271,** 316–21.
McTaggart, J. M. E. (1927). *The nature of existence*, Vol. 2. Cambridge University Press, Cambridge.
Passmore, J. A. (1957). *A hundred years of philosophy*, p. 273. Duckworth, London.
Penrose, O. and Percival, J. C. (1962). *Proc. Phys. Soc.* **79,** 605–16.
Platt, J. R. (1956). *Am. Sci.* **44,** 183.
Pochoda, P. and Schwarzschild, M. (1964). *Astrophys. J.* **139,** 587.
Popper, K. R. (1958). *Nature (Lond.)* **181,** 402.
Price, H. H. (1937). *Proc. Aristotelian Soc.* **16** (suppl.), 211 *et seq.*
Reichenbach, H. (1953). *Ann. Inst. H. Poincaré* **13,** 154.
—— (1956). *The direction of time.* University of California Press, Berkeley, Los Angeles.
—— (1957). *The philosophy of space and time* (transl. M. Reichenbach and J. Freund), p. 145. Dover, New York.
Ritz, W. (1909). *Phys. Z.* **9,** 323.
Russell, B. (1917). *Mysticism and logic*, p. 21. Allen and Unwin, London.
Saltmarsh, H. F. (1938). *Foreknowledge*, p. 97. Bell, London.
Schrödinger, E. (1950). *Space–time structure*, p. 78. Cambridge University Press, Cambridge.
Sciama, D. W. (1963). *Proc. Roy. Soc. A* **273,** 484.
Sellars, W. (1962). Time and the world order. In *Scientific explanation, space and time* (ed. H. Feigl and G. Maxwell), pp. 527–616. University of Minnesota Press, Minneapolis.
Shapiro, I. I., Smith, W. B., Ash, M. B., Ingalls, R. P., and Pettengill, G. H. (1971). *Phys. Rev. Lett.* **26,** 27–30.
Smart, J. J. C. (1954). *Analysis* **14,** 79.
—— (1964). *Problems of space and time*, p. 13. Macmillan, London.
Smoluchowski, M. V. (1916). *Phys. Z.* **17,** 567.
Solheim, J.-E., Barnes, T. G., and Smith, H. J., (1976). *Astrophys. J.* **209,** 330–4.

STÜCKELBERG, E. C. G. (1941). *Helv. Phys. Acta* **14,** 588.

TOLMAN, R. C. (1917). *The theory of relativity of motion*, pp. 54–55. University of California Press, Berkeley.

—— (1934). *Relativity, thermodynamics, and cosmology*, p. 440. Clarendon Press, Oxford.

TYRRELL, G. N. M. (1946). *The personality of man*. Penguin Books, London.

VAN FLANDERN, T. C. (1975). *Mon. Not. r. astron. Soc.* **170,** 333–42.

WATANABE, M. S. (1953). Reversibilité contre irreversibilité en physique quantique. In *Louis de Broglie: physicien et penseur*, pp. 385–400. Albin Michel, Paris.

WEINBERG, S. (1977). *The first three minutes: a modern view of the origin of the universe.* André Deutsch, London.

WESSON, P. S. (1973). *Q. J. r. astron. Soc.* **14,** 9–64.

WEYL, H. (1949). *Philosophy of mathematics and natural science*, p. 264. Princeton University Press, Princeton, N.J.

WHEELER, J. A. and FEYNMAN, R. P. (1945). *Rev. mod. Phys.* **17,** 157.

WILLIAMS, D. C. (1951). *J. Phil.* **48,** 457–72.

ZERMELO, E. (1896). *Annln. Phys.* **57,** 485.

ZWART, P. J. (1976). *About time: a philosophical inquiry into the origin and nature of time.* North-Holland, Amsterdam.

APPENDIX

Derivation of the complete set of Lorentz formulae

In addition to the axioms set out in Chapter 5 we require the following axiom:

Axiom XII. According to both A and B, all light signals describe straight lines in Euclidean space with speed c.

Consider, with the aid of Figs. A.1 and A.2, a signal emitted by B at epoch t_2', according to B's clock, and reflected immediately on arrival at some event E not necessarily in line with A and B. Let it return to B at epoch t_3' according to B. Let A assign epoch t and spatial co-ordinates (x, y) to E, where x is in the line of A and B and y is orthogonal to it. Similarly, let B assign epoch t' and spatial co-ordinates (x', y') to E, where x' is in the line of A and B and y' orthogonal to it. Then, if A assigns epoch t_2 to the emission of the signal by B and epoch t_3 to its

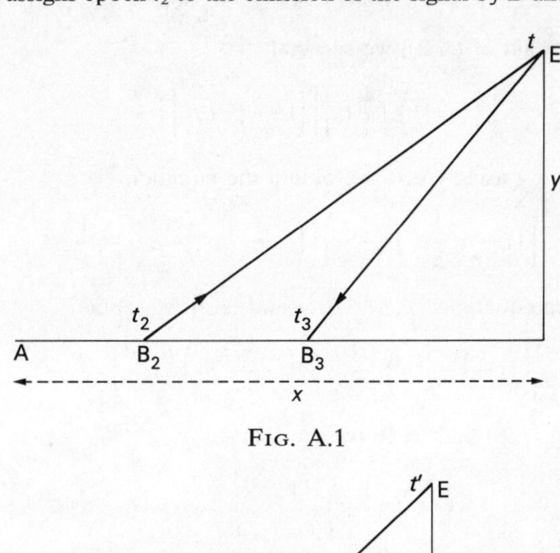

FIG. A.1

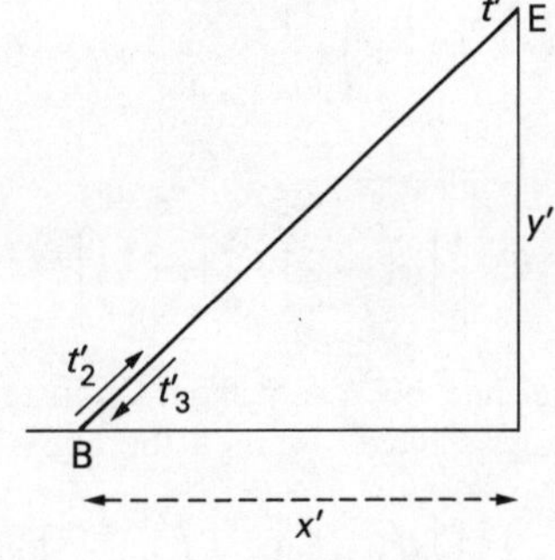

FIG. A.2

reception by B, it follows that

$$(t-t_2)^2-(x-Vt_2)^2/c^2=y^2/c^2 \tag{A.1}$$

$$t_2=\frac{t'_2}{\sqrt{(1-V^2/c^2)}} \tag{A.2}$$

$$t'_2=t'-r'/c \tag{A.3}$$

where $r'=\sqrt{(x'^2+y'^2)}$. Similarly,

$$(t_3-t)^2-(x-Vt_3)^2/c^2=y^2/c^2 \tag{A.4}$$

$$t_3=\frac{t'_3}{\sqrt{(1-V^2/c^2)}} \tag{A.5}$$

$$t'_3=t'+r'/c. \tag{A.6}$$

From equation (A.1) it follows that

$$\left\{\left(t-\frac{x}{c}\right)-t_2\left(1-\frac{V}{c}\right)\right\}\left\{\left(t+\frac{x}{c}\right)-t_2\left(1+\frac{V}{c}\right)\right\}=\frac{y^2}{c^2}$$

whence, with the aid of (A.2), we see that

$$\left\{\left(t-\frac{x}{c}\right)-\frac{1}{\alpha}t'_2\right\}\left\{\left(t+\frac{x}{c}\right)-\alpha t'_2\right\}=\frac{y^2}{c^2}. \tag{A.7}$$

Consequently, by using (A.3), we obtain the equation

$$\left\{\left(t-\frac{x}{c}-\frac{1}{\alpha}t'\right)+\frac{1}{\alpha}\frac{r'}{c}\right\}\left\{\left(t+\frac{x}{c}-\alpha t'\right)+\alpha\frac{r'}{c}\right\}=\frac{y^2}{c^2}. \tag{A.8}$$

Similarly, from equations (A.4), (A.5) and (A.6) we obtain

$$\left\{\left(t-\frac{x}{c}-\frac{1}{\alpha}t'\right)-\frac{1}{\alpha}\frac{r'}{c}\right\}\left\{\left(t+\frac{x}{c}-\alpha t'\right)-\alpha\frac{r'}{c}\right\}=\frac{y^2}{c^2} \tag{A.9}$$

On comparing (A.8) and (A.9) we see that

$$\alpha\left(t-\frac{x}{c}-\frac{1}{\alpha}t'\right)+\frac{1}{\alpha}\left(t+\frac{x}{c}-\alpha t'\right)=0$$

and consequently

$$t'=\frac{1}{2}\left\{\alpha\left(t-\frac{x}{c}\right)+\frac{1}{\alpha}\left(t+\frac{x}{c}\right)\right\}. \tag{A.10}$$

If, however, a light signal had been sent from A, instead of B, to E and back, we should by a similar analysis have obtained the equation

$$t=\frac{1}{2}\left\{\alpha'\left(t'-\frac{x'}{c}\right)+\frac{1}{\alpha'}\left(t'+\frac{x'}{c}\right)\right\}$$

where

$$\alpha' = \alpha(-V) = 1/\alpha$$

i.e. the equation

$$t = \frac{1}{2}\left\{\frac{1}{\alpha}\left(t' - \frac{x'}{c}\right) + \alpha\left(t' + \frac{x'}{c}\right)\right\}. \tag{A.11}$$

Equations (A.10) and (A.11) together are equivalent to the pair of equations

$$t' - \frac{x'}{c} = \alpha\left(t - \frac{x}{c}\right), \qquad t' + \frac{x'}{c} = \frac{1}{\alpha}\left(t + \frac{x}{c}\right) \tag{A.12}$$

whence it follows, just as in the case when E is in line with A and B, that

$$t' = \frac{t - Vx/c^2}{\sqrt{(1 - V^2/c^2)}}, \qquad x' = \frac{x - Vt}{\sqrt{(1 - V^2/c^2)}}. \tag{A.13}$$

Finally, on substituting for $(t - x/c)$ and $(t + x/c)$ by means of (A.12) in equation (A.7), we find that

$$\left\{\frac{1}{\alpha}\left(t' - \frac{x'}{c}\right) - \frac{1}{\alpha}t_2'\right\}\left\{\alpha\left(t' + \frac{x'}{c}\right) - \alpha t_2'\right\} = \frac{y^2}{c^2}$$

and hence that

$$(t' - t_2')^2 - \frac{x'^2}{c^2} = \frac{y^2}{c^2}. \tag{A.14}$$

However, equation (A.3) implies that

$$(t' - t_2')^2 - \frac{x'^2}{c^2} = \frac{y'^2}{c^2} \tag{A.15}$$

and hence that $y'^2 = y^2$. From this it is clear that $y' = y$. By cylindrical symmetry about the line AB, it follows that we can replace this equation by the pair of equations $y' = y$, $z' = z$, where y, z and y', z' are pairs of orthogonal space co-ordinates. We thus obtain the standard Lorentz formulae (5.34) on p.252, namely

$$t' = \frac{t - Vx/c^2}{\sqrt{(1 - V^2/c^2)}}, \qquad x' = \frac{x - Vt}{\sqrt{(1 - V^2/c^2)}}, \qquad y' = y, \qquad z' = z.$$

INDEX

Page numbers in italics refer to footnotes. References to authors are to the text; detailed bibliographies will be found at the end of each chapter.